STRESS-INDUCED PHENOMENA IN METALLIZATION

Previous Proceedings in the Series of Workshops on Stress-Induced Phenomena in Metallization

Year		Held in	Publisher	ISBN
1999	5th	Stuttgart, Germany	AIP Conf. Proceedings Vol. 491	1-56396-904-1
1997	4th	Tokyo, Japan	AIP Conf. Proceedings Vol. 418	1-56396-682-4
1995	3rd	Stanford, California, USA	AIP Conf. Proceedings Vol. 373	1-56396-439-2
1993	2nd	Austin, Texas, USA	AIP Conf. Proceedings Vol. 305	1-56396-251-9
1991	1st	Ithaca, New York, USA	AIP Conf. Proceedings Vol. 263	1-56396-082-6

Other Related Titles from AIP Conference Proceedings

591   Electronic Properties of Molecular Nanostructures: XV International Winterschool/Euroconference
Edited by Hans Kuzmany, Jörg Fink, Michael Mehring, and Siegmar Roth, November 2001, 0-7354-0033-4

590   Nanonetwork Materials: Fullerenes, Nanotubes, and Related Systems, ISNM 2001
Edited by Susumu Saito, Tsuneya Ando, Yoshihiro Iwasa, Koichi Kikuchi, Mototada Kobayashi, and Yahachi Saito, October 2001, 0-7354-0032-6

550   Characterization and Metrology for ULSI Technology: 2000 International Conf.
Edited by David G. Seiler, Alain C. Diebold, Thomas J. Shaffner, Robert McDonald, W. Murray Bullis, Patrick J. Smith, and Erik M. Secula, February 2001, CD-ROM included, 1-56396-967-X

544   Electronic Properties of Novel Materials—Molecular Nanostructures: XIV International Winterschool, Euroconference
Edited by Hans Kuzmany, Jörg Fink, Michael Mehring, and Siegmar Roth, November 2000, 1-56396-973-4

486   Electronic Properties of Novel Materials—Science and Technology of Molecular Nanostructures: XIII International Winterschool
Edited by Hans Kuzmany, Jörg Fink, Michael Mehring, and Siegmar Roth, September 1999, 1-56396-900-9

To learn more about these titles, or the AIP Conference Proceedings Series, please visit the webpage **http://proceedings.aip.org**

STRESS-INDUCED PHENOMENA IN METALLIZATION

Sixth International Workshop on Stress-Induced
Phenomena in Metallization

Ithaca, New York   25 –27 July 2001

EDITORS
Shefford P. Baker
Matti A. Korhonen
Cornell University, Ithaca, New York

Eduard Arzt
Max-Planck-Institut für Metallforschung, Stuttgart, Germany

Paul S. Ho
University of Texas at Austin, Austin, Texas

AMERICAN INSTITUTE of PHYSICS

Melville, New York, 2002
AIP CONFERENCE PROCEEDINGS ■ VOLUME 612

Editors:

Shefford P. Baker
Matti A. Korhonen

Cornell University
Department of Materials Science and Engineering
Bard Hall
Ithaca, NY 14853-1501
USA

E-mail: shefford.baker@cornell.edu
E-mail: mak30@cornell.edu

Eduard Arzt
Max-Planck-Institut für Metallforschung
Heisenbergstrasse 3
70569 Stuttgart
GERMANY

E-mail: arzt@mf.mpg.de

Paul S. Ho
University of Texas at Austin
Interconnect and Packaging Group
10100 Burnett Rd.
Austin, TX 78712-1100
USA

E-mail: paulho@mail.utexas.edu

L.C. Catalog Card No. 2002102851
ISBN 0-7354-0058-X
ISSN 0094-243X
Printed in the United States of America

CONTENTS

PART I: ELECTROMIGRATION STRESSES AND MECHANISMS

PART II: CYCLIC LOADING EFFECTS

PART III: STRESS VOIDING

PART IV: DEFORMATION AND STRESSES

PREFACE

The International Workshop on Stress-Induced Phenomena in Metallization provides a forum for authoritative presentations and high-level discussions on issues related to stresses in metallizations. The first workshop was held at Cornell University in Ithaca, New York, in 1991. Subsequent workshops have been held in Austin, Texas (University of Texas) in 1993, Palo Alto, California (Stanford University) in 1995, Tokyo, Japan (University of Tokyo) in 1997, and Stuttgart, Germany (Max-Planck-Institut für Metallforschung) in 1999. The workshop series returned to Ithaca and the Cornell University campus on July 25-27, 2001.

Overall, the focus of this workshop series has been to identify critical issues and to present fundamental results relating to stress induced phenomena in metallizations. The format is intended to foster high-level discussions and collaborations. Industry and academe are roughly equally represented, with active participation by a wide range of researchers from graduate students to industrial research and development and production managers. This workshop series has traditionally focused on problems related to reliability in microelectronic devices, but the scope of the series has expanded with each workshop. The first workshop focused primarily on stress voiding, and electromigration phenomena were added in the second. Atomic transport processes and metallizations in displays and acoustic devices were included in the fourth workshop and the fifth featured new damascene and Cu-based technologies.

This book contains papers from the sixth workshop. Despite a severe downturn in the microelectronics industry, the workshop was well attended, indicating the on-going interest in this forum. About 60 participants from Asia, Europe, and the United States attended. The scope of the workshop was expanded again to include metallization issues in micro-electro-mechanical sytems (MEMS) and cyclic stressing effects. Reports of phenomena in Cu metallizations and damascene structures were prominent, reflecting the recent rapid advances in these areas. The papers contained in this book represent the state of the art in a wide range of problems relating to stresses in metallizations.

We gratefully acknowledge the members of the program committee and the many individuals who helped to generate the workshop program: J.C. Bravman (Stanford University, Stanford, CA, USA), W.L. Brown (Agere Systems, Murray Hill, NJ, USA), M.P. de Boer (Sandia National Laboratories, Albuquerque, NM, USA), J.M.E. Harper (IBM, Yorktown Heights, NY, USA), M. Hommel (Infineon Technologies AG, München, Germany), K. Hinode, (Hitachi Central Research Lab, Tokyo, Japan), Y.C. Joo, (Seoul National University, Seoul, Korea), H. Kawasaki (Motorola, Austin, TX, USA), O. Kraft (Max-Planck-

Institut für Metallforschung, Stuttgart, Germany), K. Maex (IMEC, Leuven, Belgium), A.S. Oates (Lucent Technologies, Murray Hill, NJ, USA), R. Rosenberg (IBM, Yorktown Heights, NY, USA), J.E. Sanchez, Jr. (AMD, Sunnyvale, CA, USA), S. Shingubara (Hiroshima University, Hiroshima, Japan), C.V. Thompson (Massachusetts Institute of Technology, Cambridge, MA, USA), and C.A. Volkert (Max-Planck-Institut für Metallforschung, Stuttgart, Germany)

We would also like to thank the companies and institutions that provided financial support for the workshop. The workshop would not have been possible without their generous support: Agere Systems (Allentown, PA, USA; www.agere.com), Cornell Center for Materials Research (Ithaca, NY, USA; www.ccmr.cornell. edu), Frontier Semiconductor (San Jose, CA, USA; www.frontiersemi.com), Hamamatsu Photonics Systems (Bridgewater, NJ, USA; usa.hamamatsu.com), Hysitron Incorporated (Minneapolis, MN, USA; www.hysitron.com), and Infineon Technologies (München, Germany; www.infineon.com)

Finally, we would like to express our thanks to P. Baranski, D.E. Nowak, P. Pant, and especially to J.B. Shu (all of Cornell University, Ithaca, NY, USA) for assistance organizing and running the workshop, as well as the production of these proceedings.

Shefford P. Baker
Matti A. Korhonen
Eduard Arzt
Paul S. Ho

World Leaders In Nanomechanical Test Instruments
Hysitron Incorporated

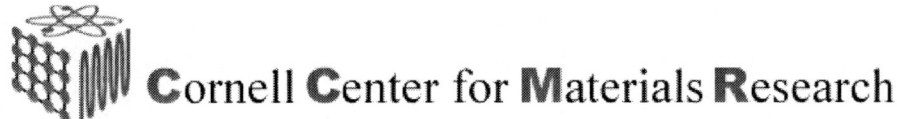

Cornell Center for Materials Research

HAMAMATSU
PHOTONIC SYSTEMS

A DIVISION OF HAMAMATSU CORPORATION

agere systems

Infineon
technologies

PART I: ELECTROMIGRATION STRESSES AND MECHANISMS

Electromigration in the Nineties

J.R. Lloyd
IBM TJ Watson Research Center
Yorktown Heights NY 10598

Abstract

There has been considerable progress in the past decade in understanding electromigration failure. Some of this understanding has come from a belated appreciation of concepts introduced years earlier, but not fully incorporated into the "world picture" of electromigration until just recently. This paper will outline the progress made in the past 10 years from a personal viewpoint. Hopefully, it will reflect the opinions of the general worker in the field, but it must be viewed with that fact that there is inevitable personal bias in mind.

Introduction

In 1990, the picture in the mind of most workers of electromigration failure was something like the following. Vacancies diffusing along grain boundaries in aluminum alloy thin film conductors, pushed along by the electron wind, gathered at regions of negative flux divergence (where more material is leaving than is arriving) and formed voids that produce failures that (for some unknown reason) followed a lognormal failure distribution. In the minds of many, everything was reasonably well understood and all that was required in the future was to take measurements. Jim Black's "Black Equation" (1) was all we needed and the world was a happy place.

$$t_{50} = Aj^{-n} \exp\left(\frac{\Delta H}{kT}\right) \qquad (1)$$

However, there were nagging inconsistencies in this quiet world picture. As dimensions shrank, Black's law was not obeyed by as many experiments as before. Activation energies for failure changed as continuous grain boundary diffusion pathways disappeared and the failure distributions were becoming distinctly non-lognormal.

Most importantly, the metal of choice for conductors in the microelectronic industry changed from Al to Cu. Al and Cu share the same crystal structure (FCC) and are both good conductors, but the similarities end there. The differences in the properties of these two metals are reflected in their substantially different behaviors in response to electromigration as a stimulus.

CP612, *Stress-Induced Phenomena in Metallization:* Sixth Int'l. Workshop, edited by S. P. Baker et al.
© 2002 American Institute of Physics 0-7354-0058-X/02/$19.00

Vacancies and Stress

One of the most important features of the nineties is that a much better appreciation of the role of mechanical stress in electromigration failure was gained. Actually, the seeds of this understanding were planted years earlier, but for some reason, they didn't take all that well. (2,3) Since 1990, however, we have been able to put stress in its proper place. Before this time, the rather naïve picture of vacancies cruising along grain boundaries until stopped by a blocking boundary was the accepted mode of viewing electromigration failure. After this, the correspondence with creep was better appreciated.

Three papers written in the nineties have substantially increased our understanding in this area. (4-6) In all three, it is realized that the naïve concept of vacancies accumulating at blocking boundaries leading to failure (7) was not reasonable in its previously understood form. Vacancies are rather ethereal entities and annihilate readily at the many convenient sinks available in a conductor. When a vacancy annihilates, either the metal changes volume, or, in a constrained system, a tensile stress is generated. Since in conductor constrained by interlevel dielectric, the stress generated is primarily hydrostatic, the stress levels can become very large. The stress possible in these circumstances is much higher than the traditional engineering yield stress. In the presence of boundary conditions blocking the flow of vacancies, induced gradients in the hydrostatic stress become important driving forces for mass transport and are instrumental in determining the behavior of a metal conductor when carrying current.

So, in the nineties, we began looking at electromigration failure not as the condensation of excess vacancies, but as the interaction between the driving forces caused by the electron/atom momentum transfer and the induced stress gradients. In this context, failure can be described as a three-stage process.

In the first stage, a stress gradient is generated by the interaction of the electromigration driving force and the boundary conditions for mass transport. The electromigration force is working in the direction of electron flow and the stress gradient is acting in the opposite direction. It is useful to examine the steady state stress profile for a particular geometry to get an idea as to how things will eventually develop. (8) If the steady state is achieved without failure, we have an immortal conductor. If it cannot be achieved because of materials limitations, at some point in time and space, the stress will increase until it is high enough to nucleate damage in the form of a void or an extrusion. At this time, (second stage) the boundary conditions change markedly, altering the steady state stress profile, sometimes significantly. In the case of a void, there will be a short time when the stress gradient and electromigration will be working in the same direction (instead of in opposition as in the first stage) and an inflationary void growth phase will result until an opposing stress gradient can be re-established. Something similar happens with an extrusion, promoting an initially rapid growth. After this rapid growth period, (third stage), the boundary conditions contain at least one stress free boundary, permitting continuous growth of damage until another steady state stress profile is achieved or until failure occurs due to an open circuit, a short circuit, or when the resistance reaches an unacceptable level.

The role of stress has also been understood in its relationship with microstructure, especially in wide lines where the probability of finding grain clusters mixed with "bamboo" regions can prove interesting. (9,10) In addition, regions where mechanical damage had presumably altered the grain structure of narrow lines showed similar behavior. (11) The effect of thermal stresses to shorten lifetime and the calculated Blech Length are readily accommodated in this view. In addition, the effects of different strength passivation layers are easily understood. In addition, direct observations of electromigration-induced stress profiles have reinforced our view. (12-14)

Perhaps this understanding of the role of stress is the single most important development of the nineties in electromigration science. It has increased our understanding of conductor behavior in situations where electromigration is significant and has pointed the way towards more reliable product through design rules and materials choices.

Al/Cu vs Cu

Perhaps the most important change in conductor technology over the past decade has been the incorporation of Cu metallization in integrated circuits. Prior to the nineties, virtually all Si based integrated circuits were made with Al or Al alloy conductors. Al is a very forgiving material. It comes in a close 4th in the conductivity sweepstakes and has other properties that make it very attractive for semiconductor processing. The presence of a thin tenacious self-limiting native oxide was a blessing we did not really appreciate until we started using Cu.

As the performance of integrated circuits increased, the limitations imposed by the resistivity of Al alloys became important. Pure Al could not be used due to its poor electromigration performance, so in order to withstand the rigors of high performance use, the Al had to be alloyed, primarily with Cu. This increased the resistivity still further (from 2.7 to well over 3 $\mu\Omega$–cm) and the designers complained loudly. Also, even with the addition of Cu, electromigration was becoming a limiting factor in the design of the most advanced high speed integrated circuits. Something had to be done if Moore's Law was to be obeyed.

There aren't too many choices if one wants I to increase conductivity over Al alloys. There is pure Al, but this has reliability problems, and there are the noble metals, Cu, Ag and Au. All the noble metals have lower resistivity than Al, but also have other problems that Al does not have. For one, the lack of the native oxide means that the noble metals require a diffusion barrier to obviate inter and intra-level shorting and leakage problems, not present with Al metallization. The other issue was somewhat of a surprise to us, but in retrospect, it shouldn't have been. (15)

The diffusion coefficient of any metal is thermally activated, and the activation energy for diffusion, to first order, is a function of the melting temperature and the crystal structure. Therefore, metals with a higher melting point should diffuse more slowly than those with a lower melting point. The noble metals and Al share the same crystal structure (FCC) so it was expected that at any temperature the electromigration lifetime, dependent on diffusion, would go roughly exponentially as the difference in the melting temperatures. With the activation energy for Cu grain boundary diffusion being measured at around 1.1 eV and that for Al alloys being ~0.7 eV, (16) it was expected that at operational temperatures (~100C) Cu electromigration lifetime would be more than an order of magnitude better than Al.

What happened, was that Cu lifetimes were somewhat longer, but not as much as one might think and not all the time. In addition, the activation energy for failure varied significantly, sometimes within the same laboratory, from a high of ~ 1.2 eV as expected to less than 0.3 eV. At the same time, Al alloy electromigration performance was improving markedly as bamboo structure Al was becoming the norm in narrow lines. (17) Suddenly, the reliability improvement we were expecting with Cu began to evaporate. The resistivity advantage was still there but there was the question of whether we would be able to take advantage of it if the electromigration limitations were to preclude higher current densities.

These results puzzled us, but really shouldn't have. We soon realized that the source of this behavior was the difference in the way the metals oxidize. Al oxide forms a self-limiting refractory shell that is an excellent diffusion barrier, and also adheres extremely well to Al. Thus, the Al surface, when oxide is present ,is effectively shut off as a diffusion pathway. This is why electromigration in Al alloys had always been via grain boundary transport, even when geometrically there was plenty of surface available. Since the surface is not available for diffusion, only the grain boundaries were left as diffusion pathways. In the case where Al was deposited onto other oxides, the high free energy of formation of aluminum oxide would ensure that Al would reduce the existing oxide, forming Al_2O_3 at the interface, once again cutting off this interface as a diffusion pathway. If Al were deposited onto a barrier metal, the diffusion would take place at this interface since it wouldn't be as strong as the oxide interface. Al, however is pretty reactive, and many Al/metal interfaces allow slower diffusion than in pure Al grain boundaries.

Cu, on the other hand, does not form such a protective oxide, and does not adhere well to either most oxides or to many metals. Therefore, the Cu interface and surface are available for diffusion and are effectively always the most important diffusion pathways, whether grain boundaries are present or not. So, instead of comparing Al and Cu grain boundary diffusion to assess their relative electromigration performance, we need to compare Cu interfacial and surface diffusion to Al interfacial diffusion. Interfacial diffusion is not so much correlated with the melting temperature as it is with the reactivity of the metal at the interface. Al is generally more reactive than Cu, so the diffusion is usually more difficult and slower, providing better electromigration reliability.

Electromigration interest in the past decade changed from diffusion along Al alloy grain boundaries to diffusion along Cu interfaces and surfaces. Microstructure is less important since grain boundaries don't take part in the diffusion process in Cu most of the time and the status of the Cu/barrier interface is absolutely critical.

Failure Distribution

Electromigration failure has been traditionally assumed to follow a lognormal failure distribution. The reason for adopting this is obscure, legend having it that Jim Black just happened to have this kind of paper in his desk when he first plotted electromigration failures in the 1960s.

From time to time this assumption has been challenged, one reason being that the lognormal distribution is not extendable. What this means is that if we have a series of conductors in series, each with a lognormal failure distribution, the ensemble cannot fail with a lognormal distribution. It is mathematically impossible. This presents a problem in extrapolating test structure data. Test structures are simplified circuits that can be easily tested, whereas product chips are very complicated with meters of line and millions of circuits. It was shown experimentally (18) and argued theoretically that some form of extreme value lognormal distribution is more appropriate. (19) The best analogy is that of a chain made up of individual links. The chain fails when the weakest of the links fails. If we assume that each link fails lognormally, something that can be justified theoretically, one can estimate the reliability of more complex systems.

In addition, careful testing of large numbers of samples revealed that the assumption of a single mode failure distribution was probably incorrect. It has recently become clear that electromigration failures typically follow at least a bimodal if not a multi-modal failure distribution. New test structures and techniques have shed light on what is an important feature of electromigration failure when relating test structure data to integrated circuits. (20)

Additional work in this area is needed, since it is not clear that, if there are two failure modes, they would necessarily follow the same failure kinetics. This may mean that the failure distribution is a sensitive function of the testing conditions, complicating the extrapolations to use conditions necessary for realistic reliability evaluations.

Low k Dielectrics

The choice of interlevel or passivating dielectric has been shown to affect electromigration lifetime significantly. This has been known since the earliest days of electromigration research. (21) Up until recently, virtually all integrated circuits were constructed such that they were covered by relatively thick, strong passivation layers composed of oxides or nitrides of silicon. These layers encapsulated the conductor in "pipes" that constrained the conductor volume and allowed for high compressive stresses. Furthermore, the main stress component was hydrostatic, allowing for much higher stresses in the conductor than would normally be possible in bulk materials.

7

Low-k dielectrics, on the other hand, are extremely weak, often with elastic moduli more than an order of magnitude lower than the conductor materials. Therefore the advantages we may have incurred with the constraining effect of the dielectric are no longer there. The hydrostatic stresses possible with the weaker dielectric are much less than if covered by oxide and the possibility of intra and interlevel extrusions is much greater.

In addition, the dielectric constant appears to track closely with thermal conductivity. Those materials that are good choices for high performance based on the dielectric constant are the worst choices for thermal management.

These are areas that will be of great interest in the coming years as we migrate in high performance microcircuits to these exotic materials.

Final Comments

This level of understanding came about from many papers by many investigators, many discussions at conferences and dinners and many arguments behind the scenes. Not everything that has happened can be encompassed in this short review. Workshops like this one were instrumental in furthering our understanding of this most important failure mechanism. Often it is the unpublished work, the conversations over snacks and coffee cups that further science the most. These contribution cannot be quantified, yet cannot be overlooked. Journal articles are good, essential and the way science is done, but arguments in the hallways and in the queue for free pastries are equally important. We need to keep the dialogues coming freely and often.

References

1) J.R. Black , Proc. 6th Ann. Reliab. Physics Symp., 148 (1967)
2) I.A. Blech, J. Appl. Phys., 47, 1203 (1976)
3) I.A. Blech and C. Herring, Appl. Phys. Lett., 29, 131 (1976)
4) R. Kircheim, Acta Metall.Mater. 40, 309 (1992)
5) M.A. Korhonen, P. Borgesen, K.N. Tu, and C.-Y. Li, J. Appl. Phys., 73, 3790 (1993)
6) J.J. Clement, J. Appl. Phys., 82, 5991 (1997)
7) M. Shatzkes and J.R. Lloyd, J. Appl. Phys., 59, 3890 (1986)
8) J.R. Lloyd, Microelectronic Engineering 49, 51 (1999)
9) B.D. Knowlton, J.J. Clement, R.I. Frank and C.V. Thompson, Mat. Res. Soc. Symp. Proc. Vol. 391, 189 (1995)
10) D.D. Brown, J.E. Sanchez, Jr., P.R. Besser, M.A. Korhonen and C.-Y. Li, Mat. Res. Soc. Symp. Proc. Vol. 391, 197 (1995)
11) Y.-C. Joo, S.P. Baker, M.P. Knauss and E. Arzt Mat. Res. Soc. Symp. Proc. Vol. 428, 225 (1997)
12) H.-K. Kao, S. Cargill III and C.-K. Hu, Mat. Res. Soc. Symp. Proc. Vol. 612, D1.8.1 (2000)
13) P.-C. Wang, I.C. Noyan, S.K. Kaldor, J.L.Jordan-Sweet, E.G. Liniger and C.-K. Hu , Appl. Phys. Lett., 76, 1 (2000)
14) S. Chiras and D.R. Clarke, J. Appl. Phys., 88, 6302 (2000)
15) J.R. Lloyd and J.J. Clement, Thin Solid Films, 262, 135 (1995) and references therein
16) I. Kauer, W. Gust and L. Kozma, *Handbook of Grain and Interphase Boundary Diffusion Data,* Ziegler Press, Stuttgart (1989)
17) M.J.C. van den Hamberg, P.F.A Alkamede, A.H/ Verbruggen, A.G. Dirks, E. Ochs and S. Redelaar, Mat. Res. Soc. Symp. Proc. Vol. 473, 211 (1997)
18) L.P. Murray, L.C. Rathbun and E.D. Wolf, Appl. Phys/ Lett., 53, 1414 (1988)
19) J.R. Lloyd and J. Kitchin, J. Mater. Res., 9, 563 (1994)
20) E.T. Ogawa, V.A. Blaschke, A. Bierwig, K.-D. Lee, H. Matsuhashi, D. Griffiths,A. Ramamurthi, P.R. Justison, R.H. Havemann and P.S. Ho, Mat. Res. Soc. Symp. Proc. Vol. 612, D2.3.1 (2000)
21) S.M. Spitzer and S. Schwartz, J. Electrochem. Soc., 116, 1368 (1969)

Electromigration in Epitaxial Cu(001) lines

G. Ramanath[§], H. Kim[§], H. S. Goindi[§], M. J. Frederick[§], C.-S. Shin[†],
R. Goswami[§], I. Petrov[†], and J. E. Greene[†]

[§]*Department of Materials Science & Engineering, Rensselaer Polytechnic Institute, Troy, NY 12180.*

[†]*Frederick Seitz Materials Research Laboratory and Department of Materials Science & Engineering,
University of Illinois, Urbana, IL 61801.*

We report the electromigration (EM) response of single-domain epitaxial Cu(001) lines on layers of Ta, TaN, and TiN. Epitaxial Cu(001) lines on nitride layers exhibit nearly two orders of magnitude higher mean-time-to-failure (MTTF) values than those on Ta, indicating the strong influence of the underlayer. The activation energy of EM for Cu on the nitrides is ~0.8-1.2 eV, and that of Cu on Ta is ~ 0.2 eV, for 200-300 °C. Our results also indicate that the MTTF values correlate inversely to the crystal quality of the Cu layers measured by X-ray diffraction. The EM resistance of epitaxial Cu lines with different crystal quality on TaN were measured to separate the effects of interface chemistry and crystal quality. While higher quality epitaxial films reveal a higher EM resistance, the magnitude of the change is smaller than that obtained by changing the interface chemistry. Epitaxial lines exhibit more than 3-4 orders of magnitude higher MTTF than polycrystalline lines on the same underlayer. Based upon our results, we propose that the Cu/underlayer interface chemistry and presence of grain boundary diffusion play important roles in unpassivated Cu films.

1. INTRODUCTION

Electromigration (EM) is current-induced displacement of atoms caused by momentum transfer from mobile electrons to the atoms in the conductor [1], which results in the formation of voids and hillocks, leading to interconnect failure. Current-induced material transport can take place by different diffusion paths, e.g., surfaces, grain boundaries, interfaces and line defects. Each path is associated with a characteristics activation barrier. EM has been investigated in Al lines over the last three decades [2,3]. It is generally understood that the major EM failure mechanism in Al lines is grain boundary diffusion. Al lines with large grains [4,5] of a bamboo-like structure [6] and having a (111) preferred orientation [7] show the highest resistance to EM.

Recently, Cu has replaced Al as the interconnect metal because of its higher electrical conductivity and greater resistance to EM. While the higher EM resistance of Cu can be explained by its higher melting point [8], the mechanism of EM-induced failures in Cu lines is not well understood. Conflicting results are reported by various groups. For instance, Ryu et al [9] have shown that Cu films with (111) preferred

CP612, *Stress-Induced Phenomena in Metallization:* Sixth Int'l. Workshop, edited by S. P. Baker et al.
© 2002 American Institute of Physics 0-7354-0058-X/02/$19.00

orientation show superior electromigration resistance while Vanasupa and co-workers [10], and Jo et al [11] have shown that the underlayer on which the Cu film is deposited is of primary importance. Other studies[12] have reported on the importance of surface-diffusion by comparing behavior of passivated and unpassivated Cu lines. All these studies are based on results obtained from polycrystalline films where the effects of grain size, orientation and distribution are difficult to separate.

Here, we present the results of an investigation aimed to eliminate the effects of grain boundaries, film orientation and interface chemistry by examining the behavior of single-domain epitaxial Cu(001) films. This will allow us to isolate the effects of different underlayers by keeping the film microstructure constant. Moreover, understanding the EM in epitaxial films is important because future Cu interconnects in sub-100-nm vias are likely to be single-crystal-like due to the bottoms of growth of large-grained Cu films during electrodeposition[13]. Our results on unpassivated Cu lines demonstrate that grain boundaries decrease EM resistance and the inclusion of a nitride underlayer improves it. Both the Cu/underlayer interface chemistry and the quality of epitaxial film—determined by the orientation inheritance from the underlayer—affect the EM behavior.

2. EXPERIMENTAL DETAILS

Polished $10 \times 10 \times 0.5$ mm^3 MgO(001) single-crystal substrates were used for growing epitaxial Cu films and oxidized Si(001) wafers with a 400-nm-thick thermal SiO$_2$ were used for growing polycrystalline films. The substrates were rinsed successively in baths of trichloroethylene, acetone, isopropanol and DI water, and inserted into a load-locked ultra-high vacuum (base pressure = 5×10^{-10} Torr) dc magnetron sputter deposition system, described elsewhere[14]. For this study, an additional sputtering source was added in the view-port flange opposing the original source. The target-to-substrate separations for the two magnetrons of size 6.35 and 7.5 cm dia, were 6.5 and 10 cm, respectively.

Inside the chamber, the Si(001) and MgO(001) substrates were thermally degassed at 500 and 800 °C, respectively, for 1 hour. The MgO crystals gave rise to sharp 1x1 reflection high-energy electron diffraction spots after the procedure. Ta, TaN, TiN and Cu layers were grown by sputtering 99.999% pure Cu, Ti and Ta targets either in flowing Ar or Ar/N$_2$ mixture. The sputtering gas pressures (99.999% pure), substrate temperatures T_s, discharge powers, and growth rates are listed in Table 1.

A 500-nm-thick epitaxial Cu layer was grown by DC magnetron sputter-deposition on MgO(001) crystals with, and without, a 20-nm-thick epitaxial interfacial layer of Ta, TaN, or TiN. In order to obtain high quality interfacial layers, a uniform axial magnetic field of $60 \leq B_{ext} \leq 180$ G—created by a pair of external Helmholtz coils with Fe pole pieces—was utilized to shape the discharge and provide a high ion flux to growing film surface[14,15].

Polycrystalline Cu films with (001) and (111) preferred orientations were grown using texture inheritance effects on dense textured polycrystalline layers of TiN(001) and TiN(111), respectively, for comparing their EM response with that of epitaxial Cu/TiN(001) structures. The details of texture control in dense TiN layers on oxidized

Si substrates using high-flux, low-energy ion irradiation are described elsewhere[16-18]. In order to prevent oxidation during handling and storage, the Cu film surface was protected by benzoyl triazine (BTA)[19] and samples were stored in a desiccator.

Table 1. Deposition conditions

Films	Pressure(mTorr)/Gas	T_s (°C)	Power (W)	Rate (nm/min)
TiN/MgO(001)	20 /N$_2$	700	170	8
TiN/SiO$_2$	20 /N$_2$	350	170	8
Ta/MgO(001)	20 /Ar	600	100	43
TaN/ MgO(001)	20 /85%Ar + 15%N$_2$	600	150	43
Cu	5 /Ar	100	60	70

X-ray diffraction (XRD) was used to measure the quality of the epitaxial Cu films. Pole figures were obtained using a SCINTAG XDS 2000™ system, and the line width of the out-of-plane (002) Bragg reflection was measured by θ-2θ scans in a Rigaku D-Max diffractometer with Cu K$_\alpha$ radiation.   Plan-view and cross-section transmission electron microscopy (TEM) were carried out in a 120 kV Philips CM 12 microscope to characterize film microstructure, and orientation relationships of the epitaxial layers with the substrate.

Arrays of 2-mm-long, 2-μm-wide lines were created from the 500-nm-thick blanket films by photolithography and ion etching. The MgO crystals were then bonded to a package and mounted on a sample holder and placed inside the furnace of a computer-interfaced EM test station. The samples were tested in the constant voltage mode at 240-300 °C in an Ar ambient to obtain a preset nominal current density of ≈ 3.5 MA cm^{-2}, flowing through each line.

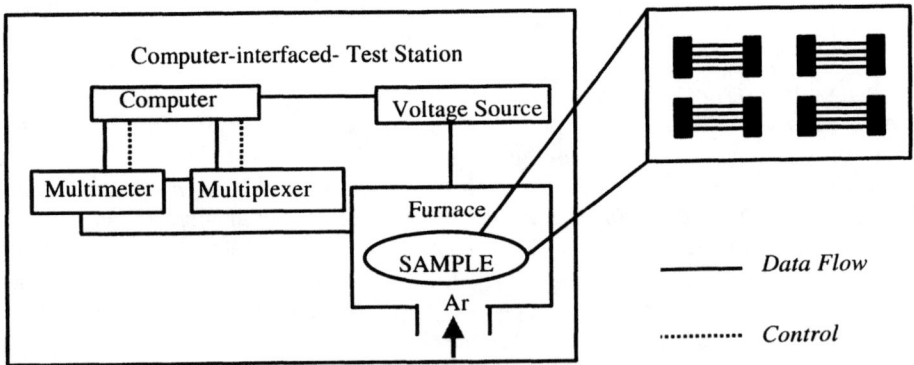

FIGURE 1. A schematic sketch of the electromigration test station.

The failures of lines appear as a current drop steps in a current vs. time plot. The failure time values thus determined were used to calculate the mean time to failure (MTTF) for each type of sample. The MTTF of different samples were compared after normalizing them to a common current density value. Scanning electron microscopy (SEM) measurements of the lines were carried out, both prior to and after EM testing, in a JEOL JSM-6330F microscope with a field emission electron gun operated at 20 kV to characterize the failure morphology.

3. RESULTS AND DISCUSSION

3.1. Epitaxial layer quality

3.1.1. Interfacial layers

High-quality epitaxial interfacial layers of TiN, TaN and Ta are essential to examine their effect on the EM behavior of epitaxial Cu films deposited on them. Figure 2 shows representative cross-section and plan-view TEM micrographs depicting the film microstructure of epitaxial Ta and TaN films grown on MgO(001).

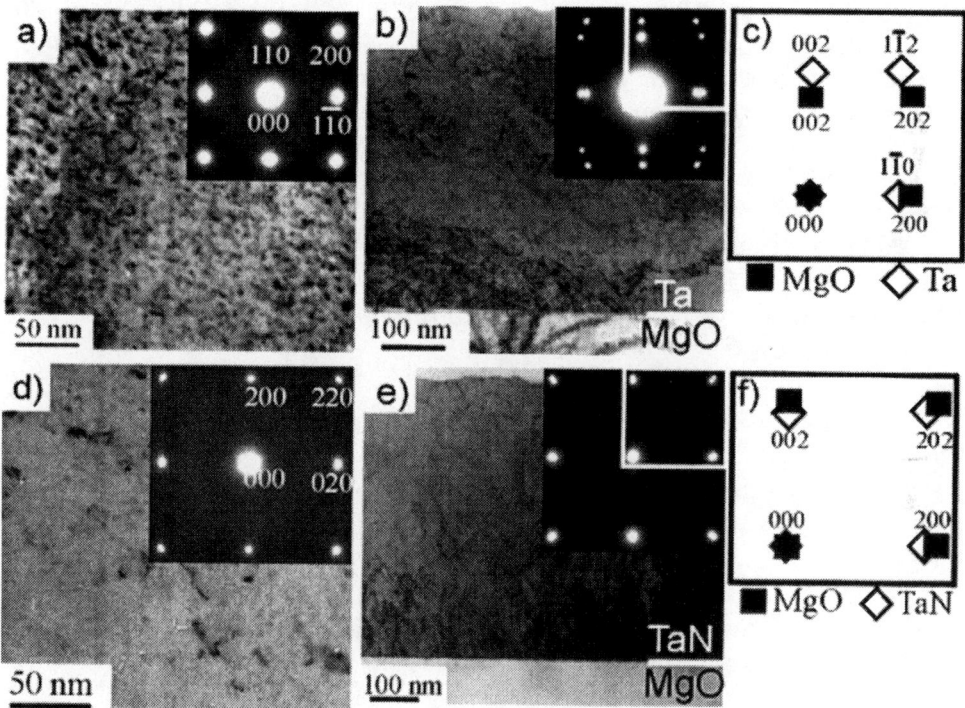

FIGURE 2 Bright field (a) plan view and (b) cross-section TEM micrographs of epitaxial Ta grown on MgO(001). The insets in (a) and (b) are corresponding selected area electron diffraction (SAED) patterns (c) A calculated SAED pattern corresponding to (b) indicating that the bcc Ta lattice is rotated 45° about the [001] growth axis with respect to MgO(001). Bright field (d) plan view and (e) cross-section TEM micrographs of epitaxial TaN grown on MgO(001). The insets in (d) and (e) are corresponding SAED patterns. (f) A calculated SAED pattern corresponding to (e), showing cube-on-cube epitaxy.

All the nitride films show a single-domain microstructure without any evidence of grain boundaries. However, crystallographic defects such as threading dislocations, dislocation loops, and nanopipes[20] (in the case of TaN) are observed. In Ta films, while the (001) planes are aligned to (001) planes of MgO (in the direction normal to the substrate), the Ta unit cell is rotated by 45° in the plane of the substrate such that Ta [110] // MgO [100] (see Figure 2c). TaN films have a cube-on-cube orientation relationship with the MgO substrate—i.e., TaN(001)//MgO(001) and TaN[100]// MgO[100] (see Figure 2f). TEM micrographs obtained from TiN films (not shown) reveal a microstructure similar to that of TaN.

3.1.2. Cu(001) layers

The orientation distribution of planes in the Cu(001) films was measured by XRD pole-figures by aligning the diffractometer to the Cu(111) Bragg reflection angle, $2\theta =42.9°$.   An example pole-figure obtained from epitaxial Cu on MgO(001) substrate, without any interfacial layer, is shown in Figure 3a.   The four Cu(111) Bragg peaks seen at $\psi = 54.7°$, $\phi = 45, 135, 225$ and 315°, and the lack of any significant intensity along the annulus at the given ψ angle are characteristics of (001) epitaxial films with a single-domain structure.   The central peak arises from the MgO(002) peak whose Bragg angle is almost identical to that of Cu(111) at $\psi = 0°$, $\phi = 0°$.   The linewidth of the Cu(002) ω rocking curves reveal that Cu films on TaN and TiN have a lower mosaicity (better epitaxial quality) than the ones grown on MgO substrates without any overlayers. Cu(001) films with the poorest epitaxial quality are obtained on a Ta layer. These results are summarized in Figure 3b.

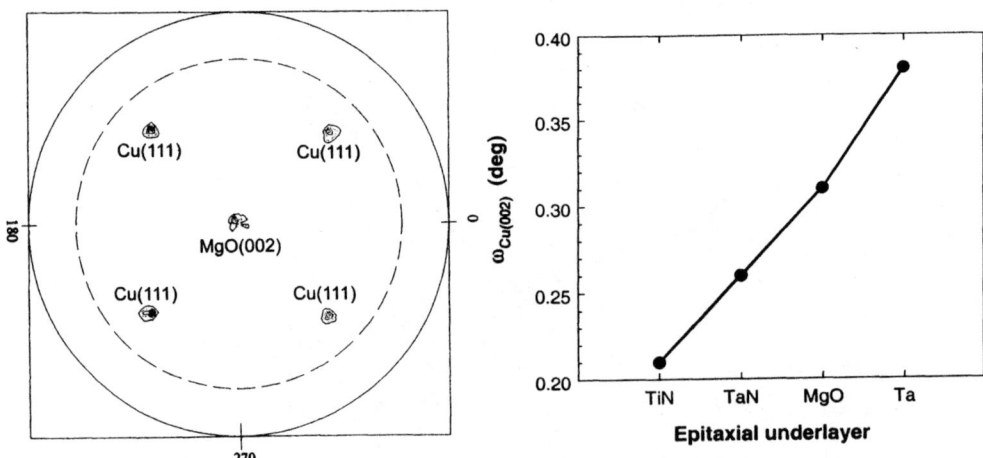

FIGURE 3 (a) A representative (111) XRD pole-figure obtained from epitaxial Cu(001) films. The central peak arises from the MgO substrate. (b) The linewidth (FWHM) of the Cu(002) ω rocking curves from epitaxial Cu films grown on different underlayers.

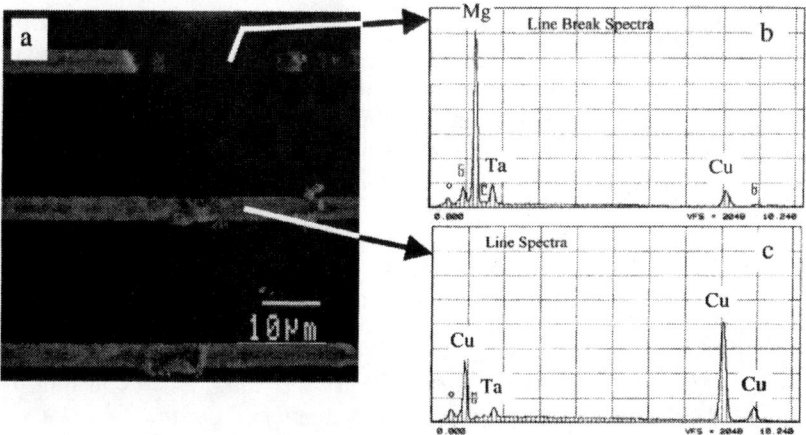

FIGURE 4. (a) SEM micrograph showing typical EM-induced failure in epitaxial Cu(001) on TaN underlayer. EDX spectra obtained from (b) broken portion of the line, and (c) an unfailed line.

3.2. Failure morphology

EM-induced failure results in line breakage and hillock formation in both epitaxial and polycrystalline Cu lines. In Cu lines on different underlayers, line breakage is observed only in the Cu layer and the underlayer is intact, acting as a shunt. This is illustrated in Figure 4, which shows a representative SEM micrograph and energy dispersive x-ray spectra obtained from the different regions of epitaxial Cu lines on TaN, after EM failure.

FIGURE 5. High-resolution SEM micrographs of failed Cu lines (a & b) with, and (c & d) without a nitride underlayer.

The morphology of the failed epitaxial Cu(001) lines is dependent on the underlayer used. A large fraction ($\approx$70%) of the hillocks and voids observed in failed Cu(001) lines on TaN and TiN underlayers exhibit a smooth morphology with faceted voids and ball-like monolithic hillocks (see Figures 5a & 5b). This morphology is similar to that observed in single-crystal Al lines [21]. In contrast, polycrystalline Cu lines show rounded edges, with only occasional presence of facets. These results

15

suggest that failure in epitaxial lines initiates at planar/volume defects and proceeds along crystallographic planes to minimize energy and area associated with the voids. In contrast, epitaxial Cu lines *without* a nitride underlayer—i.e., on MgO and on Ta(001)— exhibit a very rough morphology consisting of ≈ 0.05-1 μm fragments (see Figures 5c & 5d). The lines appear as though they have exploded due to electrical and/or thermal stresses. The reason for the formation of such explosive fragmentation is not clear at present.

3.3. Effect of Grain boundaries

Figure 6 compares the EM characteristics of epitaxial Cu(001) lines with that of (111) and (001) textured polycrystalline Cu lines to examine the effect of grain boundaries on EM failure. The interface chemistry was maintained constant by using TiN underlayer for this set of samples, as described in the experimental details section. All three curves show a knee, indicating deviation from lognormal behavior, probably due to early failures[22]. It is seen that the EM resistance of polycrystalline lines is three to four orders of magnitude lower than epitaxial lines. This clearly indicates that grain boundaries play a significant in accelerating EM in Cu films, probably by providing faster diffusion paths.

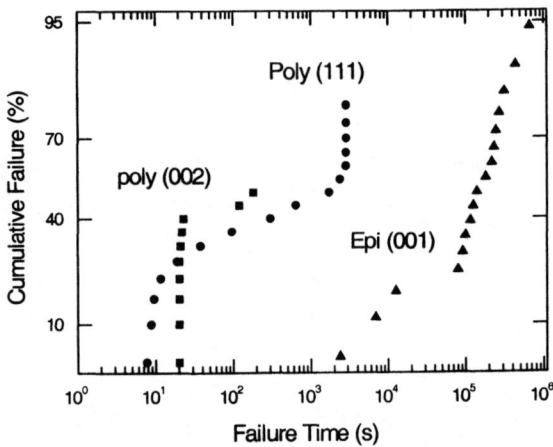

FIGURE 6. Failure statistics of epitaxial Cu(001) lines(triangles) and polycrystalline Cu lines with (111) [circles] and a (001) [squares] preferred orientation during accelerated EM tests at 240 °C. In all three cases, the Cu film was grown on a TiN layer to keep the interface chemistry constant.

Lines with (111) texture show a higher lifetime than the (001) line, suggesting a trend similar to that observed in Al interconnects. However, since the grain sizes may be different (not measured) in the polycrystalline lines, EM measurements of lines with different textures, but the same grain size, on the same underlayer are needed to conclusively determine the effect of texture.

16

3.4. Effect of underlayers

Figure 7 shows the EM characteristics (at 240 °C) of epitaxial Cu(001) lines grown on MgO(001) single crystals with different interfacial layers. The results from Cu lines deposited directly on pristine MgO substrates, without any interfacial layers, is also shown for comparison. Cu(001) lines on TaN and TiN have ≈ a factor-of 4-to-8 higher electromigration lifetime, respectively, when compared with lines on MgO. In contrast, Cu(001) lines on Ta exhibit more than an order of magnitude lower EM lifetime. Measurements at higher temperatures (not shown), e.g., 280 and 300 °C, also reveal similar overall trends, indicating that the underlayer influences the EM behavior of Cu lines. The mean time to failure (MTTF) of Cu lines on different underlayers bear an inverse correlation with the epitaxial quality of the Cu(001) lines—i.e., Cu films with a better crystal quality (lower $\omega_{Cu(002)}$) have a superior EM resistance (see Figure 8).

Interface chemistry and the quality of orientation inheritance of the epitaxial Cu(001) layer are two factors to be considered in order to understand the nature of the interfacial layer influence on the EM response of the Cu lines. Since the epitaxial films contain only a single domain, the contribution from the second factor would be primarily due to point and line defects, and possibly low angle grain boundaries. In order to separate the relative effects of epitaxial quality and interface chemistry, we tested three Cu(001) lines on epitaxial TaN(001)[15] (same interface chemistry) but different $\omega_{Cu(002)}$ (epitaxial quality). The data show that the lines with smaller linewidth (better epitaxial quality) show a higher MTTF. However, this alone, does not account for the decrease in MTTF with increasing $\omega_{Cu(002)}$ observed when the underlayer is changed. This difference can be qualitatively inferred from the difference in the MTTF values indicated by the bold and dashed lines in Figure 8. Our results thus indicate that while both interface chemistry and of epitaxial crystal quality of Cu(001) affect EM behavior, the effect of the former appears to be greater.

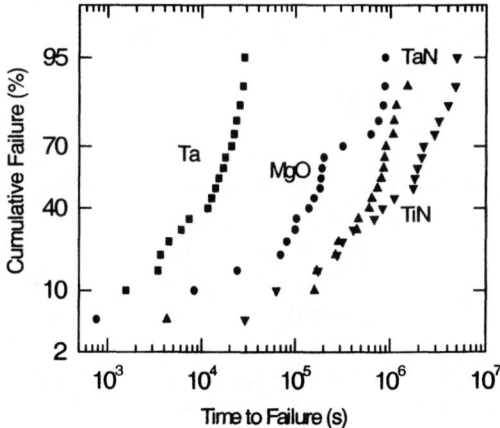

FIGURE 7. Failure statistics of Cu(001) on MgO(001) with different interfacial layers: Ta(solid squares), none(solid circles), TaN (upward triangles), TiN(downward triangles).

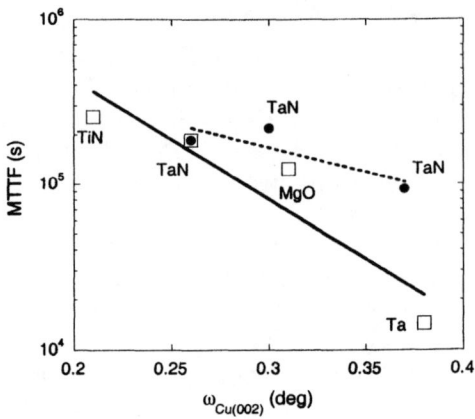

FIGURE 8. MTTF vs. linewidth of the Cu(002) peak on epitaxial films deposited on various underlayers (closed squares). Closed circles and dashed lines correspond to MTTF of epitaxial Cu(001) lines with different epitaxial quality grown on TaN underlayers.

3.5. Failure Kinetics

The activation energy E_a of EM in epitaxial Cu on the different underlayers was estimated from Arrhenius plots of MTTF vs. $1/T$ values obtained from accelerated EM tests between 200-300 °C. Since the EM behavior exhibits deviation from lognormal characteristics, both mean-time-to-failure and median-time-to-failure were used for our analysis (see Figure 9).

Cu(001) lines on nitride layers have E_a values in the ~0.8-1.2 eV range. While these E_a values are consistent with those reported by several researchers[12,23-27], they have been attributed varyingly to grain boundary diffusion and/or surface diffusion. Cu lines on MgO and Ta show E_a between ~0.2-0.4 eV, a factor of 3-4 smaller than obtained for Cu on TiN and TaN. Jo and Wook [11,28] have reported similar E_a values from *in situ* EM measurements of Cu lines in ultra-high vacuum, and have attributed them to surface diffusion.

The reasons for such large variations in E_a are not yet clear. Nonetheless, our E_a values cannot be attributed to grain boundary diffusion because single-domain epitaxial Cu lines—established by X-ray pole-figure results—are used in our experiments. Surface diffusion is also unlikely to be the only mechanism because *all* the epitaxial lines have three exposed surfaces, and the distinguishing feature is the underlayer. If surface transport was the only dominant factor, altering the underlayer from Ta to TaN would not have resulted in an eight-fold increase in EM lifetime observed in our experiments. Changing the underlayer can, however, alter the defect concentrations and surface morphology of the Cu layer, in addition to interface chemistry. Based on these arguments, we suggest that the E_a values measured in our experiments correspond to a combination surface and interfacial diffusion related phenomena. Further studies are needed to relate the E_a values to atomic-level mechanisms.

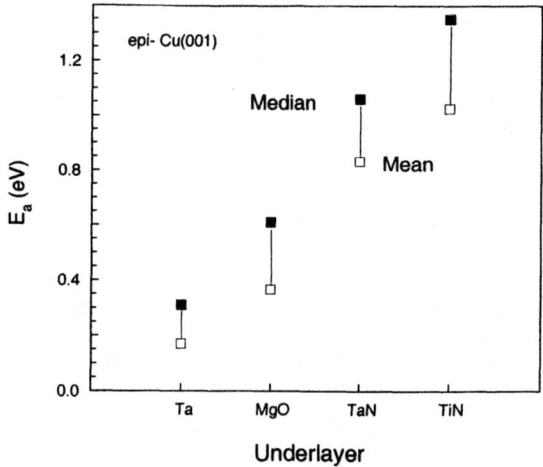

FIGURE 9. E_a plotted for Cu(001) films on different underlayers. Solid squares are activation energy based on median time to failure, open squares are activation energy based on mean time to failure.

4. SUMMARY

We have successfully grown single-domain epitaxial Cu(001) films onto epitaxial layers of Ta, TaN, and TiN. Cu films grown on TiN and TaN undelayers have a better epitaxial quality than those grown on Ta underlayer.

EM-induced failure results in line breakage and hillock formation in both epitaxial and polycrystalline Cu lines. Line breakage is observed only in the Cu layer, and the undelayer is intact. Failed epitaxial Cu(001) lines with nitride underlayers exhibit faceted voids, whereas polycrylstalline Cu lines show rounded edges. Epitaxial Cu lines without nitride underlayer exhibit an explosive failure morphology.

Epitaxial Cu(001) lines show superior EM resistance than polycrystalline lines with (111) and (001) texture, indicating that grain boundaries play an important role in the EM behavior of Cu lines. The Cu lines on TaN and TiN show up to eight-fold greater EM lifetimes compared with those on Ta. Higher quality epitaxial lines exhibit a higher EM resistance, but the magnitude of the change is smaller than that obtained by changing the interface chemistry. Characteristic activation energies estimated from line-failure lifetimes are in the ~0.8-1.2 eV range for epitaxial Cu on nitride layers; MgO and Ta is in the ~0.2-0.4eV range.

ACKNOWLEDGMENTS

We gratefully acknowledge funding support from NSF under CAREER Award DMR 9984478, and Novellus Systems Inc., San Jose, CA. We also acknowledge the Cornell Nanofabrication Facility, and the Center of Microanalysis of Materials, University of Illinois.

REFERENCES

1. Hummel, R. E., *Electro- and Thermo- Transport in Metals and Alloys* (AIME, New York, 1977).
2. Pierce, D. G. and Brusius, P. G., Microelectronics and Reliability **37** (7), 1053-1072 (1997).
3. Sorbello, R. S., Sol. Stat. Phys. **51**, 159-231 (1998).
4. Wada, J.-I., Suguro, K., Hayasaka, N., and Okano, H., Jap. J. Appl. Phys. **32** (6B), 3094-3098 (1993).
5. d'Heurle, F. M. and Ames, I., Appl. Phys. Lett. **16** (2), 80-81 (1970).
6. Cho, J. and Thompson, C. V., Appl. Phys. Lett. **54** (25), 2577-2579 (1989).
7. Vaidya, S. and Sinha, A. K., Thin Solid Films **75**, 253-259 (1981).
8. Renz, T., Ph.D. thesis, Rensselaer Polytechnic Institute, 1997.
9. Ryu, C., Kwon, K.-W., Loke, A. L. S., Lee, H., Nogami, T., Dubin, V. M., Kavari, R. A., Ray, G. W., and Wong, S. S., IEEE Trans. Electron. Dev. **46** (6), 1113-1120 (1999).
10. Vanasupa, L., Joo, Y.-C., Besser, P. R., and Pramanick, S., J. Appl. Phys. **85** (5), 2583-2590 (1999).
11. Jo, B. H. and Vook, R. W., Thin Solid Films **262**, 129-134 (1995).
12. Hu, C.-K., Rosenberg, R., and Lee, K. Y., Appl. Phys. Lett. **74**, 2945-47 (1999).
13. Lingk, C., Gross, M. E., and Brown, W. L., J. Appl. Phys. **87** (5), 2232-6 (2000).
14. Petrov, I., Adibi, F., Greene, J. E., Sproul, W. D., and Munz, W., J. Vac. Sci. Technol. A **10**, 3283-87 (1992).
15. Shin, C.-S., Gall, D., Desjardins, P., Vailionis, A., Kim, H., Petrov, I., and Greene, J. E., Appl. Phys. Lett. **75**, 3808 (1999).
16. Greene, J. E., Sundgren, J.-E., Hultman, L., Petrov, I., and Bergstrom, D. B., Appl. Phys. Lett. 67 **67** (20), 2928-30 (1995).
17. Hultman, L., Sundgren, J.-E., Greene, J. E., Bergstrom, D. B., and Petrov, I., J. Appl. Phys. **78** (9), 5395-403 (1995).
18. Chun, J.-S., Petrov, I., and Greene, J. E., J. Appl. Phys. **86**, 3633 (1999).
19. Brusic, V., Frisch, M. A., Eldridge, B. N., Novak, F. P., Kaufman, F. B., Rush, B. M., and Frankel, G. S., J. Electrochem. Soc. **138**, 2253 (1991).
20. Ramanath, G., Carlsson, J., Greene, J. E., Allen, L. H., Hornback, V. C., and Allman, D. J., Appl. Phys. Lett. **61**, 3179 (1996) and references therein.
21. Joo, Y.-C. and Thompson, C. V., J. Appl. Phys. **81** (9), 6062-6072 (1997), and references therein.
22. Gall, M., Capasso, C., Jawarani, D., Hernandez, R., Kawasaki, H., and Ho, P. S., Appl. Phys. Lett. **76**, 843-845 (2000).
23. Hu, C.-K., Lee, K. Y., Gignac, L., and Carruthers, R., Thin Solid Films **308-309**, 443-47 (1997).
24. Nitta, T., Ohmi, T., Hoshi, T., Sakai, S., Sakaibara, K., Imai, S., and Shibata, T., J. Electrochem. Soc. **140**, 1131-37 (1993).
25. Lee, K. Y., Hu, C.-K., and Tu, K. N., J. Appl. Phys. **78**, 4428-37 (1995).
26. Frankovic, R. and Bernstein, G. H., IEEE Trans. Electron. Dev. **43**, 2233-39 (1996).
27. Gladikh, A., Lereah, Y., Karpovski, M., Palevski, A., and Kaganovski, Y. S., Mat. Res. Soc. Symp. Proc. **427**, 121-126 (1996).
28. Vook, R. W. and Jo, B. H., Mat. Res. Soc. Symp. Proc. **428**, 31-41 (1996).

Immortality of Cu Damascene Interconnects

Stefan P. Hau-Riege

Intel Corporation, Hillsboro, OR 97124

Abstract. We have studied short-line effects in fully-integrated Cu damascene interconnects through electromigration experiments on lines of various lengths and embedded in different dielectric materials. We compare these results with results from analogous experiments on subtractively-etched Al-based interconnects. It is known that Al-based interconnects exhibit three different behaviors, depending on the magnitude of the product of current density, j, and line length, L: For small values of (jL), no void nucleation occurs, and the line is immortal. For intermediate values, voids nucleate, but the line does not fail because the current can flow through the higher-resistivity refractory-metal-based shunt layers. Here, the resistance of the line increases but eventually saturates, and the relative resistance increase is proportional to (jL/B), where B is the effective elastic modulus of the metallization system. For large values of (jL/B), voiding leads to an unacceptably high resistance increase, and the line is considered failed. By contrast, we observed only two regimes for Cu-based interconnects: Either the resistance of the line stays constant during the duration of the experiment, and the line is considered immortal, or the line fails due to an abrupt open-circuit failure. The absence of an intermediate regime in which the resistance saturates is due to the absence of a shunt layer that is able to support a large amount of current once voiding occurs. Since voids nucleate much more easily in Cu- than in Al-based interconnects, a small fraction of short Cu lines fails even at low current densities. It is therefore more appropriate to consider the probability of immortality in the case of Cu rather than assuming a sharp boundary between mortality and immortality. The probability of immortality decreases with increasing amount of material depleted from the cathode, which is proportional to (jL^2/B) at steady state. By contrast, the immortality of Al-based interconnects is described by (jL) if no voids nucleate, and (jL/B) if voids nucleate.

INTRODUCTION

In today's Si-based integrated circuit (IC) technology, several meters of metal interconnects are required to build a single high-performance circuit, so that many millions of metal segments exist in each IC. These metallic circuit elements are a significant reliability concern owing mainly to electromigration. This concern increases with each new generation of Si technology, which requires the use of a larger number of narrower interconnects, stressed at ever-higher current densities.

During operation, atomic redistribution due to electromigration leads to the build up of mechanical stresses at sites of an atomic flux divergence. This leads to possible failure of the circuit, since high tensile stresses lead to voiding, and high compressive stresses lead to the formation of metallic extrusions. If the stresses stay small enough, however, void nucleation does not occur. If the line is bound by zero-flux boundaries, such as contacts and vias lined with an impermeable diffusion barrier, the stress along the length of the line will evolve until there is a uniform stress gradient, at which point the back force due to this gradient balances the electron wind force [1]. It has been

CP612, *Stress-Induced Phenomena in Metallization:* Sixth Int'l. Workshop, edited by S. P. Baker et al.
© 2002 American Institute of Physics 0-7354-0058-X/02/$19.00

shown for Al-based metallization schemes that this saturation effect or immortality occurs when the product of line length, L, and current density, j, stays below a critical product [2]. When sufficiently thick refractory metal cladding-layers are present, voids can form without causing immediate failure, because current can shunt through the refractory layers that do not electromigrate. However, continued electromigration leads to void growth and can eventually lead to unacceptably high resistances [2]. If the resistance increase can be tolerated, void growth will saturate, and again a steady state develops; the line will not fail and is immortal. The relative resistance increase at steady state is proportional to (jL) [3].

The described Blech-length effects have been used extensively in the design and layout of integrated circuits with Al-based metallization schemes with W plugs, which are zero-atomic-flux boundaries. Thick Ti, TiN, or Al_3Ti under- and overlayers provide an alternative current path and allow voids to nucleate and grow without significant resistance increase [4]. Similar effects have been observed in more complex interconnect trees [5].

The increasing ratio of wiring delay to the intrinsic transistor delay has motivated the IC industry to move from aluminum-based interconnects embedded in SiO_2 to copper-based metallization systems with inter-level dielectrics (ILD) having lower dielectric constants, k, than SiO_2 [6]. This change in materials and consequent metallization scheme has significant impact on electromigration. For example, electromigration experiments on Cu interconnects have shown that diffusion takes place primarily along the Cu/passivation interface, regardless of the grain boundary structure [7]. In addition, low-k ILD's are often polymer-based, and are mechanically much more compliant than SiO_2 [8], affecting the electromigration-induced stress evolution in the lines. The goal of this paper is to discuss if the principles leading to short-line effects in Al-based interconnects still apply to the newer metallization technology based on Cu. We will analyze whether immortality in Cu can still be categorized by (jL) [9-10], or if it is necessary to introduce a new figure of merit for immortality. Finally, we will determine the effects of different dielectrics on immortality.

BACKGROUND: MODELING OF ELECTROMIGRATION

The electromigration force originates from scattering events with flowing electrons. A more compressive stress develops in volumes in which atoms accumulate, whereas the stress becomes more tensile in volumes in which atoms are depleted. The relationship between the change in the concentration of atomic lattice sites, C, and the hydrostatic stress, σ, is described by the effective bulk modulus, B [11], through

$$\frac{dC}{C} = -\frac{d\sigma}{B}.$$
(1)

Stress gradients result in gradients in the chemical potential, leading to back diffusion, which opposes the electromigration wind force.

In case where void nucleation does not occur, the stress will evolve until there is a uniform stress gradient along the length of the line, and the back force due to this gradient balances the electron wind force [1]. Under this condition,

$$z * epj = \Omega \frac{\Delta\sigma_{max}}{L},$$ (2)

where z* is the effective charge of the migrating metal atoms, e is the fundamental electronic charge, ρ is the resistivity of the conducting metal, Ω is the atomic volume, L is the length of the line, and $\Delta\sigma_{max}$ is the difference in the stress at the anode and the cathode. Equation (2) shows that the maximum stress in the line is related to the product of the current density and the line length, (jL). If the stress required for void nucleation is less than the maximum stress, the line will not fail and will be immortal. There is a critical line-length current-density product that defines the condition for immortality,

$$jL < \frac{\Omega\Delta\sigma_{crit}}{z * e\rho} \equiv (jL)_{crit},$$ (3)

where $\Delta\sigma_{crit}$ is the minimum stress difference that leads to void nucleation. Therefore, short lines, and/or lines tested at low currents, are more likely to be immortal. Interestingly, the critical current-density line-length product is independent of the effective bulk modulus, B.

If (jL) exceeds the critical current-density line-length product, void nucleation does occur but is not necessarily fatal if the void does not pinch off the line from the via. Often a shunt layer is present that connects the via to the metal, thereby serving as an alternative current path. The stress at the void surface will fall to zero, and the stress in the nearby metal will quickly decrease as the void grows. Eventually, a force balance will develop where Equation (2) again applies. The line will be immortal if the maximum compressive stress does not cause yielding or fracture of the dielectric, given that the refractory-metal based liner is able to support the current, which is typically the case for Al-based interconnects. The void volume, and thereby line resistance, will change over time. Void growth will eventually saturate and the line will be immortal. The immortality condition is

$$\left(\frac{jL}{B}\right) < \frac{\rho/_A}{\rho_1/_{A_1}} \frac{\Delta R_{crit}^{max}}{R} \frac{2\Omega}{e\rho z *} \equiv \left(\frac{jL}{B}\right)_{crit},$$ (4)

where ρ and ρ_1 are the resistivity of the high-conductivity metal and the shunt layers, respectively, R is the initial resistance of the line, and ΔR_{crit}^{max} is the maximum acceptable resistance increase of the line. This line-length and current-density dependent saturation of the resistance increase has been demonstrated in experiments on Al-based interconnects by Filippi and co-workers [4].

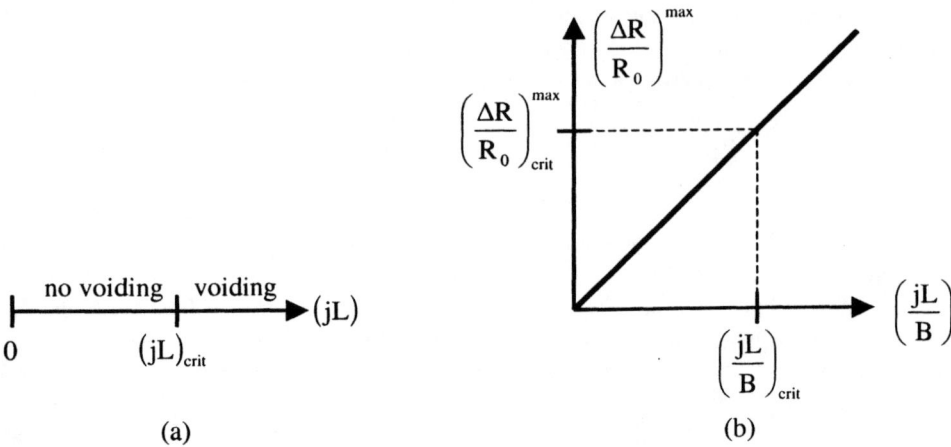

(a)   (b)

FIGURE 1. Immortality conditions for Al-based interconnects with shunt layers. According to Equation (3), if (jL) < (jL)$_{crit}$, no void formation occurs and the line is immortal, as indicated in (a). If (jL) ≥ (jL)$_{crit}$, voids do form, leading to a maximum relative resistance increase, $(\Delta R/R_0)^{max}$, that is proportional to (jL/B), as shown in Equation (4) and sketched in (b). The line is immortal if (jL/B) < (jL/B)$_{crit}$.

In summary, Al-based interconnects with shunt layers exhibit three different modes of short-line effects, depending on the magnitude of (jL) [4] and (jL/B) [12]. For (jL) < (jL)$_{crit}$, no void nucleation occurs, and the line is immortal, as depicted in Figure 1 (a) (mode 1). For (jL) > (jL)$_{crit}$, voids nucleate, and the current must flow through the higher-resistivity refractory-metal-based shunt layer. The resistance of the line increases but eventually saturates due to back-stress effects, and the maximum relative resistance increase at steady state is proportional to (jL/B) [13], as depicted in Figure 1 (b). If the resistance increase is acceptable, the line is considered to be immortal (mode 2), otherwise it is mortal (mode 3).

FIGURE 2. Schematic diagram of the Cu interconnects for the electromigration experiments. The electrons are passed from the via above the line.

EXPERIMENTS

We performed electromigration experiments on electroplated damascene Cu interconnects. A schematic diagram of the test configuration is shown in Figure 2. The interconnects are contacted by Cu vias, that are lined with a Ta-based diffusion barrier layer and located above the line ends. The thickness of the liner is less than 10% that of the line width. It was confirmed through transmission electron microscopy that the barrier is continuous in the vias, so that the interconnects are bound by zero-atomic-flux boundaries. The lines were 0.21 μm wide and 10.5, 21, 42, 70, and 210 μm long. Electromigration testing was conducted on Cu lines that were embedded in dielectrics of different mechanical properties. The effective bulk moduli, B, for the different types of samples were obtained through finite element calculations [12] and are listed in Table 1.

TABLE 1. Effective bulk modulus, B, for the different dielectric materials used in this study, calculated using finite-element methods as described in reference [12].

Interlevel Dielectric	B (GPa)
ILD #1	23
ILD #2	19
ILD #3	4

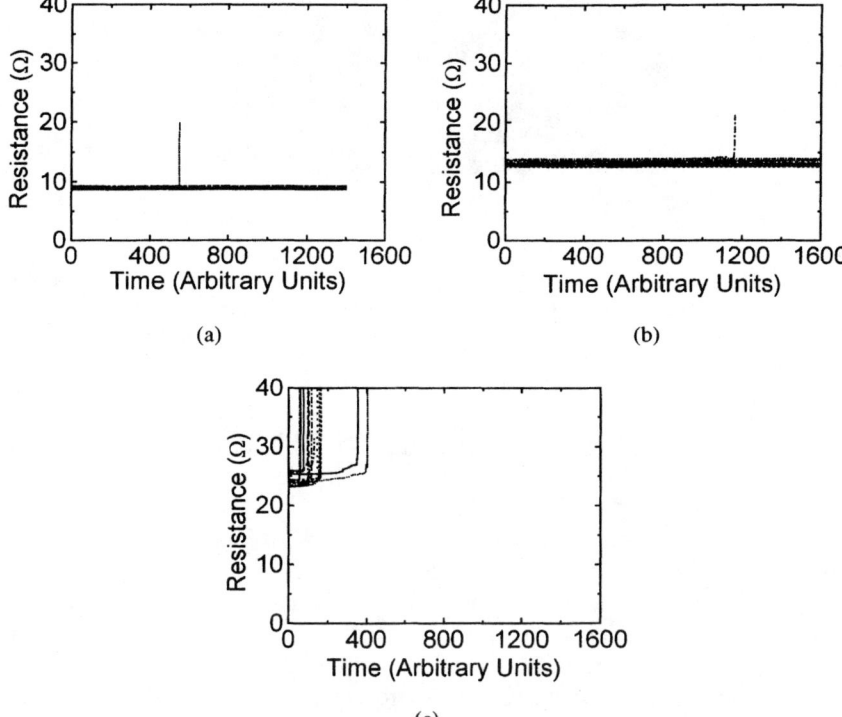

(a)   (b)

(c)

FIGURE 3. (a) to (c) Evolution of the electrical resistance of 18 interconnects of different lengths, L. The interconnects were tested at 300°C and passing a current density of 2 MA/cm^2 from the via above. The line lengths were (a) L = 10.5 μm, (b) L = 21 μm, and (c) L = 42 μm.

The samples were tested in samples sizes of approximately 20 at a temperatures of 300°C for ILD #1 and #2 and 250°C for ILD #3, by forcing a constant current density, j, ranging from 1 – 4 MA/cm^2. The electrical resistance across the interconnects was measured in situ to monitor electromigration-induced damage. After electromigration testing, the samples were analyzed using focus-ion-beam (FIB) microscopy.

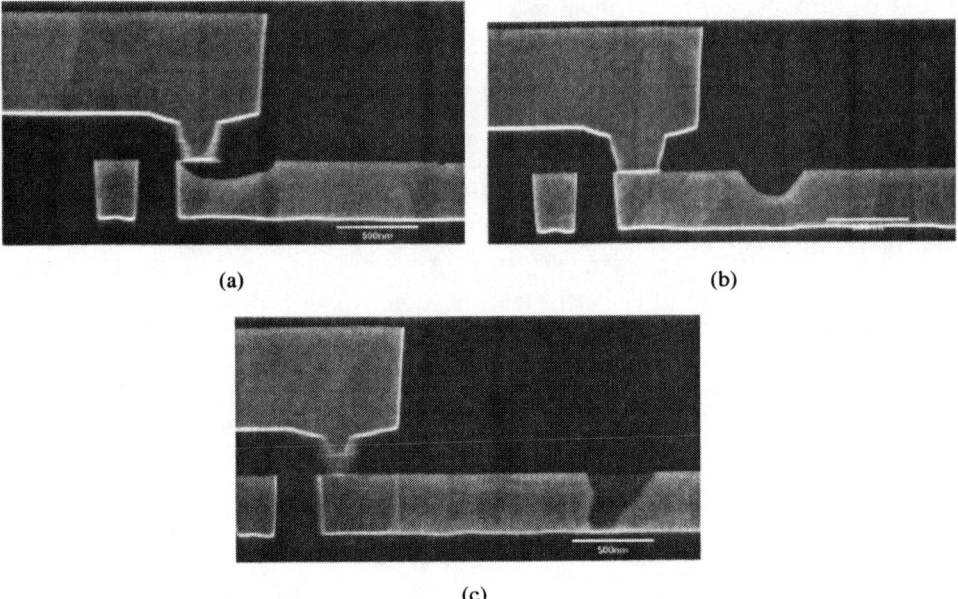

(a)   (b)

(c)

FIGURE 4. (a) to (c) FIB cross sections of the interconnects with the resistance signature shown in Figure 5 (a), cut along the length of the line after electromigration testing. The line shown in (a) corresponds to the only failed line in Figure 5 (a). Here, the bottom of the via is exposed and the line failed by open-circuit failure. For the cases shown in (b) and (c), a void has formed which lead to an only minor resistance increase.

EXPERIMENTAL RESULTS

Figures 3 (a) to (c) show the evolution of the electrical resistance of interconnects tested at 300°C while passing a current density of 2 MA/cm^2. The samples were embedded in ILD #1, and the line length, L, was 10.5 µm, 21 µm, and 42 µm, respectively. As depicted in Figure 3, we observed either no resistance increase, or an abrupt increase in resistance indicating open-circuit failure. We did not observe a failure mode for which the resistance increases and eventually saturates. The samples were analyzed post-mortem through FIB microscopy, and voids were detected in all samples. The voids were always located in the proximity of the cathode end of the line at the Cu/Si$_3$N$_4$ interface. Voids were not observed further downstream. We also did not observe metallic extrusions in any of the lines tested. Figures 4 (a) through (c) are examples of cross sections at the cathode end along the length of three of the 10.5 µm-long lines. The interconnect shown in Figure 4 (a) corresponds to the only failed line in Figure 3 (a). Here, the void has exposed the bottom of the via, leading to open-circuit failure. For the lines shown in Figures 4 (b) and (c), the voids led to only small resistance increases that is within the noise of the resistance measurements.

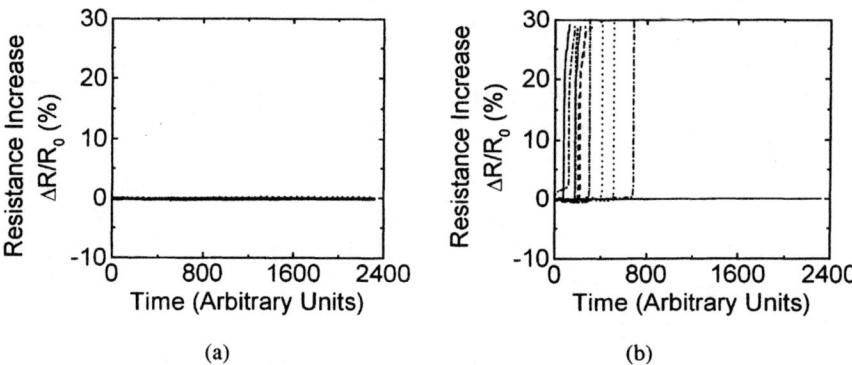

FIGURE 5. (a) and (b) Evolution of the electrical resistance of 18 interconnects of different lengths, L. The interconnects were tested at 300°C passing a current density, j, from the via above. For both cases, $(j*L) = 3780$ A/cm, and (a) L = 10.5 μm and j=3.6 MA/cm^2 $(j*L^2 = 4.0$ A) and (b) L = 21 μm and j = 1.8 MA/cm^2 $(j*L^2 = 7.9$ A).

Two sets of experiments were conducted with the same magnitude of (jL) (i.e., 3780 A/cm) but with different line lengths and current densities. The evolutions of the electrical resistance are shown in Figures 5 (a) and (b). The tests were performed at 300°C. All interconnects were embedded in ILD #2. In case (a), the lines were 10.5 μm long and stressed passing a current density of 3.6 MA/cm^2, and no failures were observed for the duration of the test. In case (b), the lines were 21 μm long and stressed passing a current density of 1.8 MA/cm^2, and, in contrast to case (a), the majority of the lines failed by open-circuit failure.

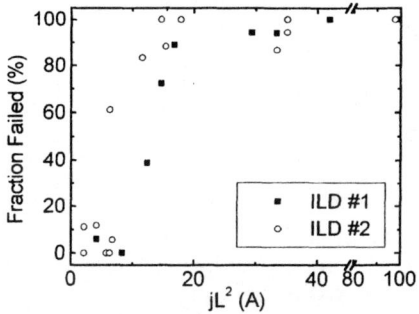

FIGURE 6. Fraction of interconnects failed as a function of (jL^2) for two different inter-level dielectrics (ILD). The effective bulk modulus, B, is lower for ILD #2 than for ILD #1. All tests were terminated after the same amount of time.

All electromigration tests were terminated after the same time, t_{term}, which is at least three times longer than the failure times of very long lines, where "very long lines" are considered to be those for which the failure distribution does not change if the line length is increased further. As shown in Figure 6, the fraction of lines failed after time t_{term} depends monotonically on (jL^2). Rather than having a critical (jL^2)

27

below which all lines are immortal and above which all lines are mortal, the fraction of lines failed varies *continuously* with (jL^2).

We also found that the fraction of failed lines increases as the ILD becomes more compliant, characterized by a lower B: We conducted electromigration experiments for sets of lines embedded in ILD #1, ILD #2, and ILD #3 with the configuration sketched in Figure 2. We tested lines of different lengths and passing different current densities. The fraction of lines that failed by open-circuit failure at time t_{term} is plotted as a function of (jL^2) for ILD #1 and #2 in Figure 6. We did not observe immortality for samples embedded in ILD #3 even at test conditions as low as $(jL^2) = 1$ A.

MODELING RESULTS

To determine the total void volume at steady state, we needed to consider the hydrostatic stress profile, σ, at steady state. The maximum compressive stress, $\Delta\sigma$, is proportional to (jL) and is independent of B. To the first order, σ depends linearly on the concentration of atomic lattice sites, C, through

$$C = C_0 \exp\left(-\frac{\sigma}{B}\right) \approx C_0\left(1 - \frac{\sigma}{B}\right), \tag{5}$$

with $C_0 = C(0)$. At the anode end, $\Delta C = C - C_0 \approx C_0\Delta\sigma/B$, so that the total amount of material depleted from the cathode is

$$C_0\frac{\Delta\sigma}{2B}L = C_0\frac{z^*e\rho j}{2B\Omega}L^2, \tag{6}$$

which is proportional to (jL^2/B).

DISCUSSION

We first discuss the characteristics of voiding in Cu-based interconnects. We then analyze the immortality behavior of Cu-based interconnects, and contrast it with the immortality behavior of Al-based interconnects with shunt layers.

Voiding and Failure in Cu Interconnects

We observed that voids always nucleated at the top Si_3N_4/Cu interface, as indicated in Figure 4. This observation can be attributed to the fact that the Si_3N_4/Cu interface energy is lower than the barrier/Cu interface and the grain boundary energies [14]. We also found that voids formed even under the mildest test conditions of $j = 2$ MA/cm^2 and $L = 10.5$ μm. Using Equation (3) and $\sigma_{crit} = \Delta\sigma_{crit}/2$, and by allowing $z^* = 1$ [7] and $\rho = 4$ μΩ×cm at 300°C, we found that $\sigma_{crit} \leq 41$ MPa for the Cu interconnects studied in this paper. This value represents the tensile stress needed to nucleate a void in the metal, and is much lower than the critical stress of 600 MPa observed for Al-based interconnect systems [15], which indicates that voids form much more readily in Cu- than in Al-based interconnects.

It can also be seen in Figure 4 that the void location, although always near the cathode end where the tensile stresses are largest, varies from line to line. This local variation is significant because only certain void locations lead to failure. Specifically, open-circuit failure only occurs in cases in which the void pinches off the electron flow from the via. To cause failure in the "via-above" case, the void has to nucleate near the via and grow large enough to expose the bottom of the via.

Immortality In Cu-Based Interconnects

The short-line effects in Cu interconnects, as reported in this paper, greatly contrast the short-line effects previously observed in Al interconnects. These differences can be demonstrated by considering the three modes of short-line effects found in Al-based interconnects with shunt layers, which are categorized by (jL) [4] and (jL/B) [12]. For (jL) < (jL)$_{crit}$, no voids are observed to nucleate in Al-based interconnects, as depicted in Figure 1 (a), so that the lines are immortal [2] (mode 1). This mode of immortality was not observed for Cu interconnects for (jL) values as small as 2100 A/cm, even though it can be expected that for much smaller (jL) values Cu lines will eventually be immortal due to the absence of void nucleation. For (jL) > (jL)$_{crit}$, voids nucleate both in Al and Cu interconnects. In the case of Al interconnects with shunts, void growth eventually saturates, leading to a maximum relative resistance increase at steady state that is proportional to (jL/B), as shown in Figure 1 (b). If this resistance increase is acceptable, the line is immortal (mode 2). Like the first mode, mode 2 of immortality was also not observed for Cu interconnects. Instead, the resistance of the Cu lines either remained constant for the duration of the test, or the resistance increased rapidly, leading to abrupt open-circuit failure. Clearly, (jL/B) is not the correct figure of merit to describe the immortality of Cu lines, which is exemplified by Figure 3 showing that lines with the same (jL) but different current densities and line lengths can show very different immortality behavior. Finally, whereas the immortality of Al interconnects is deterministic, in the sense that critical (jL) and (jL/B) values exist below which Al interconnects are immortal, the immortality of Cu interconnects is probabilistic since a small fraction of lines fails even for low current densities and line lengths.

However, even though the phenomenology is very different for Al- and Cu-based interconnects, the physical models that explain electromigration and short-line effects as observed in Al still apply to Cu. For instance, immortality is still a result of a force balance between the electromigration-driving force and the back stress. In fact, all observed differences in Cu immortality as reported in this paper can be attributed solely to differences in materials and metallization architecture. These differences are: (i) The critical stress for void nucleation is much lower in Cu than in Al, so that voids nucleate much more readily in Cu than in Al metallizations. (ii) The Cu samples described in this paper do not have an effective shunt layer, so that in the case of voiding, the thin and highly-resistive Ta-based barrier would melt due to Joule heating rather than support a significant amount of current. (iii) Finally, the transition to Cu as the interconnect material of choice coincides with the introduction of novel dielectrics.

For these more compliant dielectrics, the back-stress force is reduced because a larger change in the number of atoms per unit volume is required to induce changes in the hydrostatic stresses. We will now discuss the consequences of these differences on the immortality of Cu-based interconnects.

As a consequence of a missing shunt, the resistance is not observed to first increase and then saturate, as in the case for Al-based interconnects, since no alternative current path is present that bridges a void that disconnects the Cu line from the via. Instead, Cu lines are observed to either fail under test by sudden open circuit, or maintain a nearly constant resistance during the duration of the test due to the presence or absence of a *fatal* void, respectively.

Secondly, even though immortality is likely to occur for short lines tested at low current densities, a small fraction of lines still fails, as depicted in Figures 3 (a) and (b). This is due to the low critical stress for void nucleation and the missing shunt layer. Since voids nucleated in all lines tested, there is always a small but finite probability that a void is located and shaped in a way that it cuts off the electron flow from the via to the line. When this occurs, the line will fail by open-circuit failure because, once again, there is no shunt layer. Therefore, immortality is less robust for Cu- than for Al-based interconnects, and the immortality concept for Cu-based interconnects without shunt layers is probabilistic.

Finally, (jL) is not the correct figure of merit to describe the likelihood of immortality of Cu lines for the current-density and line-length values studied in this paper, since lines tested at the same (jL) product can exhibit very different electromigration behavior, as shown in Figures 5 (a) and (b). We will now propose a new figure of merit that is capable of characterizing the immortality of Cu interconnects.

If the stresses in the line stay below σ_{crit}, no void will nucleate. Independent of metallization system, this condition can be expressed by a critical (jL) product, $(jL)_{crit}$, below which no void nucleation occurs (Equation (3)). (jL) has been used successfully as a figure of merit for the immortality of Al-based interconnects [1]. However, because σ_{crit} for Cu interconnects is more than ten times smaller than σ_{crit} for Al interconnects, the $(jL)_{crit}$ we found in our experiments is smaller than 2100 A/cm. Therefore, an immortality condition based on $(jL)_{crit}$ is not useful for the design of IC's with Cu interconnects.

For the much more relevant case of $(jL) > (jL)_{crit}$, voids will nucleate, as shown in this study. Given that the void does not lead to open-circuit failure, it will grow under continued electromigration stressing, and eventually void growth will saturate due to back stress. According to Equation (7), the void size at steady state is proportional to (jL^2/B). If shunt layers are present, the maximum relative resistance increase at steady state, $\Delta R/R_0$, is proportional to (jL/B), which is the figure of merit for immortality of Al-based interconnects with shunt layers in case voiding occurs. However, if shunt layers are not present, there is no gradual increase in resistance during electromigration test, so that the concept of $\Delta R/R_0 \propto (jL/B)$ is not applicable. Rather, voiding occurs either at places with little resistance impact, or voids induce open-circuit failure. Immortality is therefore determined by the probability that a void pinches off the via from the line. For a given void volume and void shape, the likelihood for immortality decreases with increasing total void *volume* at steady state, since a larger void it is more likely to expose the bottom of a via or span the full cross section of the interconnect. (jL^2/B) is therefore the correct figure of merit to describe the immortality of Cu-based interconnects without shunt layers, since it is proportional to the void volume at steady state. In this model, the probability of immortality decreases monotonically with (jL^2/B), as sketched in Figure 7.

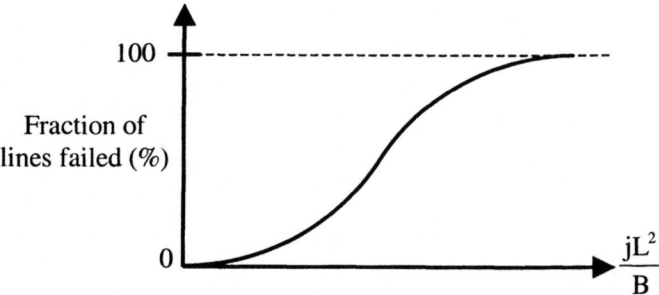

FIGURE 7. Probabilistic immortality conditions for Cu-based interconnects. The fraction of lines failed increases monotonically with (jL^2/B).

Figure 6 (a) shows the fraction of failed lines as a function of (jL^2) at time t_{term}. The fraction of failed lines increases with increasing (jL^2), and the fraction of failed lines is generally larger for ILD #2 than for ILD #1, because ILD #2 has a lower B value and is therefore more compliant. For the most compliant ILD #3, about 100% failure was observed even for the lowest value of (jL^2) of 1 A. This agrees with the trend that immortality is less likely for lower values of B. Here it can be reasoned that very large void volumes are necessary to induce back stresses that balances the electromigration driving force, and that these voids almost always lead to open-circuit failure.

SUMMARY AND CONCLUSIONS

We have studied short-line effects in Cu-based interconnects, and compared it with short-line effects in interconnects based on Al. For both metallization schemes, lines with $(jL) < (jL)_{crit}$ are immune to electromigration-induced failure, because the hydrostatic stresses are small, and void nucleation does not occur. This phenomena is independent of the presence of shunt layers. For $(jL) > (jL)_{crit}$, void nucleation does occur in both metallization schemes. Given that shunt layers are present, or voiding occurs in way that a continuous current path remains, void growth will eventually saturate. The void size at steady state is proportional to (jL^2/B).

For Al-based metallization schemes, shunt layers are typically present that serve as an alternative current path in case of voiding, making the interconnects insensitive to defects and small voids. The relative resistance increase at steady state is proportional to (jL/B). If the maximum relative resistance increase is acceptable, the line is considered immortal. For these reasons, (jL/B) can be used as the figure of merit for immortality when voiding occurs and shunt layers are present. In the case when only one dielectric material is considered, the figure of merit can be further reduced to the familiar (jL) [2].

For state-of-the-art Cu interconnects, the barrier layers are too thin and resistive to serve as an effective shunt for the electron current, so that resistance saturation is not observed: Cu interconnects either fail by abrupt open-circuit failure, or the lines are immortal. Furthermore, voids nucleate more readily than in Al-based interconnects due to a much lower critical stress for void nucleation. For technologically relevant test conditions (jL) typically exceeds (jL)$_{crit}$, and voids nucleate. Due to voiding, a small fraction of lines always fails, even when short lines were tested at low current densities, so that immortality is not deterministic for (jL) > (jL)$_{crit}$. Rather, the immortality of shunt-less interconnects with a low barrier for void nucleation, as in the case of most Cu-based interconnects, is probabilistic and more accurately described by (jL^2/B) than by (jL) or (jL/B).

To regain a deterministic immortality, the metallization scheme described in this paper needs to be modified. The barrier for void nucleation must be improved by, for example, replacing the Si_3N_4 etch stop layer with a layer that adheres better to Cu in order to increase σ_{crit}. Also, the introduction of a thick, low-resistivity shunt layer as a continuous current path will make Cu lines insensitive to defects and smaller voids, so that a resistance saturation regime would once again be observed.

ACKNOWLEDGEMENTS

The author thanks Christine Hau-Riege and Carl Thompson for valuable discussions, Thomas Marieb and Jose Maiz for continuous support, and Michael McKeag for the failure analysis. The process community of Intel's Portland Technology Development is acknowledged for sample fabrication and testing.

REFERENCES

[1] I.A. Blech and H. Sello, *Physics of Failures in Electronics* 5, ed. T.S. Shilliday and J. Vacarro, Rome Air Development Center, 496 (1967).

[2] I. A. Blech and C. Herring, *Appl. Phys. Lett.* **29**, 131 (1976).

[3] Z. Suo. *Acta mater.* **46**, 3725 (1998).

[4] R.G. Filippi, R.A. Wachnik, H. Aochi, J.R. Lloyd, and M.A. Korhonen. *Appl. Phys. Lett.* **69**, 2350. (1996).

[5] S.P. Hau-Riege and C.V. Thompson, *J. Appl. Phys.* **88**, 2382 (2000).

[6] D. Edelstein, J. Heidenreich, R. Goldblatt, W. Cote, C. Uzoh, N. Lustig, P. Roper, T. McDevitt, W. Motsiff, A. Simon, J. Dukovic, R. Wachnik, H. Rathore, R. Schulz, L. Su, S. Luce, and J. Slattery, *IEEE Intl. Electron Devices Meeting Digest*, 773 (1997).

[7] C.-K. Hu, R. Rosenberg, and K.Y. Lee, *Appl. Phys. Lett.* **74**, 2945 (1999).

[8] D.T. Price, R.J. Gutmann, and S.P. Murarka, *Thin Solid Films* **308-309**, 523 (1997).

[9] P.C. Wang and R.G. Filippi, *Appl. Phys. Lett.* **78**, 3598 (2001).

[10] E.T. Ogawa, A.J. Bierwag, Ki-Don-Lee, H. Matsuhashi, P.R. Justison, A.N. Ramamurthi, P.S. Ho, V.A. Blaschke, D. Griffiths, A. Nelsen, M. Breen, R.H. Havemann, *Appl. Phys. Lett.* **78**, 2652 (2001).

[11] M. A. Korhonen, P. Boergesen, K.N. Tu, and Che-Yu Li, *J. Appl. Phys.* **73**, 3790 (1993).

[12] S.P. Hau-Riege and C.V. Thompson, *J. Mater. Res.* **15**, 1797 (2000).

[13] J.C. Doan, S. Lee, S.H. Lee, P.A. Flinn, J.C. Bravman, and T.N. Marieb, *J. Appl. Phys.* **89**, 7797 (2001).

[14] Jessica Xu, Intel Corporation, Hillsboro, OR, *private communication*.

[15] C.S. Hau-Riege and C.V. Thompson, *J. Appl. Phys.* **87**, 8467 (2000).

The Electromigration Short-Length Effect in AlCu and Cu Interconnects

R. G. Filippi[a], P.-C. Wang[a], R. A. Wachnik[a], D. Chidambarrao[a],
M. A. Korhonen[b], T. M. Shaw[c], R. Rosenberg[c], T. D. Sullivan[d]

[a] IBM Microelectronics Division, 2070 Route 52, Hopewell Junction, New York 12533-6531
[b] Department of Materials Science and Engineering, Cornell University, Ithaca, New York 14853
[c] IBM Research Division, T. J. Watson Research Center, Yorktown Heights, New York 10589
[d] IBM Microelectronics Division, Essex Junction, Vermont 05452

Abstract. The electromigration short-length effect is investigated for AlCu and Cu interconnects in SiO$_2$ dielectrics. Simple models based on first principles are shown to accurately describe the length-dependent electromigration behavior of both metallization systems. The model for AlCu describes resistance saturation while the model for Cu describes the incubation time. Experimental results for different combinations of current density and stripe length are in good agreement with the predictions of both models. The effect of line width and temperature on the electromigration threshold are also investigated. In addition, an electromigration resistant power grid is proposed as a practical application of the short-length effect.

I. INTRODUCTION

It has been over twenty years since electromigration phenomena in the presence of a diffusion barrier were first reported for aluminum (Al) interconnects [1]. Since that time, the so-called "Blech Effect" or "Short-Length Effect" has captured the interests of many investigators [2,3,4,5,6,7,8]. The short-length effect has been traditionally characterized for Al-based interconnects, although recent studies have also demonstrated the electromigration threshold in copper (Cu) interconnects [9,10]. As the microelectronics industry favors Cu as the interconnect material for advanced integrated circuits (IC), it will become increasingly important to characterize the short-length effect in Cu-based metallization systems. In this paper, we quantify the short-length effect for Al-based and Cu-based interconnects using simple, first principle models. In addition, we construct an electromigration resistant power grid as a practical application of the short-length effect.

During electromigration, the electron wind applies a force that results in an atomic flux, J [11]

$$J = nv_e = n\left(\frac{D}{kT}\right)j\rho eZ^* \qquad (1)$$

CP612, *Stress-Induced Phenomena in Metallization:* Sixth Int'l. Workshop, edited by S. P. Baker et al.
© 2002 American Institute of Physics 0-7354-0058-X/02/$19.00

where n is the atomic density, v_e is the average atomic drift velocity, D is the diffusion coefficient, k is Boltzmann's constant, T is the ambient temperature plus Joule heating, j is the current density, ρ is the resistivity of the conductor and eZ^* is the effective ion charge. In the presence of a diffusion barrier, however, Blech reported that the accumulation of conductor atoms at the anode end of the line results in a stress gradient which opposes the electromigration driving force [1]. The stress gradient produces a back-flow atomic drift velocity, v_b, such that the net atomic flux, J_{eff}, is given by

$$J_{eff} = n(v_e - v_b) = \frac{nD}{kT}\left(j\rho eZ^* - \frac{\Delta\sigma\,\Omega}{L}\right)$$ (2)

where Δσ is the back-flow stress, Ω is the atomic volume and L is the conductor length. In Equation (2), Δσ/L is the electromigration-induced compressive stress gradient along the length of the interconnect. Blech showed, by observing void motion in unpassivated Al lines, that J_{eff} becomes zero for a critical product of current density and conductor length, $(jL)_{th}$, referred to as the threshold product.

During electromigration of multilevel interconnects with redundant refractory layers, a void will nucleate and grow at the cathode end of the line. The resistance of the line increases as the void grows beyond the front edge of the diffusion blocking stud or via [12]. As reported for Al-based multilayered interconnects with blocking studs or vias at both ends of the line, the stress gradient limits void growth and thus limits the increase in resistance of the metal structure [7,8].

The short-length effect is generally regarded as beneficial but of limited use in practice because of the difficulty of systematically constructing nets with short lengths. In addition, effective use requires developing new models and introducing stochastic design rule methodologies which are substantially different than those in customary use today.

II. EXPERIMENT

Multilayered metallization systems are evaluated in this study for Al-based and Cu-based interconnects. The Al-based interconnects are composed of aluminum-copper (AlCu) sandwiched between redundant layers of titanium (Ti). Due to anneals and oxide deposition temperature excursions (350-400 °C), Ti and Al react to form the $TiAl_3$ intermetallic [13,14]. A titanium nitride (TiN) cap layer is also present. The Cu concentration is 0.5 weight percent (wt %). Metal 1 (M1) is a 0.54 μm stack of Ti/AlCu/Ti/TiN. The AlCu lines are reactively ion etched (RIE) and passivated with silicon dioxide (SiO_2). The Cu-based interconnects are composed of electroplated Cu, where M1 is a single-damascene interconnect (0.31 μm thick) with tantalum-based conductive liners at the bottom and sidewalls. The Cu lines are passivated with silicon nitride (Si_3N_4) and SiO_2.

Two-level test structures are used in this study, shown schematically in Figure 1(a) for the AlCu interconnects and in Figure 1(b) for the Cu interconnects. In the AlCu case, tungsten (W) studs, V1 and V2, are located at both ends of M1. The stripe length is 30, 50, 70 or 100 μm, while the stripe width is either 0.33 or 1.50 μm. The 0.33 μm-wide structure has one W stud at the cathode end, while the 1.50 μm-wide structure has two W studs across the width of the line at the cathode end. In the Cu case, a W stud, V1, is located at the anode end, while a damascene Cu via, V2, is located at the cathode end. The V2 via includes tantalum-based conductive liners at the bottom and sidewalls. The stripe length is 30, 50, 70 or 100 μm, while the stripe width is 0.245 μm.

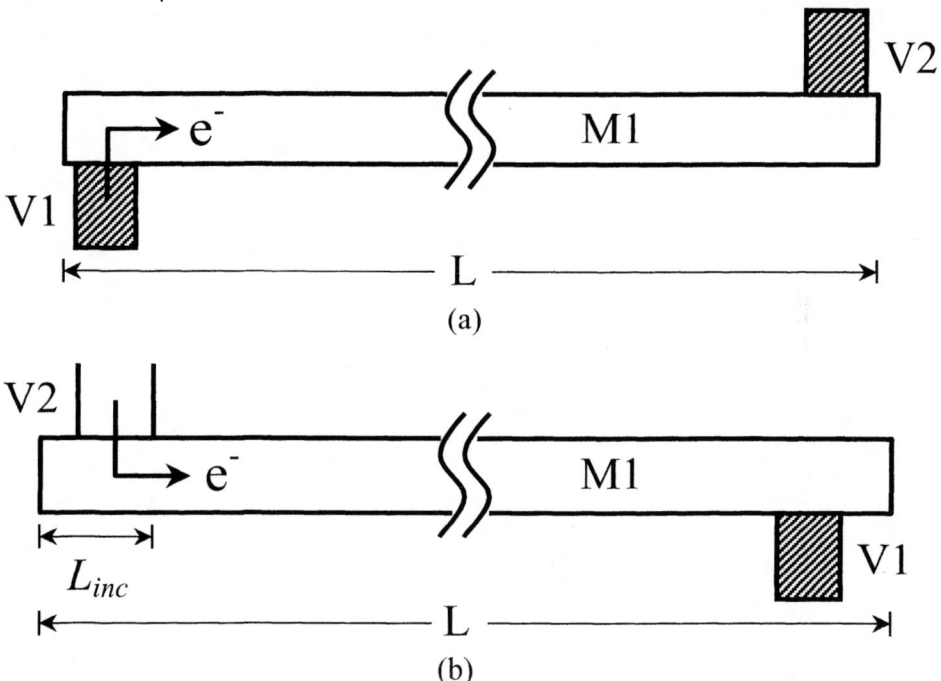

Figure 1. Schematic cross-sectional view of the two-level electromigration test structures used for (a) the AlCu interconnects and (b) the Cu interconnects. In both cases, L is the total length of the interconnect, while for the Cu case, L_{inc} is the length depleted during the incubation period.

Electromigration testing is conducted at various combinations of current density (AlCu: 0.4-2.0 MA/cm^2; Cu: 0.6-1.5 MA/cm^2) and temperature (0.33 μm-wide AlCu: 170, 210, 250 °C; 1.50 μm-wide AlCu: 250 °C; Cu: 295, 350, 400 °C). The resistance across the interconnects is measured in real time with four point measurements to monitor the electromigration damage. Each stress cell for the AlCu interconnects, which is described by a current density, stripe length, stripe width and temperature, consists of 9-12 samples. Each stress cell for the Cu interconnects, which is described by a current density, stripe length and temperature, consists of 16-24 samples.

III. RESULTS AND DISCUSSION

A. Short-Length Effects in AlCu Interconnects

In this section, short-length effects in AlCu interconnects are described. (See Reference [15] for a more detailed discussion.) Figure 2 shows a plot of the fractional resistance increase versus time for a few of the 50 μm-long AlCu samples tested in this study. The saturation of the resistance increase with time indicates the suppression of further electromigration-induced damage. A simple and intuitive argument can be used to derive the following equation, which links resistance increase, current density and stripe length for the case of diffusion blocking boundaries at both ends of the interconnect [4,8]:

$$L(\Delta R / R)_{sat} = \left(\frac{K\rho e Z^*}{2\Omega B} \right) jL^2 - K(\Delta L_0) \tag{3}$$

In Equation (3), $(\Delta R/R)_{sat}$ is the fractional resistance increase at saturation, K is a proportionality constant that relates void length to resistance increase, B is an effective bulk modulus and ΔL_0 represents the void length that does not contribute to a resistance increase. The derivation of Equation (3), which is based on the Blech electromigration model (see Equation (2)), assumes that the short-length effect in AlCu is caused by a mechanical stress gradient and that the steady-state stress gradient develops after long times.

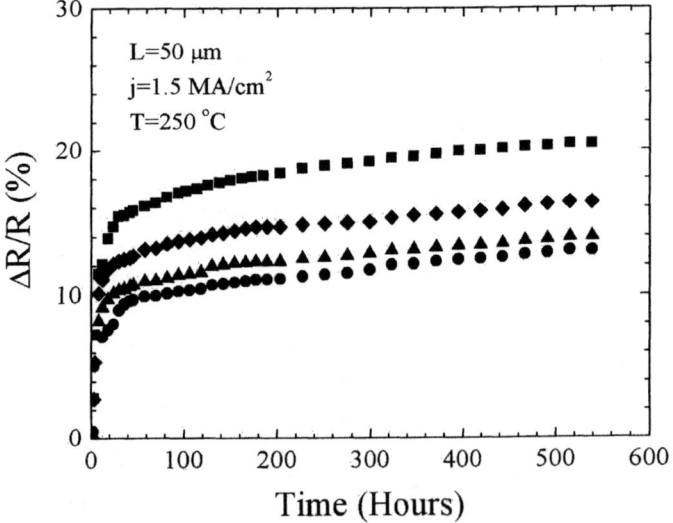

Figure 2. Resistance shift (%) versus time for the 50 μm-long AlCu samples. The stripe width is 0.33 μm. The test conditions are 1.5 MA/cm² and 250 °C. Only four out of twelve samples are shown.

1. Narrow-Line Electromigration Behavior

According to Equation (3) a plot of $L(\Delta R/R)_{sat}$ versus jL^2 should yield a straight line. This is demonstrated in Figure 3 for the 0.33 µm-wide AlCu samples tested at 250 °C. The figure shows good agreement between the least-squares fit, given by the solid line, and the actual data. A plot similar to Figure 3 was recently published in terms of the absolute resistance change at saturation [8].

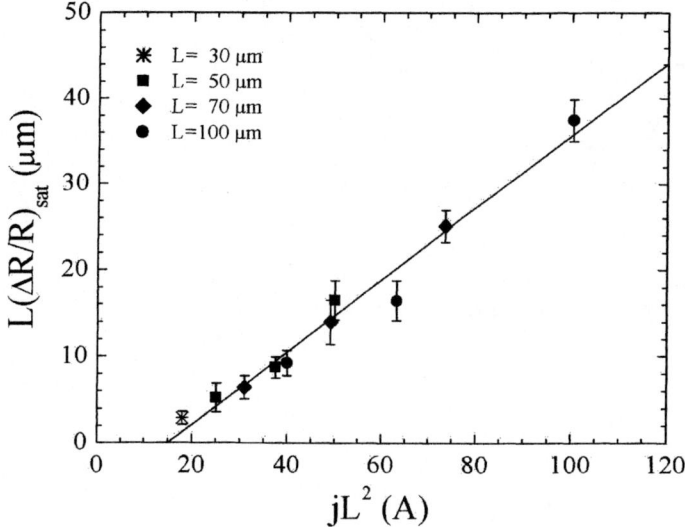

Figure 3. $L(\Delta R/R)_{sat}$ versus jL^2 for the 0.33 µm-wide AlCu samples tested at 250 °C. The symbols correspond to the average values and the error bars represent the 95% confidence interval. The solid line is the least-squares fit.

As suggested by the error bars in Figure 3, there is a statistical distribution for $(\Delta R/R)_{sat}$. The statistical nature of $(\Delta R/R)_{sat}$ is illustrated by the lognormal cumulative distribution function (CDF) plot shown in Figure 4. The shape of the distribution, or the lognormal sigma (σ), is typically found to lie between 0.2-0.3 for the 0.33 µm-wide samples, and is independent of current density, stripe length and temperature.

The temperature dependence of $L(\Delta R/R)_{sat}$ is an important consideration for determining the resistance saturation behavior at operating conditions. Figure 5 indicates that the 0.33 µm-wide AlCu samples tested at 170, 210 and 250 °C generally obey the same linear relationship independent of temperature. Based on the resistance saturation model, this result is not surprising since Equation (3) describes an equilibrium situation where most of the parameters, such as B and Ω, depend weakly on temperature. Furthermore, the product ρZ^* is insensitive to temperature [11]. The lack of a strong temperature dependence also suggests that the level of resistance saturation is not significantly affected by thermal strain.

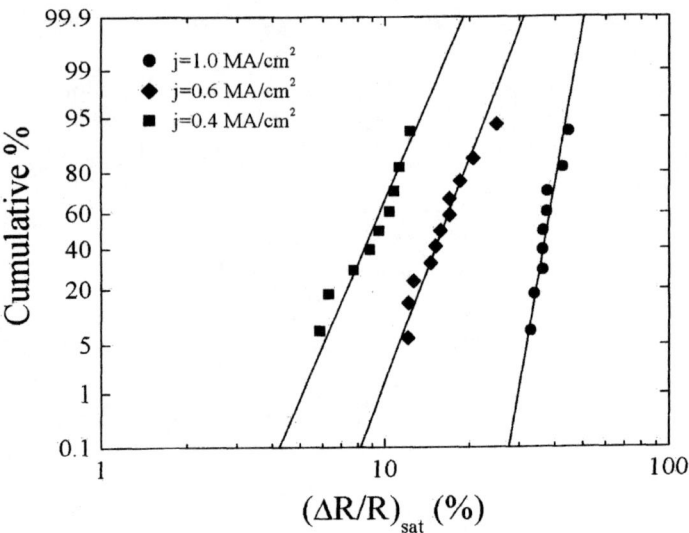

Figure 4. Lognormal CDF plot for the AlCu samples in terms of $(\Delta R/R)_{sat}$ at 250 °C and different current densities. The stripe width is 0.33 μm and the stripe length is 100 μm.

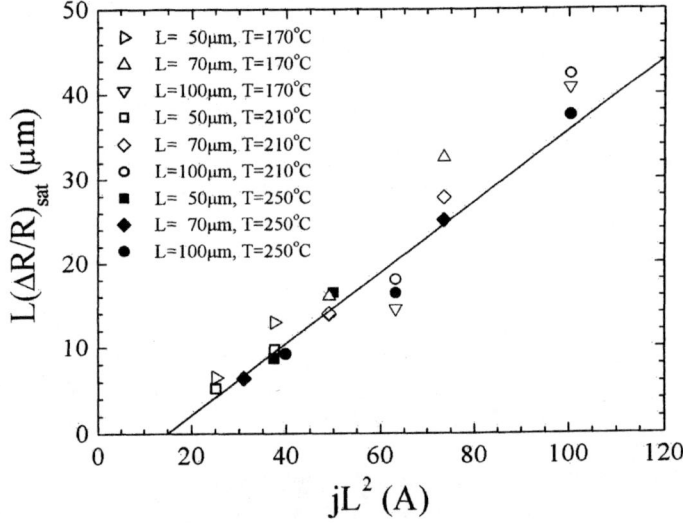

Figure 5. $L(\Delta R/R)_{sat}$ versus jL^2 for the 0.33 μm-wide AlCu samples tested at 170, 210 and 250 °C. The symbols correspond to the average values. The solid line is the least-squares fit based on the 250 °C data.

It should be appreciated that resistance saturation allows higher current densities to be used at operating conditions for short interconnects [16]. Implementation of a design rule (i.e., allowed current density) would depend on a stochastic treatment of $(\Delta R/R)_{sat}$ as opposed to a failure time, t_f (see Figure 4) [16]. A worst case value for $(\Delta R/R)_{sat}$ would correspond to a pre-defined CDF target (i.e., 99%). Figure 6 demonstrates that higher current densities are obtained for shorter lines and/or higher resistance changes at saturation. The curves in Figure 6 are generated from Equation (3) and the regression parameters for the data shown in Figure 3. Compared to a conventional time-to-failure (at a pre-defined criterion) model, a parametric relationship as shown in Figure 6 offers additional flexibility in design.

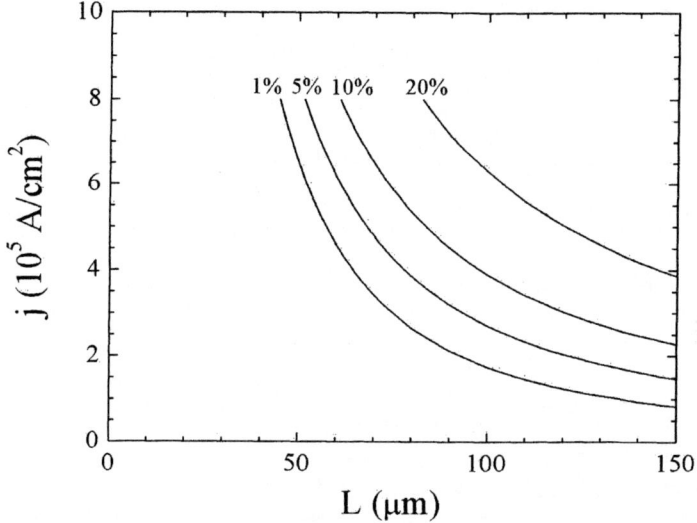

Figure 6. Current density, j, versus stripe length, L, for the AlCu samples at different levels of resistance saturation. The stripe width is 0.33 µm.

2. Wide-Line Electromigration Behavior

The 1.50 µm-wide AlCu samples are tested in order to check for a width dependence for resistance saturation, even though Equation (3) does not explicitly depend on width. Figure 7 shows a plot of $L(\Delta R/R)_{sat}$ versus jL^2 for the 0.33 and 1.50 µm-wide AlCu samples tested at 250 °C. The plot indicates a higher slope for the wide lines. The slope for the narrow and wide lines is found to equal 0.42 and 1.40 µm/A, respectively.

The wide lines also exhibit a larger variation in resistance saturation than the narrow lines. The lognormal σ for $(\Delta R/R)_{sat}$ is typically found to lie between 0.4-0.5 for the 1.50 µm-wide AlCu samples compared to 0.2-0.3 for the 0.33 µm-wide AlCu samples. The fact that the statistical distribution for $(\Delta R/R)_{sat}$ is wider for the 1.50 µm-wide lines as compared to the 0.33 µm-wide lines can be explained in terms of the final void shape in the AlCu lines [15].

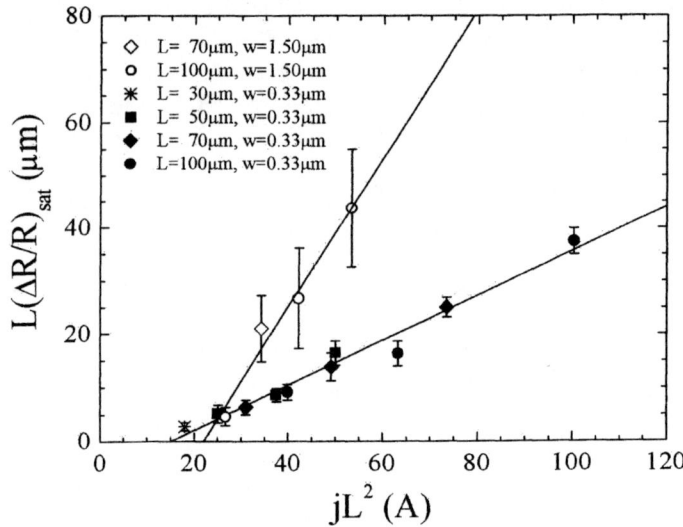

Figure 7. $L(\Delta R/R)_{sat}$ versus jL^2 for the 0.33 and 1.50 μm-wide AlCu samples tested at 250 °C. The symbols correspond to the average values and the error bars represent the 95% confidence interval. The solid lines are the least-squares fits.

3. Effect of B on Resistance Saturation

In Equation (3), B can be considered an effective bulk modulus. In studies of void growth by stress or electromigration in confined interconnects, we are often interested in the number or volume of atoms depleted from a void. As inserted further away from the void into the interconnect line, these atoms provide for a volumetric eigenstrain (analogous to thermal expansion) that gives rise to stresses.

Experiments show that the stresses in Al-based interconnect lines relax towards an equi-axial hydrostatic stress state (i.e., all principal stress components become equal) [17]. When the stresses relax, there is an elastic interaction with the surrounding matrix (refractory layers and passivations) such that the volume of the interconnect metallization changes. In order to calculate the relaxed bulk modulus, it is necessary to use the Eshelby theory of inclusions [18] (which automatically couples the deformation of the interconnect metallization and surounding matrix) or, equivalently, to use finite element modeling (FEM), where the interaction between interconnect and matrix must be taken into account. It is obvious that the composite, relaxed bulk modulus must depend on both the moduli of the metallization and the matrix, as well as the geometrical situation (i.e., aspect ratio of line and the extent of passivation).

It is known that for longer times at typical accelerated testing temperatures the electromigration-induced stress in Al-based interconnect lines becomes closely hydrostatic [19,20]. Obviously, under these conditions, the practical definition of the "relaxed" bulk modulus must be that of the relaxed hydrostatic stress, $\sigma_{hydrostatic}$, divided by the volumetric eigenstrain, ε_{vol} (proportional to the number of atoms extracted from the void) [21]:

$$B_{relaxed} = \sigma_{hydrostatic} / \varepsilon_{vol} \qquad (4)$$

Figure 8 depicts the differently defined bulk moduli as a function of the line aspect ratio as calculated from the Eshelby theory of inclusions for the case of an Al line embedded in an infinite SiO_2-matrix. Since the stresses are non-uniform in lines of rectangular cross section, we estimated their average stresses from elliptic lines (where stresses are uniform) of the same aspect ratio [21]. (See Reference [15] for more details concerning this analysis.) $B_{elastic}$ gives simply the elastic bulk modulus of pure Al, which of course does not depend on the aspect ratio, while $B_{composite}$(unrelaxed) is the composite modulus accounting for partitioning of stresses between the Al metallization and matrix before shear stress relaxation. Finally, $B_{composite}$(relaxed) shows the result after complete shear stress relaxation to an equi-axial hydrostatic stress state.

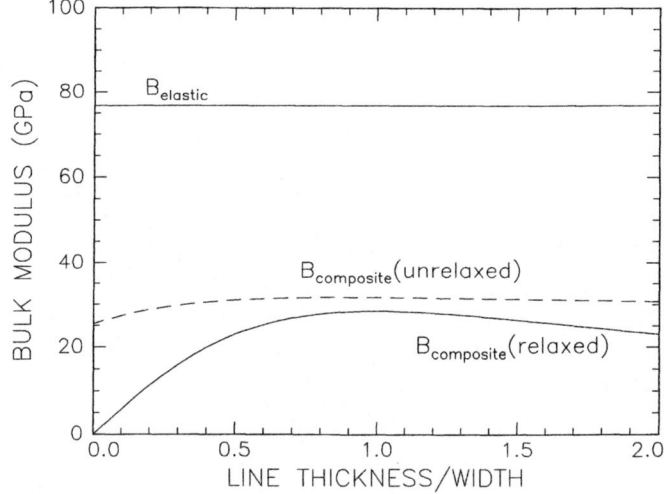

Figure 8. The elastic bulk modulus of pure Al, $B_{elastic}$, and the effective composite bulk moduli before, $B_{composite}$(unrelaxed), and after, $B_{composite}$(relaxed), shear stress relaxation for different aspect ratios of interconnect lines. The composite moduli apply for an Al line embedded in an infinite SiO_2-matrix.

Resistance saturation is strongly dependent on line width, as evidenced by the fact that a plot of $L(\Delta R/R)_{sat}$ versus jL^2 shows a higher slope for the 1.50 µm-wide AlCu samples compared to the 0.33 µm-wide AlCu samples (see Figure 7). The slope for the narrow and wide lines is found to equal 0.42 and 1.40 µm/A, respectively (a factor of 3.3 difference). According to Equation (3), the slope of $L(\Delta R/R)_{sat}$ versus jL^2 is inversely proportional to the effective bulk modulus, B. The difference in slope for the narrow and wide lines is qualitatively explained in terms of a "relaxed" composite bulk modulus, $B_{composite}$(relaxed), which depends strongly on the interconnect aspect ratio. The Al thickness (after $TiAl_3$ formation) is approximately equal to 0.40 μm. According to Figure 8, $B_{composite}$(relaxed) is equal to 27.9 GPa for the 0.33 μm-wide AlCu lines (aspect ratio = 0.40/0.33 = 1.21) and 14.7 GPa for the 1.50 μm-wide AlCu lines (aspect ratio = 0.40/1.50 = 0.27). While this does not completely account for the factor of 3.3 difference for the data shown in Figure 7, it at least shows why B (and therefore line width) should play a significant role in resistance saturation.

B. Short-Length Effects in Cu Interconnects

In this section, short-length effects in Cu interconnects are described. (See Reference [10] for a more detailed discussion.) Figure 9 shows a plot of the absolute resistance increase versus time for a few of the Cu samples tested at the same conditions, where the short-length effect in Cu interconnects is clearly demonstrated. The 50 and 70 μm-long Cu samples show massive electromigration damage as evidenced by the large resistance increase, while, on the other hand, the 30 μm-long Cu samples show only a slight resistance shift during the entire experiment. For the samples with massive degradation, the resistance change can generally be separated into three regimes: (1) an incubation period when the resistance remained unchanged at the early stage of electromigration, (2) a sudden resistance jump where the resistance increased abruptly by more than 10 Ω, and (3) a steady resistance increase due to the gradual electromigration damage.

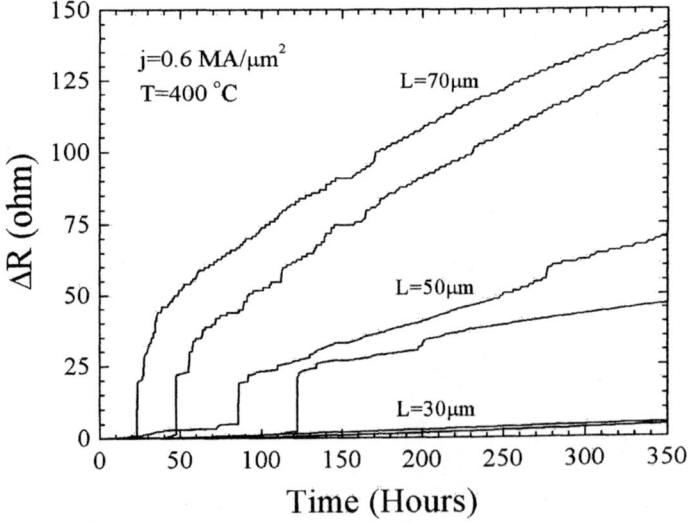

Figure 9. Resistance shift (ohm) versus time for different line lengths of the Cu samples. The test conditions are 0.6 MA/cm² and 400 °C. Only two samples per line length are shown.

During the initial incubation period, t_{inc}, a void nucleates and grows in the M1 interconnect below the V2 via (see Figure 1(b)). The resistance remains unchanged until the electromigration-induced void grows beyond the front edge of the V2 via [12]. The abrupt resistance increase, which signals the end of the incubation period, can be attributed to excessive Joule heating caused by the more resistive M1 liner shunting current at the voided region. Note that t_{inc} corresponds to the time for depleting L_{inc}=0.4 μm of material from the cathode end of the interconnects (see Figure 1(b)). Also, the void sizes, L_{inc}, formed during the incubation period are identical, regardless of the line length.

In general, as observed in Figure 9, the incubation times are shorter for longer interconnects. A simple argument can be used to derive the following equation, which links the incubation time, current density and stripe length for the case of diffusion blocking boundaries at both ends of the interconnect [10]:

$$\frac{L}{t_{inc}} = C_1 jL - D_1 \tag{5}$$

where

$$C_1 \propto n \frac{D}{kT} \rho e Z^* \tag{6}$$

and

$$D_1 \propto n \frac{D}{kT} \Delta\sigma_{max} \Omega \tag{7}$$

In the above equations, $\Delta\sigma_{max}$ is the maximum stress difference between the anode and cathode ends generated by electromigration, and C_1 and D_1 are constants at a given temperature. The derivation of Equation (5), which is based on the Blech electromigration model (see Equation (2)), makes the following three assumptions. First, the short-length effect in Cu is caused by a mechanical stress gradient. Second, a maximum stress exists that limits the stress gradient attainable in Cu. Third, the time to develop the steady-state stress gradient is very short compared to the incubation time. This last assumption allows the net electromigration flux to be treated as inversely proportional to the incubation time ($J_{eff} \propto 1/t_{inc}$). All three assumptions have been experimentally justified for pure Al by x-ray microdiffraction [22].

According to Equation (5), a plot of L/t_{inc} versus jL should yield a straight line at a given temperature. This is demonstrated in Figure 10 for the Cu samples tested at 295, 350 and 400 °C. The figure shows good agreement between the least-squares fits, given by the solid lines, and the actual data.

As suggested by the error bars in Figure 10, there is a statistical distribution for t_{inc}. The statistical nature of t_{inc} is illustrated by the lognormal CDF plot shown in Figure 11. The lognormal σ is found to increase with decreasing current density and/or decreasing stripe length.

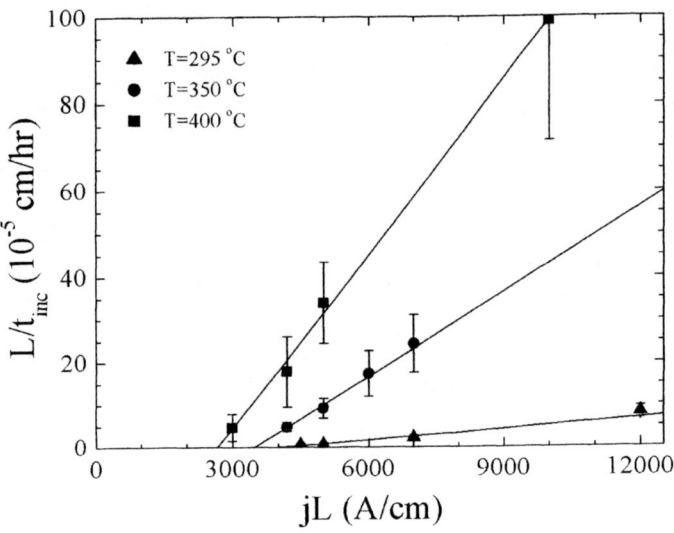

Figure 10. L/t_{inc} versus jL for the Cu samples tested at 295, 350 and 400 °C under various combinations of current density and stripe length. The symbols correspond to the median values and the error bars represent the 95% confidence interval. The solid lines are the least-squares fits.

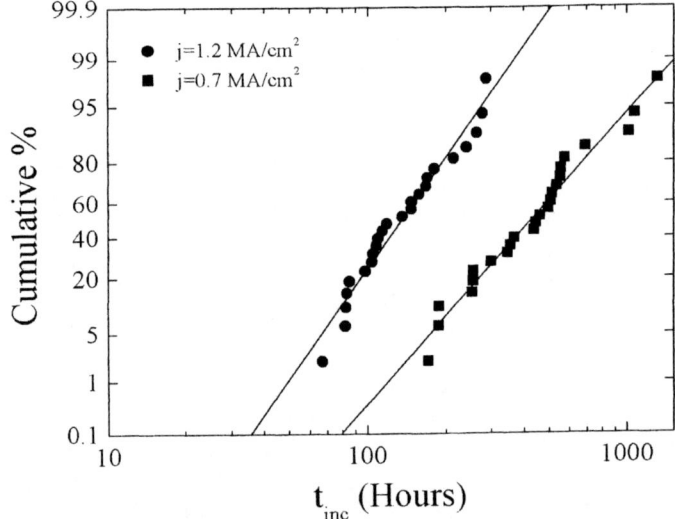

Figure 11. Lognormal CDF plot for the Cu samples in terms of t_{inc} at 295 °C and different current densities. The stripe length is 100 μm.

In Figure 10, the intercepts with the x axis determined from the linear fits are the jL values that yield infinite incubation time. Therefore, these intercepts determine the threshold products, $(jL)_{th}$, below which there is no electromigration-induced resistance degradation at their corresponding temperatures. The $(jL)_{th}$ values determined from Figure 10 are 3940 A/cm at 295 °C, 3470 A/cm at 350 °C and 2660 A/cm at 400 °C. To justify these $(jL)_{th}$ values, several experiments were conducted with testing conditions below the $(jL)_{th}$ values at their respective temperatures (i.e., 30 μm-long lines tested at 1.0 MA/cm^2 and 295 °C). As expected, there is no major resistance degradation in a practically long period of time (i.e., ≈ 2000 hours at 295 °C).

The $(jL)_{th}$ values for Cu lines are significantly higher than the values for Al lines determined previously [1]. This can be attributed to the higher mechanical strength of Cu thin films rigidly confined by surrounding Si_3N_4 and SiO_2 dielectrics [23,24]. Also, the $(jL)_{th}$ value of Cu appears to decrease with increasing temperature, which can be partly attributed to the temperature-dependent mechanical properties of the system [1].

The short-length effect in Cu lines allows higher current densities to be used for short interconnects. Implementation of a design rule (i.e., allowed current density) would depend on a stochastic treatment of t_{inc} (see Figure 11). A worst case value for t_{inc} would correspond to a pre-defined CDF target (i.e., 1%). Figure 12 demonstrates that higher threshold current densities, j_{th}, are obtained for shorter lines. The curves in Figure 12 are generated from the $(jL)_{th}$ values for the data shown in Figure 10.

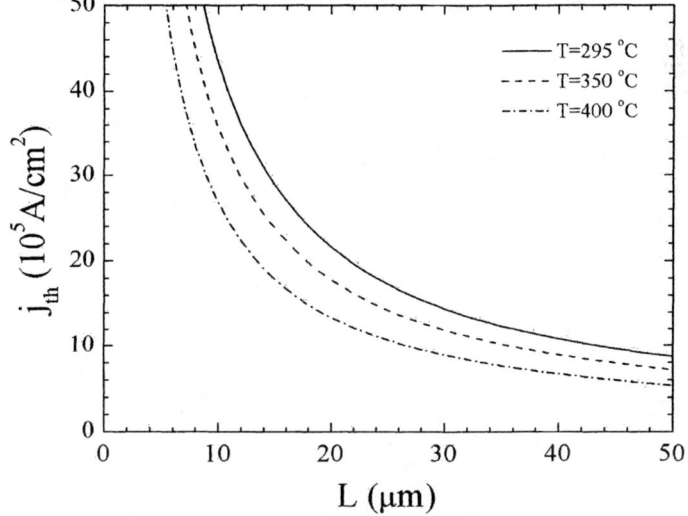

Figure 12. Threshold current density, j_{th}, versus stripe length, L, for the Cu samples at different temperatures.

Finally, it should be explained why different models are used to describe short-length effects in AlCu and Cu metallization systems. As seen in Figure 9, the resistance change with time does not saturate for the Cu samples, which prevents a resistance saturation model from being applied. The fact that the resistance continues to increase after long times may be a result of extrusion formation, which has been observed in the Cu samples after stress. In addition, the third assumption listed above for the incubation time model may not be applicable to alloyed metallizations like AlCu since the stress evolution is more complicated due to the preferential diffusion of solute atoms before the development of the steady-state stress gradient.

C. Electromigration Resistant Power Grid

In this section, an electromigration resistant power grid is described as a practical application of the short-length effect. (See Reference [25] for a more detailed discussion.) Given the models of the short-length effect described previously for AlCu and Cu metallization systems, the question still remains as to whether there is a way to systematically take advantage of the short-length effect. One area where this can be addressed is with regard to the distribution of power on an IC chip. A class of meshes has been discovered that allows the construction of a power distribution grid for IC chips with nearly arbitrary electromigration resistance. The meshes can be viewed as an extension of common grid layouts in which gaps are systematically introduced. Figure 13 illustrates a common grid layout while Figure 14 illustrates the layout of an electromigration resistant power grid. Power supply and ground interconnects are referred to as V_{dd} and V_{ss}, respectively. The grid in Figure 14 is formed from orthogonal segments on adjacent interconnect levels such that the projection of both levels is continuous. In an actual IC design, line segment portions lying outside the chip (or within macro) boundaries are deleted. The length of each gap must be longer than the lithography limited space and must be spaced such that the resulting interconnect line segment has improved electromigration resistance. As a result of systematic phase shifts within the layout, the overall grid has multiple redundant paths connecting any two points. The phase shift is seen in Figure 14, where V_{ss} segment #2 is shifted along the y axis with respect to V_{ss} segment #1 by a fraction of the segment length. Such a grid has the additional benefit of improved local wireability (i.e., signal lines can run through the gaps).

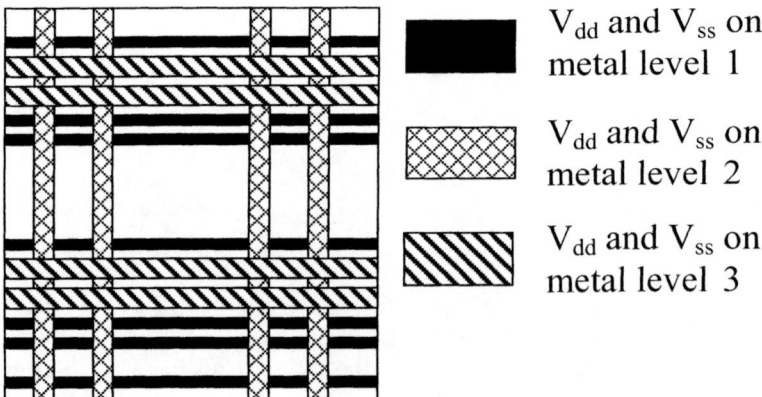

V_{dd} and V_{ss} on metal level 1

V_{dd} and V_{ss} on metal level 2

V_{dd} and V_{ss} on metal level 3

Figure 13. Top view of a continuous power grid formed from orthogonal sets of lines on three different wiring levels.

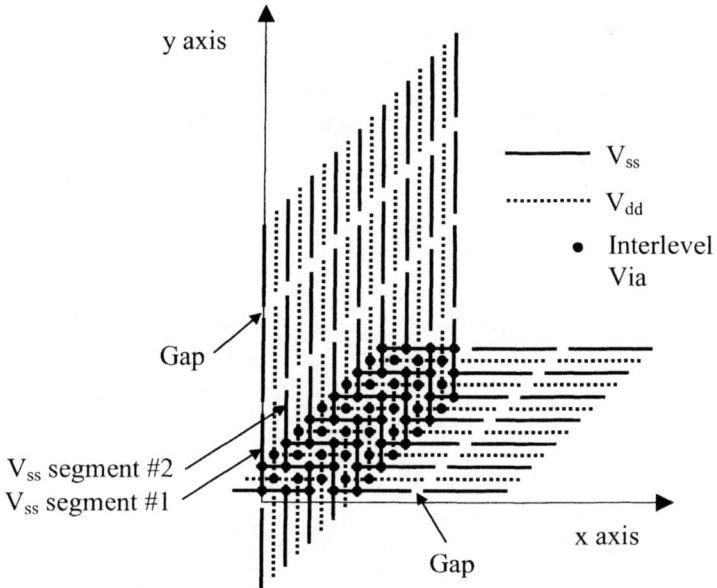

Figure 14. Top view of a partially constructed electromigration resistant power grid with V_{ss} and V_{dd} segments on 1st (along x axis) and 2nd (along y axis) metal levels. Interlevel vias (studs) connect segments from each metal level.

Since the electromigration models described previously for AlCu and Cu interconnects are based on first principles, the grid layout is extendible to many multilevel integration schemes, including low-k dielectrics if other failure mechanisms do not occur. In addition, since the class of meshes is general, one can vary the pitch of V_{dd} and V_{ss}, vary the gap size to enhance wireability, and trade off segment length and redundancy within the network.

IV. CONCLUSIONS

The electromigration short-length effect has been demonstrated for AlCu and Cu interconnects in SiO_2 dielectrics. Simple models based on first principles are shown to accurately relate resistance saturation (AlCu case) and the incubation time (Cu case) to the current density and stripe length. For AlCu, resistance saturation is relatively independent of temperature in the range 170-250 °C, but is strongly dependent on line width. The width effect is qualitatively explained by a relaxed bulk modulus that depends on the aspect ratio of the interconnect lines. For Cu, the incubation time model allows one to estimate the threshold product, $(jL)_{th}$, at a certain temperature. The resulting $(jL)_{th}$ value appears to be temperature dependent, decreasing with increasing temperature in the range 295-400 °C. For both AlCu and Cu, the short-length effect allows higher current densities to be used at operating conditions for short interconnects. An electromigration resistant power grid may be constructed as a way to systematically take advantage of the short-length effect on an IC chip.

V. ACKNOWLEDGEMENTS

The authors would like to thank R. Raviart, W. Malkus, N. Lowitz and R. Edwards for sample preparation, C.-K. Hu for valuable discussions, and the process community of IBM Microelectronics Division for processing the samples used in this study.

REFERENCES

1. I. A. Blech, J. Appl. Phys. **47**, 1203 (1976).
2. E. Kinsbron, I. A. Blech, Y. Komem, Thin Solid Films **46**, 139 (1977).
3. H.-U. Schreiber, Solid-State Electronics **28**, 617 (1985).
4. M. A. Korhonen, P. Borgesen, D. D. Brown, C.-Y. Li, J. Appl. Phys. **74**, 4995 (1993).
5. M. A. Korhonen, P. Borgesen, D. D. Brown, C.-Y. Li, T. D. Sullivan, P. A. Totta, Second International Workshop on Stress-Induced Phenomena in Metallization (AIP, New York, 1994), Conf. Proc. 305, p. 15.
6. J. J. Clement, J. R. Lloyd, C. V. Thompson, Mat. Res. Soc. Symp. Proc. **391**, 423 (1995).
7. R. G. Filippi, G. A. Biery, R. A. Wachnik, J. Appl. Phys. **78**, 3756 (1995).
8. R. G. Filippi, R. A. Wachnik, H. Aochi, J. R. Lloyd, M. A. Korhonen, Appl. Phys. Lett. **69**, 2350 (1996).
9. E. T. Ogawa, A. J. Bierwag, K.-D. Lee, H. Matsuhashi, P. R. Justison, A. N. Ramamurthi, P. S. Ho, V. A. Blaschke, D. Griffiths, A. Nelsen, M. Breen, R. H. Havemann, Appl. Phys. Lett. **78**, 2652 (2001).
10. P.-C. Wang, R. G. Filippi, Appl. Phys. Lett. **78**, 3598 (2001).
11. H. B. Huntington, A. R. Grone, J. Phys. Chem. Solids **20**, 76 (1961).
12. C-K. Hu, M. B. Small, P. S. Ho, J. Appl. Phys. **74**, 969 (1993).
13. R. W. Bower, Appl. Phys. Lett. **23**, 99 (1973).
14. E. G. Colgan, Mater. Sci. Rep. **5**, 1 (1990).
15. R. G. Filippi, R. A. Wachnik, C-P Eng, D. Chidambarrao, P.-C. Wang, J. F. White, M. A. Korhonen, T. M. Shaw, R. Rosenberg, T. D. Sullivan, submitted to J. Appl. Phys.
16. R. A. Wachnik, R. G. Filippi, T. M. Shaw, P. C. Lin, 2000 Symposium on VLSI Technology, Digest of Technical Papers, p. 220.
17. M. A. Korhonen, R. D. Black, C.-Y. Li, J. Appl. Phys. **69**, 1748 (1991).
18. J. D. Eshelby, Proc. Roy. Soc. A **241**, 376 (1957).
19. R. S. Hemmert, M. Costa, 29th Annual Proceedings of Reliability Physics 1991, Las Vegas (IEEE, New York, 1991), p. 64.
20. Q. Ma, S. Chiras, D. R. Clarke, Z. Suo, J. Appl. Phys. **78**, 1614 (1995).
21. M. A. Korhonen, P. Borgesen, K. N. Tu, C.-Y. Li, J. Appl. Phys. **73**, 3790 (1993).
22. P.-C. Wang, G. S. Cargill III, I. C. Noyan, C.-K. Hu, Appl. Phys. Lett. **72**, 1296 (1998).
23. R. P. Vinci, E. M. Zeiliski, J. C. Bravman, Thin Solid Films **262**, 142 (1995).
24. Y-L. Shen, S. Suresh, M. Y. He, A. Bagchi, O. Kienzle, M. Ruhle, A. G. Evans, J. Mater. Res. **13**, 1928 (1998).
25. R. G. Filippi, P. C. Lin, T. M. Shaw, R. A. Wachnik, U.S. Patent 6,202,191, March 2001.

A High Reliability Copper
Dual-Damascene Interconnection
With Direct-Contact Via Structure

Kazuyoshi Ueno, Mieko Suzuki, Akira Matsumoto, Koichi Motoyama,
Noriaki Oda, Hidenobu Miyamoto, and Shuichi Saito

ULSI Device Development Division, NEC Electron Devices, NEC Corporation

1120 Shimokuzawa, Sagamihara, 229-1198, Japan

Abstract. A new via technology for improving electromigration (EM) reliability of copper (Cu) dual-damascene (DD) interconnection has been developed. Early failure mode of a conventional Cu DD structure is found as void formation at the via-bottom interface, where flux divergence of Cu ions is large due to diffusion barrier-layer. In order to avoid the early failures, direct-contact via (DCV) technology whose concept is "barrier-free" at the via-bottom has been developed. The early failure mode is eliminated by the DCV technology and lower via resistance is obtained.

INTRODUCTION

Recently, copper (Cu) interconnection has been used in LSI chips due to its low resistance and high electromigration (EM) reliability. High EM reliability enhances not only reliability but also speed performance of LSI chips, since higher current density can be used for the circuit design. In order to guarantee high reliability in real LSIs, it is required to have not only a long median time to failure, T_{50}(lifetime of 50% failure), but also a long lifetime of early failure such as $T_{0.1}$ (lifetime of 0.1% failure).

Single-damascene (SD) and dual-damascene (DD) processes have been used for the fabrication of Cu interconnects. The DD process has an advantage of lower cost

CP612, *Stress-Induced Phenomena in Metallization:* Sixth Int'l. Workshop, edited by S. P. Baker et al.

because number of process step reduces. However, bimodal distributions of lifetime were observed for conventional Cu DD interconnects, and the early failures degrade $T_{0.1}$ [1-4]. In order to obtain a tight lifetime distribution and a long $T_{0.1}$, it is important to reduce the number of failure modes, especially it is important to avoid the early failure modes. In this paper, failure modes for a conventional DD interconnect are analyzed. Based on the analysis, the failure mechanism is investigated and a new via-structure and its fabrication process named direct-contact via (DCV) technology is proposed for eliminating the early failure mode. The electrical and the reliability performance of the DCV structure is described, and the EM failure mechanism for the new structure is discussed.

EXPERIMENTAL PROCEDURE

Two-level Cu DD interconnection with a conventional via structure and the new DCV structure was fabricated. Figure 1 shows the schematic cross-sections of each structure. In the conventional structure, a barrier-layer exists at the interface between the Cu via-plug and the Cu wiring (M1) underneath the via. The barrier-layer increases the via-resistance and potentially degrades the EM reliability, because Cu-flow is blocked by the barrier-layer, leading to a large Cu-flux divergence and void formation at the via-bottom. As will be described later in this paper, the voids at the via-bottom lead to early failures in a conventional structure. In order to eliminate the early failures, it is ideal to remove the barrier-layer, because Cu-flow becomes continuous. In addition, via-resistance reduces by removing the barrier-layer as we had reported previously [5].

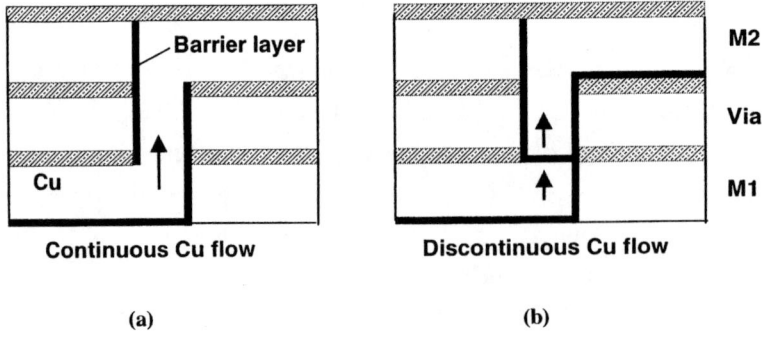

Continuous Cu flow	Discontinuous Cu flow
(a)	(b)

FIGURE 1. Schematic cross-sections of (a) the direct-contact via (DCV) structure and (b) the conventional via structure

As the first step, two-level Cu DD interconnection with the conventional via structure was fabricated and the failure modes were analyzed. The line-width and the via- diameter of the fabricated interconnection are 0.28µm, respectively, and they are contacted without border or extension. SiO_2 and SiN are used as interlayer dielectrics and cap-layer over Cu, respectively. The structures were fabricated with a conventional processing sequence. As will be shown later, via-cleaning process was optimized to improve the EM reliability. The test-structures were fabricated with and without the optimized cleaning to show that the via-bottom interface has a critical effect on the reliability of the Cu DD interconnection.

A new process sequence is proposed for fabricating the Cu DD interconnection with the DCV structure as shown in Fig. 2. A dielectric barrier-layer is deposited and is etched back to form diffusion barrier at the sidewalls after forming trenches and holes. The sidewall barrier can be formed by a plasma treatment as reported before [6]. Then, an ultra-thin adhesion layer is deposited. Since the sidewalls are already covered with the dielectric barrier, the thickness of the adhesion-layer can be minimized. Cu seed-layer is deposited after the adhesion-layer deposition without braking vacuum, leading a damage-free interface between the adhesion and the Cu seed. The trenches and the holes are filled by electroplating, followed by CMP to complete the structure.

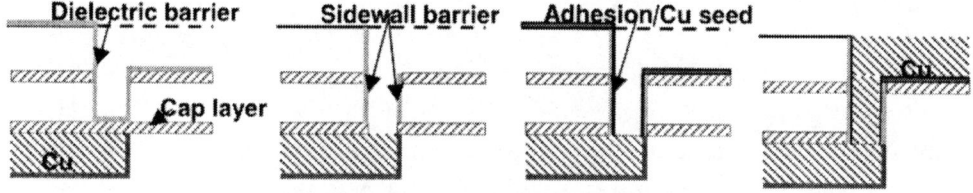

FIGURE 2. New process sequence for the DCV structure.

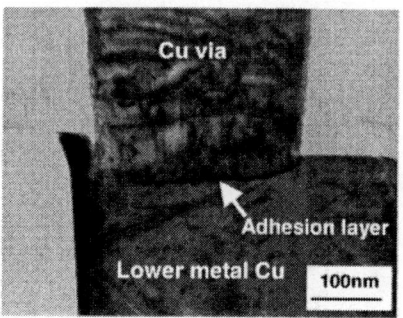

FIGURE 3. Cross-sectional TEM of the fabricated DCV.

In the previous paper, a DCV structure was fabricated by etching-back barrier-layer at via-bottom [5]. The reason why we use the new process sequence is as follows. The etch-back process potentially degrades the reliability of fabricated interconnection, because adhesion between the plasma-exposed barrier-layer and the Cu seed-layer may be degraded during the etch-back, and it is considered that the Cu/barrier interface can be a fast diffusion path for EM [7]. In order to avoid such degradation, the new process sequence is proposed in this study. In this process sequence, an ultra-thin adhesion-layer and Cu seed-layer are successively deposited without braking vacuum. Therefore, a clean interface can be formed between the adhesion-layer and the Cu seed-layer. Fig. 3 shows cross-sectional TEM of the via-bottom for the DCV structure. The dark line at the via-bottom in Fig. 3 corresponds to the adhesion-layer. The thickness of the layer is in several nano-meter range and invisible at some areas. Since thickness of the adhesion-layer is so thin at the via-bottom, Cu can punch through the layer [1, 2], and a barrier-less structure can be fabricated effectively. The new process sequence features a high compatibility with conventional DD processes and is suitable for using a Cu-electroplating process, since the process for the successive deposition of the ultra-thin adhesion layer and the Cu seed-layer is common with conventional barrier/seed deposition.

As the electrical characteristics, via-resistance was measured using 100,000 via-chain test-structures. Leakage current between the two level lines which are connected with via-plugs were measured for 0.28μm-space. EM lifetime was evaluated by the test-structure as depicted in Fig. 4. The current density was 2mA/via, which corresponds to 2.5MA/cm^2, and the typical test temperature was 300°C. Failure criteria is 2% resistance increase and the stress current is stopped at 10% increase. Activation energies for EM (E_a) were evaluated from the lifetime dependencies on the test temperature between 250°C and 300°C. EM failure modes were analyzed by FIB, SEM, and TEM.

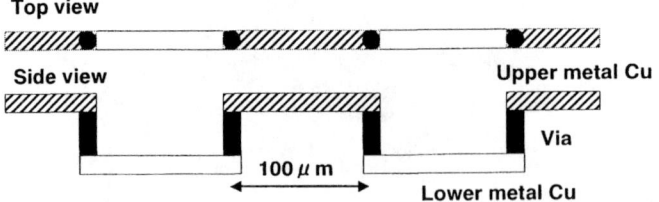

FIGURE 4. Schematic test-structure used for the EM tests.

RESULTS AND DISCUSSION

EM Failure Modes of Conventional DD Structures

Figure 5 shows the cumulative EM failure data for the conventional structures with and without using the optimized via-cleaning. Insufficient via-cleaning degraded T_{50} and σ, as shown in Fig. 5. This result indicates that the interface at the via-bottom has a critical effect on the lifetime. The lifetime was improved by the optimized cleaning, but the lifetime distribution has S-shape like a bimodal distribution [8].

Failure modes of the samples with different lifetimes are compared as shown in Fig. 6, which shows the SEM cross-sections of the failure sites for the different lifetime samples. In the short lifetime sample, voids are observed at the via-bottom in the electron flow direction from M1 to M2 and/or that from M2 to M1 as shown in Fig.6(a). In the medium lifetime sample, voids are located at the via-bottom in the direction of electron flow from M2 to M1 and at the top corner of M2, indicating that there are plural sites for the void growth. The void growth at the via-bottom is natural because Cu-flow is blocked by the barrier-layer and the flux divergence of Cu becomes large. The void at the top corner is considered due to vacancy diffusion induced by density gradient of vacancies and/or stress gradient near the corner, since electron-flow is considered to be lower than the other areas, and the electron-flow cannot be the main driving force for the Cu atomic flow. In the long lifetime sample, voids are not observed at the via-bottom. Voids are located at the top corner and in the via-plug. Hu

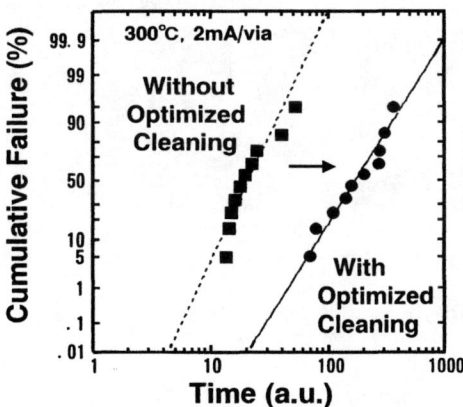

FIGURE 5. Cumulative EM failure data for the conventional via with and without the optimized via-cleaning.

53

reported similar failure mode in their long lifetime samples and they concluded that it is due to the effect of Cu atom punch through the barrier-layer at the via-bottom [1, 2]. Our results accord with their results.

EM lifetime is determined by the time when critical void at one site reach critical size. Since mean drift velocity of Cu atoms is uniquely determined by its dominant diffusion mechanism in the same test structure and process, it is considered that the total volume of vacancies is almost the same in the samples. Therefore, the difference in the lifetime depends on the position where one void reaches its critical size at first. Our observation indicates that voids at the via-bottom most easily reach its critical size and leads to early failures. The voids at the other sites such as the top corner may delay the time that the via-bottom void reaches the critical size, leading to the longer lifetime, as in the case of medium lifetime samples. In the long time samples, it is expected that vacancies at the down-stream side of the barrier-layer are compensated by Cu atoms that punch through the barrier-layer, and it avoids void formation at the via-bottom. The void at another site such as top corner reaches its critical size that is larger than that of via-bottom void in the long lifetime samples as shown in Fig. 6 (c).

Based on the above discussion, it is concluded that it is effective to remove the barrier layer at the via-bottom to eliminate the early failures.

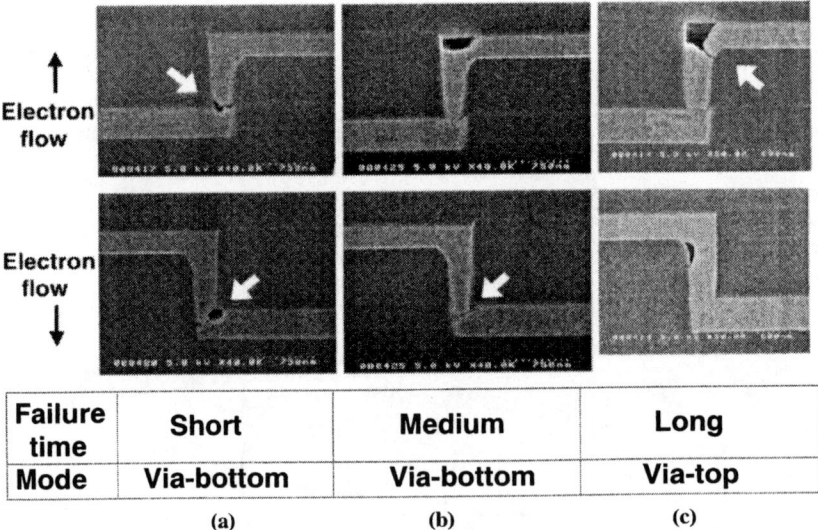

Failure time	Short	Medium	Long
Mode	Via-bottom	Via-bottom	Via-top
	(a)	(b)	(c)

FIGURE 6. Cross-sectional SEM photographs of the EM failure sites with different failure times.

Electrical Characteristics of Direct-Contact Via Structures

Good uniformity within an 8 inch wafer is obtained for the DCV structure as shown in Fig. 7, showing that the resistance distribution for the DCV structures is as tight as the conventional structures. It is considered that the developed process sequence is stable similarly to the conventional process sequence. The via-resistance of the DCV structure is 20 % lower than that of the conventional structure as expected. This is because the highly resistive barrier-layer is removed. It is concluded that the DCV structures have the advantage of lower via-resistance.

The leakage current of the DCV structures is as low as the conventional structures as shown in the dielectric breakdown characteristics (Fig. 8). The leakage current increases in proportion to the electric field up to 3MV/cm for both structures, indicating no degradation of the leakage current in using the DCV structures. The leakage current is tightly distributed within a wafer as good as the conventional structure. Further studies for the leakage characteristics such as time dependent dielectric breakdown (TDDB) are underway.

It is concluded that the DCV structures have better electrical performance than the conventional structures.

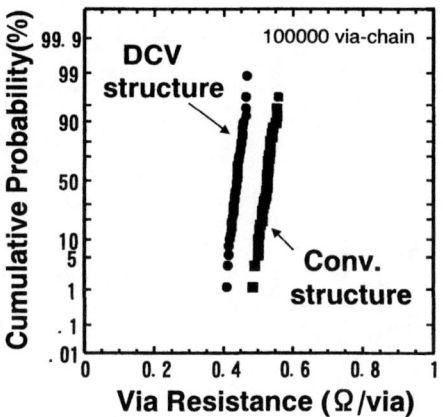

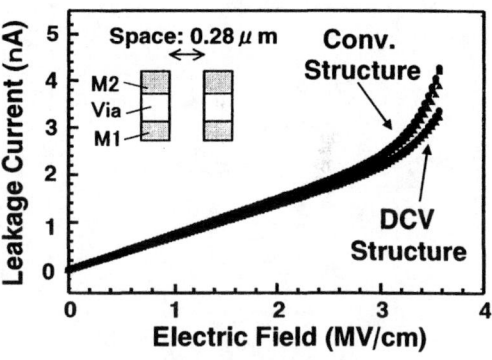

FIGURE 7. Cumulative via resistance for the DCV and the conventional via structures

FIGURE 8. Dielectric breakdown characteristics for the DCV and the conventional via structures.

EM Reliability of Direct-Contact Via Structures

Figure 9 shows a comparison of cumulative EM failure data between the DCV and the conventional structure. It is noted that early failures are eliminated by the DCV structure and the distribution, σ, is drastically reduced by the DCV structure. In this study, 4 via-chain test-structures were used for the EM test, and the lifetimes of 40 via-units are included in each distribution. Within the limit of test sensitivity for early failures, the DCV technology is effective to avoid early failures, although further tests using appropriate test-structures including more number of units, which are more sensitive to detect very early failures [3].

The activation energies (E_a) of the DCV structure and the conventional structure are 1.3eV and 1.1eV, respectively as shown in Fig. 10. The difference of the E_a values is small and the E_a values accord with the reported values for interface diffusion [9]. Since the fabricated 0.28μm interconnects have bamboo-like grain structures, it is considered that the grain-boundary is not the dominant diffusion path. Therefore, it is considered that the fast diffusion path is the Cu interfaces such as SiN/Cu and Cu/barrier-layer or adhesion-layer for the both via structures.

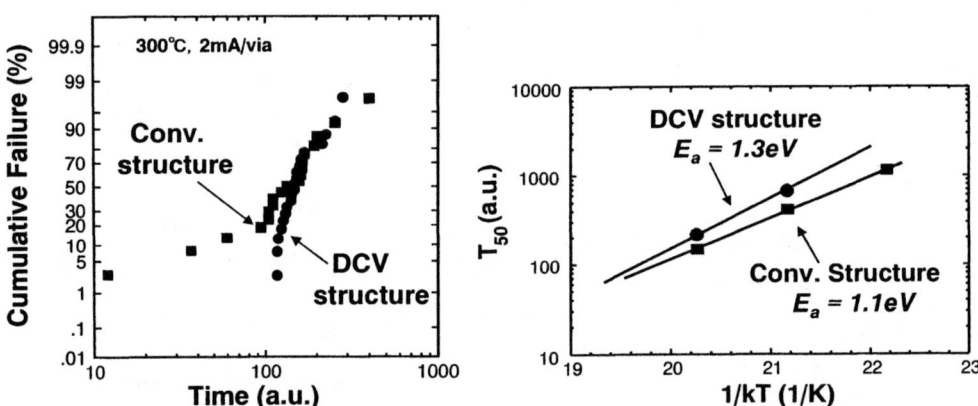

FIGURE 9.   Cumulative EM failure data for the DCV and the conventional via structures

FIGURE 10.   Activation energies for the DCV and the conventional via structures.

Figure 11 shows the typical cross-sectional SEM photographs of the DCV structure after the EM test, showing that void disappears at the via-bottom, although a small void is observed in the top corner of M2. Figure 12(a) is a cross-sectional SEM of a failure site where a critical void is observed. It is noted that the void is located at the upstream side of electron flow from the via-contact area. The void is located at the interface between the SiN cap-layer and Cu (M1). Figure 12(b) shows the mechanism of the EM failure in the DCV structure schematically. It is considered that the SiN/Cu interface is the dominant diffusion path from the activation energy and the void location at the SiN/Cu interface. Since the thickness of the adhesion layer is several nanometer-thick as shown in the TEM cross-section of the DCV structure (Fig. 3), it is considered that the Cu atoms punch through the ultra-thin adhesion layer at the via-bottom as depicted in Fig. 12(b). It makes the DCV structure like a single line without blocking area. In a such system, void nucleation and growth must take place at any points where flux divergence becomes large such as grain-boundaries. The voids grew along the grain boundaries (GB) which contact the SiN/Cu interface as shown in a cross-sectional TEM of a failure site (Fig. 13). Critical size of the void at such site is larger than that for the voids at the via-bottom. It leads to the longer lifetime. Further studies are necessary to determine which kind of grain-boundaries becomes such failure sites.

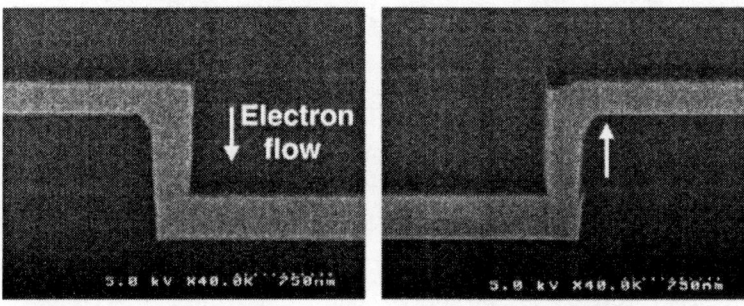

FIGURE 11. Typical cross-sectional SEM photographs for the DCV structures after the EM test.

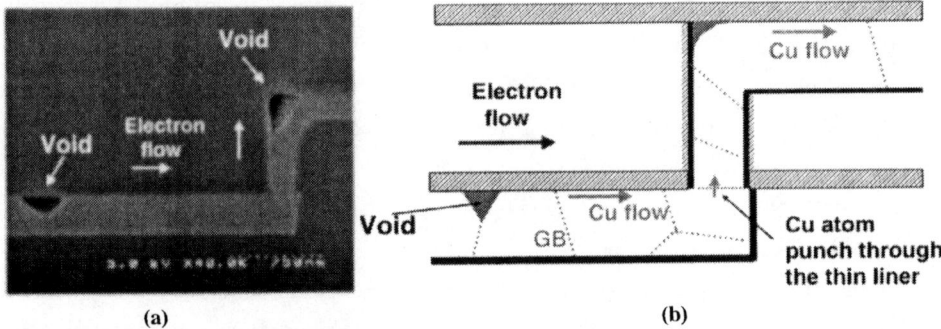

(a)   (b)

FIGURE 12. (a) Cross-sectional SEM of the failure site for the DCV structure, and (b) mechanism for the EM failure mode in the DCV structure.

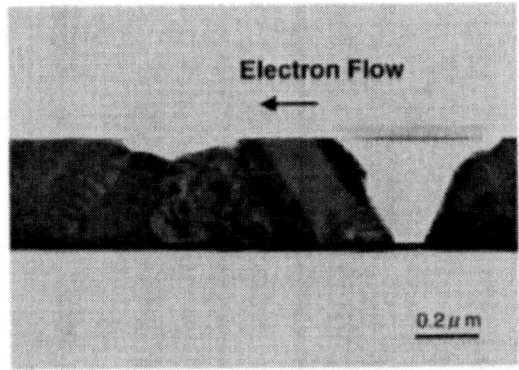

FIGURE 13. Cross-sectional TEM of a failure site for the DCV structure.

SUMMARY

Failure mechanism of the conventional Cu DD structure has been investigated. It is found that voids at the via-bottom lead to the early failures. In order to avoid the via-bottom voids, the DCV structure whose concept is barrier-free at the via-bottom. In order to realize the concept, the new process sequence using the ultra-thin adhesion layer is proposed. Two-level Cu interconnection with the DCV structure was fabricated. The via-resistance of the DCV structures has a tight distribution as good as the conventional structure, and it is 20% lower than that of the conventional structures.

Early failures are eliminated by the DCV structure, and the lifetime distribution is drastically reduced. The failure mechanism of the DCV structure was analyzed. It is considered that Cu punch through at the via-bottom effectively achieves a continuous Cu flow. The diffusion path is considered to be the SiN/Cu interface for the fabricated DCV structures and the conventional via structures. It is concluded that the developed DCV technology is essential to improve performance and reliability of LSIs with Cu DD interconnections.

ACKNOWLEDGEMENTS

The authors would like to thank O. Kudo, K. Nakamura, M. Nakamae, A. Nishizawa, T. Sakai, S. Yokogawa, and H. Tsuda for their useful discussions.

REFERENCES

1. Hu, C.-K., Rosenberg, R. , Klaasen, W. , and Stamper, A. K., "Electromigration Reliability Study of Submicron Cu Interconnection", in *Conf. Proc. XV*, edited by M. E. Gross et. al., MRS, Warrendale, 2000, pp. 691-697.

2. Hu., C.-K, Gignac, L., Malhotra, S. G., and Rosenberg, R., "Mechanisms for Very Long Electromigration Lifetime in Dual-Damascene Cu Interconnections", *Appl. Phys. Letters 78*, 904-906 (2001).

3. Ogawa, E. T., Lee, K.-D., Matsuhashi, H., Ko, K.-S., Justison, P. R., Ramamurthi, A.N.,Bierwag, A.J., Ho, P. S., Blaschke, V. A., and Havemann, R. H., "Statistics of Electromigration Early Failures in Cu/Oxide Dual-Damascene Interconnects", in *Proc. 39th Annual International Reliability Pys. Symp.*, IEEE, Piscataway, 2001, pp. 341-349.

4. Tsai, M. H., Tsai, W. J., Shue, S. L., Yu, C. H., and Liang, M. S., "Reliability of Dual Damascene Cu Metallization", in *Proc. 2000 International Interconnect Technology Conf .*, IEEE, Piscataway, 2000, pp. 214-216.

5. Tsuchiya , Y., Ueno, K., Donnelly, V. M., Kikkawa, T., Hayashi, Y., Kobayashi, A., and Sekiguchi, A., "Ultra-Low Resistance Direct Contact Cu Via Technology Using In-situ Chemical Vapor Cleaning", in *1997 Symp.VLSI Tech. Digest of Tech. Papers* , JSAP/IEEE, Tokyo/Piscataway, 1997, pp. 59-60.

6. Mikagi, K., Ishikawa, H., Usami, T., Suzuki, M., Inoue, K., Oda, N., Chikaki, S., Sakai, I., and Kikkawa, T., "Barrier Metal Free Copper Damascene Interconnection Technology Using

Atmospheric Copper Reflow and Nitrogen Doping in SiOF Film", in *1996 IEDM Tech. Digest,.,*IEEE, Piscataway, 1996, pp. 365-368.

7. Hu., C.-K, Rosenberg, R., and Lee, K. Y., "Electromigration path in Cu thin-film lines", *Appl. Phys. Letters 74*, 2945-2947 (1999).

8. Fischer, A. H., Abel, A., Lepper, M., Zitzelsberger, A. E., and von Glasow, A., "Experimental Data and Statistical Models for Bimodal EM Failures", in *Proc. 38th Annual International Reliability Phys. Symp.*, IEEE, Piscataway, 2000, pp.359-363.

9. Lloyd, J. R., Clemens, J., Snede, R., "Copper Metallization Reliability", *Microelectronics Reliability 39*, 1595-1602 (1999).

Statistical Study of Electromigration Early Failures in Dual-damascene Cu/oxide Interconnects

Ki-Don Lee, Ennis T. Ogawa, Hideki Matsuhashi, and Paul S. Ho

Interconnect and Packaging Laboratory, Microelectronics Research Center, PRC/MER, Mail Code R8650, The University of Texas at Austin, Austin, TX 78712-1000

ABSTRACT. Electromigration (EM) tests have been performed to determine early failure statistics in submicron dual-damascene Cu/oxide interconnects. Monte Carlo simulation, based on the "weakest-link" model, was developed to characterize and determine the failure modes from the cumulative distribution function (CDF) of multi-link test structures. The existence of two district failure modes (weak and strong) in our test structures was identified, where the weak mode is responsible for early failures. The weak mode is found to be void formation within the dual-damascene via, while the strong mode is associated with voiding that occurs in the dual-damascene trench. The weak mode (early failure) activation energy is found to be 1.0 ± 0.05 eV and seems consistent with void formation that is controlled by interface diffusion between the Cu metal and its Ta diffusion barrier.

Introduction

Electromigration is the phenomenon of diffusion-controlled mass transport in an interconnect, and its driving force comes from the scattering of the migrating ion with the charge carriers of electrons. Since M. Geradin first observed this phenomenon [1], it has been recognized as one of the most important failure mechanisms for on-chip interconnects. Due to continuing areal reduction of ICs, interconnects are subject to increasing current densities, which makes interconnects more vulnerable to electromigration. Therefore, implementing Cu interconnects was suggested as a replacement of Al(Cu) interconnects, because Cu has as higher melting temperature (1083°C) such that diffusion processes and electromigration are slower at a given temperature than they are for Al (Tm=660 $^{\circ}$C). Also, it helps to reduce the RC delay since Cu has the second lowest resistivity (1.67 $\mu\Omega$-cm in bulk).

In spite of those advantages, the industry hesitated to transfer from Al to Cu because Cu has several problems. Its high diffusivity into Si and difficulties in process were the tough obstacles to overcome. Recently, electromigration in Cu metallization have been investigated [2,3,4] and industry is proceeding to mass-produce advanced microprocessors having Cu dual-damascene interconnects. New techniques are required to process interconnects, including electroplating, CMP (Chemical Mechanical Polishing) and diffusion barriers. The increase in processing complexities makes it more difficult to fabricate a reliable interconnect.

Early failure becomes a crucial reliability issue in microelectronics. The lifetime of an advanced microprocessor incorporating a large number of devices and interconnects generally depends on the lifetime of the weakest component. Accordingly, the average interconnect lifetime is not as relevant as that of the weakest one. If there are intrinsic

CP612, *Stress-Induced Phenomena in Metallization:* Sixth Int'l. Workshop, edited by S. P. Baker et al.
© 2002 American Institute of Physics 0-7354-0058-X/02/$19.00

mechanism responsible for early failures, it will be important to understand the mechanism in order to improve interconnect reliability. The objective of this paper is to understand the nature and characteristics of the early failures in Cu damascene interconnects.

Cu/oxide Dual-damascene Test Structure

The samples were prepared at International Sematech using 200 mm wafers and consist of two-level interconnect structures based on a 250 Å thick Ta barrier/ 1000 Å thick low temperature PVD seed Cu/ 8000 Å thick electroplated (EP) Cu stack. Above the upper metal level (M2), there is a capping layer of 1000 Å SiN_x. Heat treatment consists of a short excursion up to roughly 400 °C during nitride passivation, a 30 min. anneal at 325 °C in forming gas after wafer processing, and a 35 min. cure at 330 °C for die attach. Consequently, the interconnects show "near bamboo" microstructure.

Fig. 1 shows the basic element of the dual-damascene structure. In this design, electromigration damage was observed at either the M2 or the via levels since they were designed to be more vulnerable to electromigration than the M1 level. M1 has broader cross section and will suffer lower current density compared with M2. Fig 1 shows two possible void growth sites. One is at cathode end of M2, and the other is at the cathode via bottom.

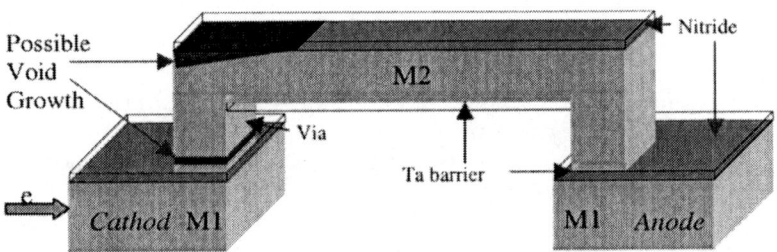

Figure 1. Dual damascene structure and possible void growth sites. The barrier covering the side wall of Cu M1 and M2 are intentionally omitted for viewing clarity.

Statistical Analysis for Early Failure Study

For this study, early failure (EF) test structures have been designed and tested. EF test structures are basically a collection of serially connected Cu/Oxide dual-damascene interconnects where N, the number of serial elements, is the statistical parameter for a given set of interconnect dimensions. The shortest lifetimes of different types of multiple interconnects are obtained from the EF test structure and analyzed to determine the characteristics of early failures.

Since EF test structure is designed according to the assumption that the weakest link in a multi-link test structure dominates the chain lifetime, this study is based on the weakest link model. [5,6,7,8] In statistical analysis, when the weakest link model is used, we have:

$$R(t) = R_1(t) \times R_2(t) \times R_3(t) \times R_4(t) \times \ldots \times R_N(t) \qquad (1)$$
$$R(t) = min(R_i(t)), \qquad i = 1, 2, 3, 4, \ldots, N, \qquad (2)$$

where $R_i(t)$ is the reliability of *ith* component at time t, $R(t)$ the system reliability, and N the number of elements in the series. Since $R(t)=1-F(t)$, one can deduce:

$$1-F(t) = (1-F_1(t)) \times (1-F_2(t)) \times (1-F_3(t)) \times \ldots \times (1-F_N(t)), \qquad (3)$$

where $F(t)$ is the unreliability function. If $F_i(t)s$ are the same regardless of i, the above equation can be simplified to:

$$1-F(t) = (1-F_i(t))^N, \qquad (4)$$
$$F(t) = 1-(1-F_i(t))^N. \qquad (5)$$

The last equation is also called the weakest link approximation. $F(t)$ is the cumulative distribution function for a series of N independent but identical elements. It has been reported and observed that the cumulative failure distribution (CFD or CDF) of single interconnects are well-fitted by a log-normal distribution. [7]
Monte Carlo simulation was developed to evaluate the lifetimes of multi-link test structures and to find the characteristics of the failure mechanism such as t_{50} and σ. The characteristics of the failure mechanism or failure mode are defined in terms of t_{50} and σ of single individual interconnect in the multi-link test structure. Since the lifetime of single interconnect is assumed to follow the log-normal distribution, it is possible to generate the lifetimes of individual interconnects of a statistical test structure from the overall distribution. Accordingly, we have:

$$\text{Lifetime, } t = F^{-1}(\text{random number as CDF, } t_{50} \text{ and } \sigma) \qquad (6)$$

where log-normal CDF, $F(t)$, is a function of t, t_{50} and σ. The lifetime of an EF/N = 10 test structure is the smallest lifetime among ten lifetimes randomly chosen from the log-normal distribution of CDF of single element. If we repeat these process several times, then the CDF of EF/N = 10 can be obtained. Also it is possible to estimate t_{50} and σ of an EF/N = 10 test ensemble from the CDF simulations. Again, the t_{50} and σ are not the characteristics of the failure mechanism but the characteristics of EF/N = 10. Fig. 2 shows that CDF plots of Monte Carlo simulation are overlapped on those derived from the weakest link approximation.
Using Monte Carlo simulation and assuming a specific CDF and known N, it is possible to deconvolute the CDF into its equivalent CDF plots for various N, the size of ensemble of the multi-link test structure. This deconvolution method can determine whether EF samples of various N were failed by the same failure mechanism. If the characteristics (t_{50} and σ) of failure mechanism are the same regardless of N (the size of ensemble), there is only one failure mechanism in the interconnects. If a discrepancy exists in failure characteristics between experimental observation and the simulation analysis, this indicates that more than one interconnect failure mechanism can exist.

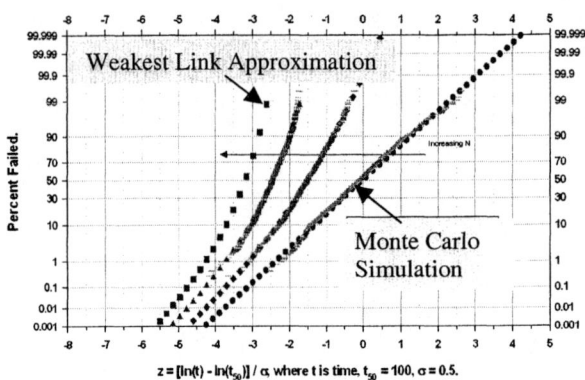

Figure 2. CDF plots of various N generated by both Monte Carlo simulation and WLA are overlapped completely. EF/N = 1 follows log-normal distribution and the t_{50} and σ are 100 hrs. and 0.5, respectively.

Depending on test conditions, process quality, and test structure types examined, systematic deviations are often observed. Two failure mechanisms, which follow different log-normal distributions, have been observed in the Cu/Oxide dual-damascene test structure. In this case, there is a race to fail in the test structure between two failure mechanisms with different statistical characteristics. The failure mechanism with a longer t_{50} is called the "strong mode," and that with a shorter t_{50} is called the "weak mode." The weak mode will fail samples earlier and it can be either a process-related or a fundamental failure mechanism.

Monte Carlo simulation for dual-mode failure mechanism is required in order to characterize and interpret the results of EM tests. If there is dual-mode failure, each interconnect can be failed by either weak or strong modes. And P, the probability that one interconnect can be failed by weak mode, is proportional to the number of samples ($N = 1$) failed by the weak mode relative to the number of total samples, where each sample has only one interconnect. The lifetime of the multi-link (N) test structure is decided by choosing the smallest lifetime among the lifetimes of N interconnects obtained by simulation. Fig. 3 (a), Fig. 3 (b) and Fig. 3 (c) show the CDF plots of various N, generated by multi-mode Monte Carlo simulation, with probabilities of weak mode (P) of 1, 10 and 50% respectively. The hatched regions on the left side of each figure show where failures are "weak" in nature while the hatched regions on the right show where failures are "strong."

Results obtained from EM tests at 400 °C and $j = 1.0$ MA/cm^2 are used to verify the validity of the Monte Carlo simulation. The solid symbols in Fig. 4 are lifetime data measured for various N. The tail of $N = 100$, which can not be fit to any mono-modal Monte Carlo simulation, represents the weak mode (early failures). The characteristics of strong mode are observed from the $N = 1$, 10 and the top half of $N = 100$. The characteristics of the strong mode deconvoluted from those three CDF plots are identical and are determined to be $t_{50} = 300$ hrs. and $\sigma = 0.45$. However, the characteristics of weak mode is hard to define due to a lack of reliable lifetime data, since at least five data points are required to obtain a reasonable precision. CDF plots created by Monte Carlo simulation can be fit using the last three lifetimes and strong mode after careful adjustments. It results in a weak mode with $t_{50} = 30$ hrs. and $\sigma = 0.3$ approximately. The error bar for t_{50} is estimated to be 10 hrs. for the weak mode and its probability is ~0.6%. In this way, the characteristics of the early failures (weak mode) can be determined from a total 28 test structures instead of using 1000 single interconnects.

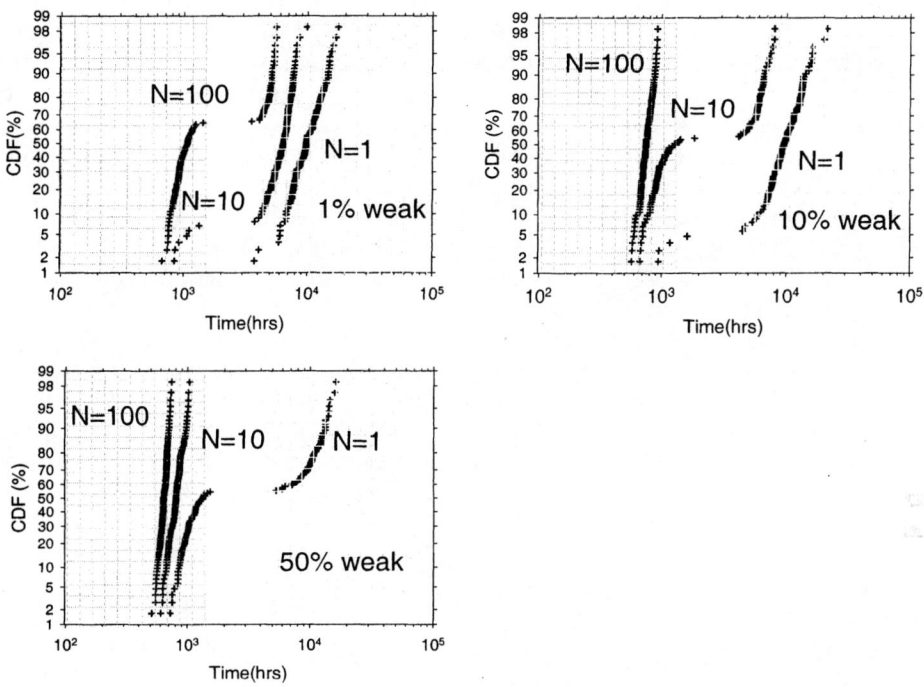

Figure 3. CDF plots of $N = 1$, 10 and 100, when strong mode has $t_{50} = 1000$ hrs., $\sigma = 0.3$; Weak mode $t_{50} = 50$ hrs., $\sigma = 0.3$; Probability of weak mode = (a) 1%; (b) 10%; (c) 50%.

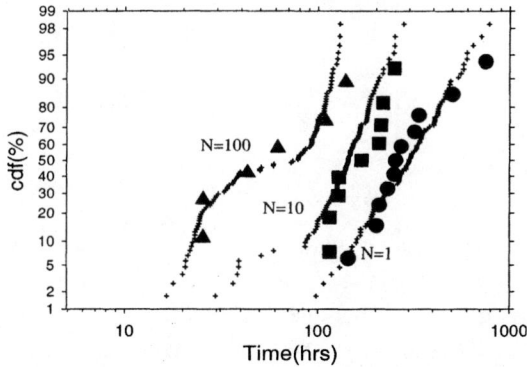

Figure 4. The real lifetime distributions (solid symbols) and CDF plots created by Monte Carlo simulation fit each other well. The strong mode has lifetime characteristics of $t_{50} = 300$ hrs. and $\sigma = 0.45$, and the weak mode has characteristics of $t_{50} = 30$ hrs. and $\sigma = 0.3$. The probability of interconnects failed by weak mode is about 0.6%.

Experimental Results of Early Failure EM test

1. TCR and Joule Heating Effect

A total of 16 sets of electromigration tests have been performed at 325, 350, 375 and 400 °C with three current densities of 1.0, 1.5 and 2.0 MA/cm^2. The early failure (EF) test structures used in the EM tests consist of a group of identical Cu/oxide dual-damascene interconnects that are serially connected to yield an interconnect chain. The controlling parameter is the number of links in the chain, N, where $N = 1, 10, 50,$ or 100. The metal interconnect height and width of all test structures are 0.5 μm and 0.5 μm, respectively. The Cu interconnects are covered by a 250 Å nominal thick Ta barrier. Above the M2, there is a capping level of 1000 Å SiN$_x$. [9] The interconnect length of the upper metal level (M2) is 300 μm, which is much longer than the critical lengths at the test current densities. Because the population of early failures can vary depending on the wafer process condition, all test structures were selected from the same wafer.

Before measuring the resistance trace at the target current and temperature, the effect of the joule heating is determined by measuring TCR. Fig. 5 shows that the curve for joule heating correlated well with the resistance change at increasing current densities. The measured amount of Joule heating is 1.5 °C at 1.0 MA/cm^2, 3.3 °C at 1.5 MA/cm^2 and 5.9 °C at 2.0 MA/cm^2. Because joule heating rises in a parabolic manner with current density, it can become significant if a higher current density is used.

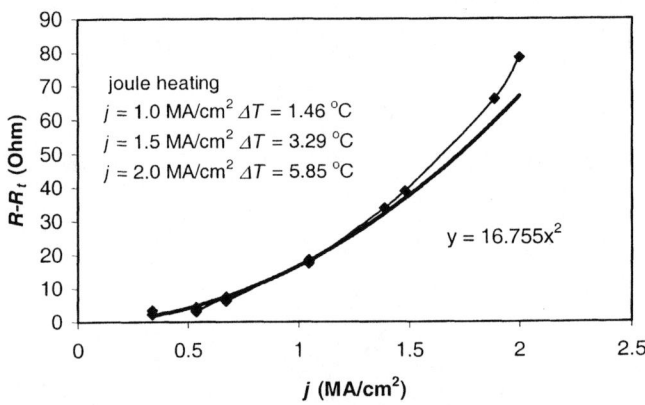

Figure 5. Joule Heating at 350 °C. The target current density is 2.0 MA/cm^2.

2. Resistance Trace for EF N = 1, 10 and 100

The damage by electromigration (EM) in our test structure was monitored from the resistance trace. Cu/oxide dual damascene interconnects show abrupt increases or steps in resistance of 200 ~ 300 Ω that represent the interconnect failures. The abrupt resistance increase (step) is the characteristic feature due to the geometry of the Cu/oxide dual-damascene structure. The circuit does not completely open after the failure because the failed interconnect is surrounded by the Ta barrier providing a secondary high resistivity current path.   Fig. 6 shows typical resistance traces of Cu/oxide dual-damascene test structures when $N = 1$, 10 and 100. For $N = 1$, failure happened after 20 hours of EM testing. For $N = 10$ and $N = 100$, multiple steps are observed, and those steps represent the failures of each individual interconnect of the test structure. Compared with $N = 10$, more steps (failures) are observed in EF $N = 100$ samples during the same period, because the chance to failure is the same for individual interconnects so the probability of observing failures increases with the number of test elements. According to the failure criterion, the time to first resistance step is considered to be the lifetime of the test structure.

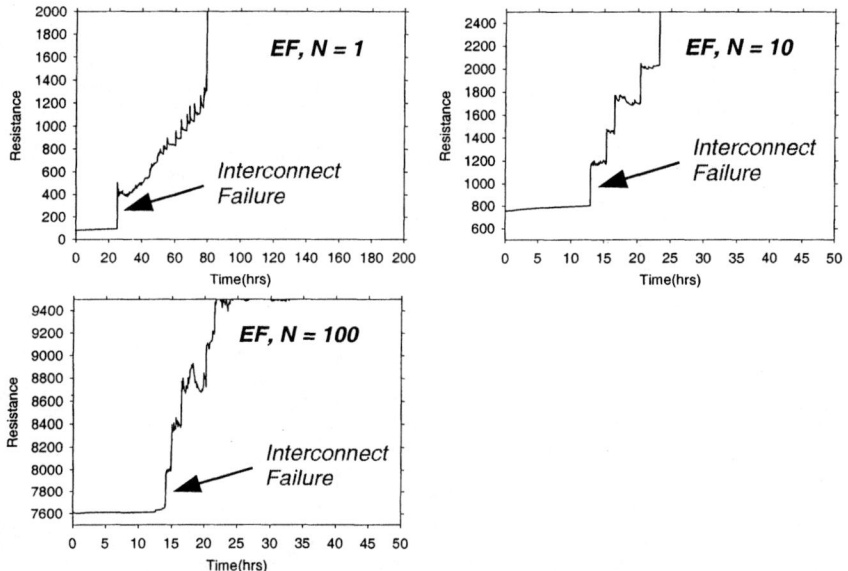

Figure 6. Resistance Trace of Early Failure Test Structure of $N = 1$, 10, and 100, where N is number of Cu/Oxide Dual-Damascene Interconnect in the test structure.

3. Activation Energy and Current Exponent for Early Failures

EM tests at 325 °C, 350 °C, 375 °C and 400 °C have been performed. The EF structures were examined to identify the failure mechanism and to build a statistical model. Fig. 7 contains the cumulative failure distributions (CFD or CDF) of early failure test structures, and they fit well to data generated using multi-mode Monte Carlo simulation. Since some of the observed CDFs have "tails" on the EF $N = 1$ CDF plots, which can not be explained by a mono-mode failure mechanism, we consider the dual-mode failure mechanism.

CDF plots of Fig. 7 (a), (b), (c) and (d) contain the actual lifetime distributions (solid symbols) observed from the EF test structures and CDF plots (small crosshairs) simulated by dual-mode Monte Carlo simulation. Overall the measured and simulated CDF plots agree well with each other. Experiments on EF with $N = 1$ at 325 and 350 °C were not done, because such experiments would take several months to obtain proper lifetime data. We are primarily interested in characterizing the early failures in this series of experiments.

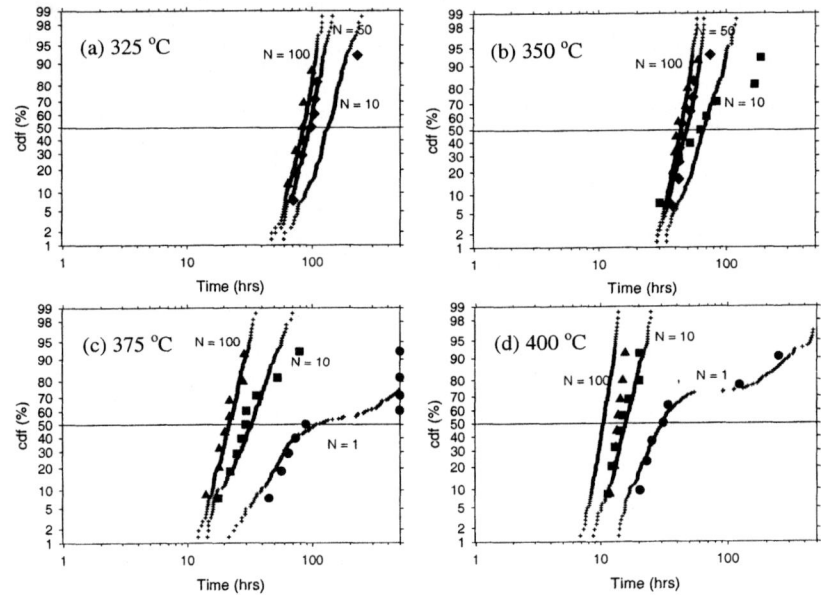

Figure 7. CDF (or CFD) plots of EF N = 1, 10, 50, 100 j = 1.0 MA/cm^2, T = (a) 325 °C (b) 350 °C (c) 375 °C (d) 400 °C Solid symbols represent the CDF of real data and small crosshair are the simulated data generated by a dual-mode Monte Carlo simulation.

Table 1 contains the characteristics (t_{50} and σ) and CDF % of weak and strong mode failure mechanisms of individual M2 line elements, which were determined using dual-mode Monte Carlo simulation. The standard deviations for both weak and strong modes seem to be about the same.

Operating Temp.	325 °C	350 °C	375 °C	400 °C
t_{50} of Weak mode	230 hrs.	105 hrs.	59 hrs.	25 hrs.
σ of Weak mode	0.45	0.4	0.5	0.35
% of Weak mode			50%	75%
T_{50} of Strong mode			~600 hrs.	250 hrs.
σ of Strong mode			~0.6	0.6
% of strong mode			50%	25%

Table 1. Descriptions of both weak and strong modes. t_{50} and σ (standard deviation) are the characteristics of the failure mode determined by the Monte Carlo simulation. Weak mode controls the early failures.

Activation energy is calculated to determine the dominating diffusion mechanism for the weak mode using the Black's equation [10]:

$$t_{50} = Aj^{-n} \, exp(Q/kT), \tag{7}$$

where A is a constant, j is the current density, n is the current density exponent, Q is activation energy, k is the Boltzmann's constant, and T is the operating temperature. Because electromigration is diffusion-controlled, activation energy (Q) depend on the diffusion path in which the metal atoms move. Fig. 8 shows the activation energies of the early failure test structures when EF $N = 1$, 10, 50 and 100. For EF $N = 10$, 50 and 100, the t_{50}s are obtained directly from the CDF plots in Fig. 8. Overall, the portion of interconnects failed by weak mode is small, therefore CDF of EF $N = 1$ is mostly dominated by the strong mode. Since Monte Carlo simulation can estimate the characteristics of the weak mode, t_{50}s of the weak mode for $N = 1$ are obtained from Table 1. According to Fig. 8, the activation energy (Q) for the weak mode (= early failures) is 1.00 ± 0.05eV, which is consistent with the value for mass transport along the interface. This suggests that the dominating diffusion for weak mode failure mechanism is interfacial diffusion.

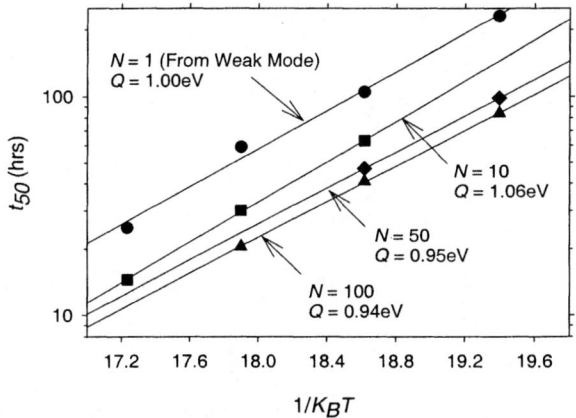

Figure 8. Activation energy from four temperatures: $T = 325, 350, 375, 400$ °C, $j = 1.0$ MA/cm^2, line width = 0.5 μm, line length = 300 μm.

In this study, the current density dependence of early failure rate has also been investigated and the results are used to determine the current density exponent, n, in Black's Equation. Fig. 9 (a) shows CDF plots of EF $N = 50$ at 350 °C and $j = 1.0$, 1.5 and 2.0 MA/cm^2. Since the CDF plots of higher current densities show transitions from the weak to the strong failure mode, only data corresponding to the weak mode failure mechanism in Fig. 9 (a) are analyzed in order to obtain a consistent t_{50}s. The t_{50} at $j = 2.0$ MA/cm^2 are determined from the extrapolation of the weak mode data to 50% CDF. The plot of t_{50} vs. j is shown in Fig. 9 (b), which is used to find the current density exponent (n). The current density exponent turns out to be 1.96, or approximately 2, consider the testing condition.

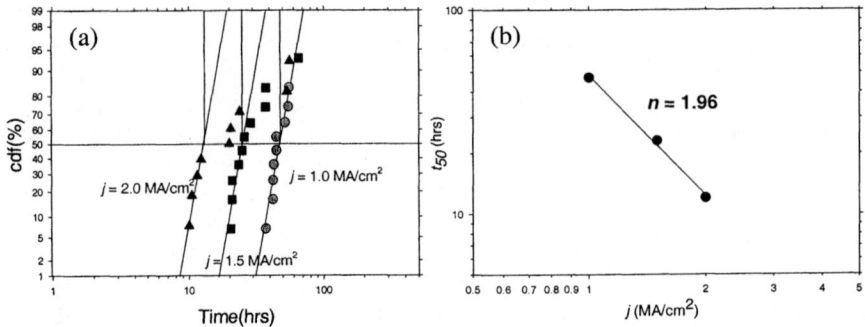

Figure 9. j dependence, EF $N = 50$, $T = 350\ °C$ and three current densities: $j = 1.0, 1.5, 2.0\ MA/cm^2$.

It has been reported that there is a transition of current density exponent (n) from 1 to 2 when j is passing over 2.5 MA/cm² at 370 °C. [11] Our recent test results also show that the current density exponent can vary between 1 and 2.

4. Failure Analysis for Weak and Strong mode

Failure analysis was performed on the EF structures after EM testing to examine the microstructure associated with the two-mode failure mechanism, which is statistically identified. Interconnects failed by weak mode (=early failures) have shorter lifetimes, which can be attributed to a small amount of mass transport that yields faster failure. Likewise, strong mode has longer lifetime and is associated with a slower process leading to the interconnect fail. Fig. 10 shows possible void sites and their failure modes. The two most probable void formation sites in the dual-damascene interconnect are the (cathode) via bottom and cathode end (of M2). While the void growth at via bottom can make the circuit open very quickly, it will take more time for interconnects to be failed by void growth at cathode end. The amount of mass depletion required to make the interconnect open for strong mode is much larger than that for weak mode in Fig. 10.

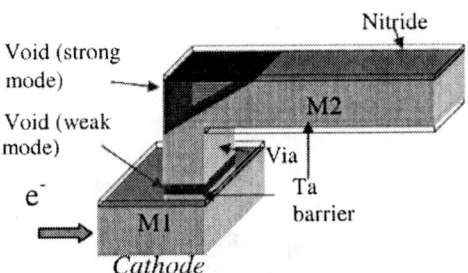

Figure 10 The possible sites for void formation. For better view, the Ta barriers along the side wall of the Cu interconnect is not depicted.

Focused Ion Beam (FIB) imaging was used to clarify multi-mode failure mechanisms and examine the void sites. Fig. 11 is the cross-sections of EF/N=10 in which two different types of void formations are identified. Interestingly, extrusions at the anode due to mass accumulation occur simultaneously along with mass depletion at cathode end. Fig. 11 (b) shows a void forming at the cathode end of M2, which can be responsible for strong mode failures. Since the failure in the interconnects is controlled by the rate of damage formation for the two failure mechanisms, the coexistence of both weak and strong mode voids can be observed in a single interconnect, as shown in Fig. 11 (a).

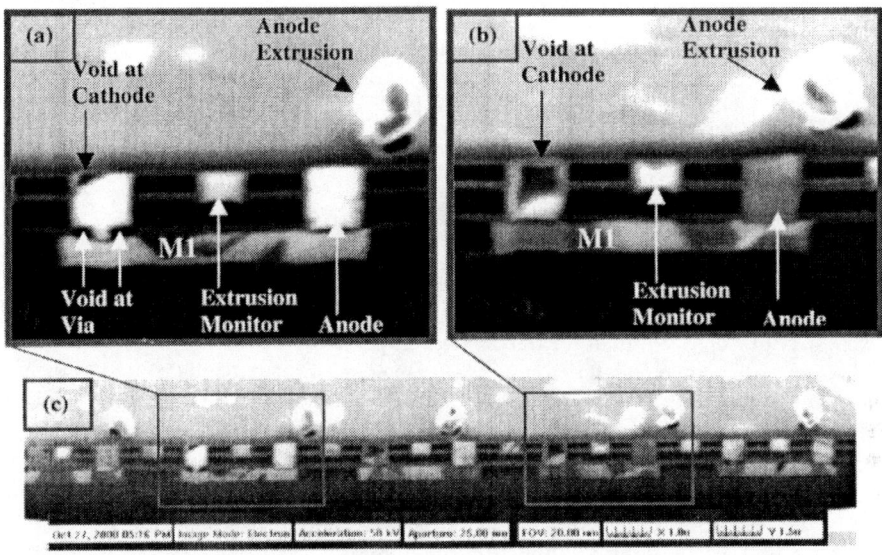

Figure 11. The FIB images show the two different modes of voiding. (a) Strong and weak mode (b) Strong mode (c) Cross-section of EF N=10 showing EM-induced damages. Weak mode is responsible for the early failures.

Discussion - Early Failure

The population of weak mode (= early failures) can vary from wafer to wafer according to the degree of process control so that it can increase if new process is developed or the feature size is decreased. Electromigration experiments have been performed to find out the relation between early failures and the process control. Table 2 summarizes the early failure populations of the test structures obtained from three wafers - A, B and C in the order of the date we received. In Wafer C, which we received the most recently, none of 0.5 µm EF test structures failed by early failures. This can be attributed to improvement in manufacturing. The population of early failure was found to increase with decreasing line width, as expected. This can be attributed to difficulty in dimensional control of interconnects. This study describes a statistical method to detect early failures, providing a quantitative basis for evaluating reliability and process control.

Wafer ID	A	B					C		
Line Width	0.5 μm	0.5 μm	0.5 μm	0.5 μm	0.5 μm	0.5 μm	0.35 μm	0.25 μm
Test Temp.	400 °C	325 °C	350 °C	375 °C	400 °C	380 °C	380 °C	380 °C
% of Weak mode	0.6%	*	*	50%	75%	0%	7.5%	15%
t_{50} of Weak mode	30 hrs.	230 hrs.	105 hrs.	59 hrs.	25 hrs.		100 hrs.	30 hrs.
σ of Weak mode	0.3	0.45	0.4	0.5	0.35		0.5	0.45
% of Strong mode	99.4%			50%	25%	100%	92.5%	85%
t_{50} of Strong mode	300 hrs.			~600hrs	250 hrs	400 hrs.	210 hrs.	130 hrs.
σ of Strong mode	0.45			~0.6	0.6	0.7	0.3	0.45
Ref.	Fig. 4	Fig. 7, Fig. 8 and Table 1						

Table 2. Descriptions of both weak and strong modes. t_{50} and σ are the characteristics of the failure mode determined by the Monte Carlo simulation. Weak mode controls the early failures. * t_{50}s of these temperatures are too long to measure properly. $j = 1.0$ MA/cm^2.

Conclusion

Electromigration study using multiply linked line/via interconnects has been performed to characterize the statistics and mechanism of early failures. For statistical analysis, Monte Carlo simulation has been developed and verified for dual-mode failures. Two distinct failure mechanisms in interconnects were observed and defined to be a weak and a strong mode with the former relating to early failures. The activation energy (Q) and the current density exponent (n) of early failures were found to be 1.0 eV and 2.0, respectively. Mass transport for weak mode (= early failures) was dominated by Cu/barrier interfacial diffusion. FIB images identified the void formation at cathode via bottom to be responsible for early failures. EM lifetime decreased with decreasing line width while the early failure population increased. The result can be attributed to difficulty in process control.

Acknowledgments

We would like to acknowledge partial support from the state of Texas, International SEMATECH and the Center for Advanced Interconnect Science and Technology of the Semiconductor Research Corporation (SRC/CAIST).

[1] M. Geradin, *Academic. Science*, 53, p. 727 (1961).
[2] C.-K. Hu, R. Rosenberg, and K.Y. Lee, Appl. Phys. Lett., Vol 74, 20, (May 1999)
[3] C.-K. Hu, R. Rosenberg, H. S. Rathore, D. B. Nguyen, and B. Agarwala, *IEEE, IITC*, 99-267
[4] C.-K. Hu, J.M.E. Harper, *Materials Chem. and Phys.*, 52 (1998) 5-16
[5] L. L. Marsh, R. D. Havens, S.-C. Wang, J. A. Malack, and H. B. Ulsh, in *Principles of Electronic Packaging*, edited by D. P. Seraphim, R. C. Lasky, and Che-Yu Li, (McGraw-Hill Book Company, New York, 1989), pp. 262-266.
[6] Wayne Nelson, *Applied Life Data Analysis*, (John Wiley and Sons, New York, 1982), pp.168-169.
[7] P. Tobias and D. C. Trindade, *Applied Reliability*, 2nd ed. (Chapman and Hall/CRC, Boca Raton, 1995), pp. 29-31; pp. 219-222.

[8] M. Gall, "Investigation of Electromigration Reliability in Al(Cu) Interconnects," Ph.D. Dissertation, The University of Texas at Austin, May, 1999.

[9] V. Blaschke, J. Mucha, B. Foran, Q. T. Jiang, K. Sidensol, A. Nelson, in *1998 Proceedings of the Advanced Metallization Conference*, pp.43-49.

[10] J. Black, *IEEE Trans. Electron. Devices*, ED-16, 338, (1969)

[11] R. Rosenberg, D. C. Edelstein, C.-K. Hu, and K. P. Rodbell, *Annu. Rev. Mater. Sci. 2000.* p229-262 (2000)

A Governing Parameter for Electromigration Damage in Passivated Polycrystalline Line and Its Verification

Kazuhiko Sasagawa[a], Masataka Hasegawa[b], Masumi Saka[b] and Hiroyuki Abé[c]

[a] Department of Intelligent Machines and System Engineering, Hirosaki University,
3 Bunkyo-cho, Hirosaki 036-8561, Japan
[b] Department of Mechanical Engineering, Tohoku University,
01 Aoba, Aramaki, Aoba-ku, Sendai 980-8579, Japan
[c] Tohoku University, 2-1-1 Katahira, Aoba-ku, Sendai 980-8577, Japan

Abstract. The atomic flux divergence due to electromigration in unpassivated metal line, AFD_{gen}, has been formulated considering two-dimensional distributions of current density and temperature and also, simply considering structures of polycrystalline and bamboo lines. The divergence AFD_{gen} has been identified as a governing parameter for electromigration damage in unpassivated polycrystalline and bamboo lines thorough experimental verification. In those studies, we have employed the uncovered metal lines as the first step in the development of a practical and universal prediction method for electromigration damage. On the other hand, it is known that in passivated lines, a mechanical stress (atomic density) gradient is induced by electromigration, and plays an important role in electromigration mechanism. In this study, a governing parameter, AFD^{*}_{gen}, is formulated for electromigration damage in passivated polycrystalline lines. The formulation is carried out by adding the effect of the atomic density gradient to AFD_{gen}. In addition, an AFD^{*}_{gen}-based method for determination of film characteristics is developed. Using this method, the characteristics included in the AFD^{*}_{gen} are derived based on experimental data of void formation. The method was applied to both covered line and uncovered one, which were made of the same Al film. It was found that the obtained characteristic constants functioned well as the constants dependent of existence of the passivation layer or as the constants independent of passivation existence. Through the showing the validity of the obtained characteristic constants, the usefulness of the AFD^{*}_{gen} and the AFD^{*}_{gen}-based method for determination of film characteristics was verified.

INTRODUCTION

The latest progress in a packaged silicon integrated circuits is remarkable. However, some problems have been raised about the deterioration of the reliability, with the advance. For instance, on interconnect metal line the higher current density due to scaling down causes electromigration. Electromigration is a phenomenon that the metallic atoms are transported by electron wind. Voids are formed as a result of depletion of metallic atoms, and hillocks are formed as a result of accumulation of atoms in the metal line. The growth of voids leads ultimately to metal line failure.

CP612, *Stress-Induced Phenomena in Metallization*: Sixth Int'l. Workshop, edited by S. P. Baker et al.
© 2002 American Institute of Physics 0-7354-0058-X/02/$19.00

Therefore, studies to predict the electromigration failure is of significance from the viewpoint of ensuring the reliability of interconnect metal lines.

A significant amount of research have been done so far to clarify the mechanism of electromigration damage. These researches have been based on experimental or analytical results obtained under specific conditions related to temperature, current density, microstructure or structure around the line (via, passivation, barrier-metal, etc.). We should point out that although the researches on the mechanism of the damage have led to some important knowledge, they could not be directly applied to the purpose of practical or universal prediction of damage due to electromigration, or to the evaluation of electromigration endurance. It is necessary to integrate the results of the researches on mechanism and to develop a practical general-purpose method for predicting the damage. Only a few works [1-3] have attempted to integrate some of the knowledge obtained from the research on damage mechanism in order to construct the electromigration failure model which is useful for predicting the lifetime. Till now, Black's empirical equation [4] has widely been used for electromigration failure prediction, though there are several problems inherent to this method[5]. On the other hand, the theoretical and precise method for prediction of electromigration failure is demanded in industry.

Sasagawa et al. have proposed an approach which is different from the above mentioned approaches and makes use of a so-called governing parameter for electromigration damage [6]. The usefulness of a parameter, AFD_{gen}, for electromigration damage in unpassivated polycrystalline and bamboo lines has been verified experimentally[6-9]. This parameter, we believe, is applicable to the lines under any condition and it corresponds directly to the actual amount of damage, i.e. the volume of void and hillock. Also, they have developed a method for failure prediction in unpassivated polycrystalline and bamboo lines [10-12]. The parameter was able to be used effectively in a numerical simulation of the failure. The lines uncovered with passivation layer have been treated in order to build up a foundation for development of practical prediction method of failure in passivated metal lines. On the other hand, the metal lines used in packaged silicon integrated circuit are covered with passivation layer. In the covered line, the hillock due to electromigration is hard to be formed because the passivation layer restrains the hillock formation. Therefore, the stress gradient (atomic density gradient) is built up in a metal line [13,14]. Consequently, the atomic diffusion due to atomic density gradient retards the progress of electromigration damage. The lifetime of the covered metal line becomes longer than that of uncovered one [15].

In this study, we formulate a governing parameter, AFD^*_{gen}, for electromigration damage in polycrystalline lines covered with a passivation film. The effect of the atomic density gradient induced by electromigration is added to the governing parameter AFD_{gen} introduced previously. A method for deriving film characteristics of covered polycrystalline lines is also developed based on AFD^*_{gen}. The AFD^*_{gen}-based method is applied to both covered metal lines and uncovered ones, which are made of the same Al film, and the film characteristics of these lines are determined experimentally. The usefulness of AFD^*_{gen} is shown through a discussion on the validity of the film characteristic constants obtained in these two kinds of lines.

A GOVERNING PARAMETER FOR ELECTROMIGRATION DAMAGE IN PASSIVATED LINE

Considering the back flow due to the stress gradient [13,14] and the effect of the stress caused in the metal line on diffusivity [16], the atomic flux in a polycrystalline line covered with passivation can be given as [17]

$$|J| = \frac{ND_0}{kT} \exp\left\{-\frac{Q_{gb} + \kappa\Omega(N - N_T)/N_0 - \sigma_T\Omega}{kT}\right\}\left(Z^* e\rho j^* - \frac{\kappa\Omega}{N_0}\frac{\partial N}{\partial l}\right)$$ (1)

where J is the atomic flux vector which corresponds with the direction of grain boundary, N - the atomic density, D_0 - a prefactor, k - Boltzman's constant, T - absolute temperature, Q_{gb} - the activation energy for grain boundary diffusion, κ - bulk modulus, Ω - atomic volume, N_0 - the atomic density under stress free condition, N_T - the atomic density under tensile thermal stress σ_T, Z^* - the effective valence, e - electronic charge, ρ is the temperature-dependent resistivity expressed as $\rho = \rho_0\{1 + \alpha(T - T_s)\}$, ρ_0 and α denoting the resistivity and the temperature coefficient at the substrate temperature T_s, respectively. Symbol j^* and $\partial N/\partial l$ are the components of the current density vector and the atomic density gradient in the direction of J.

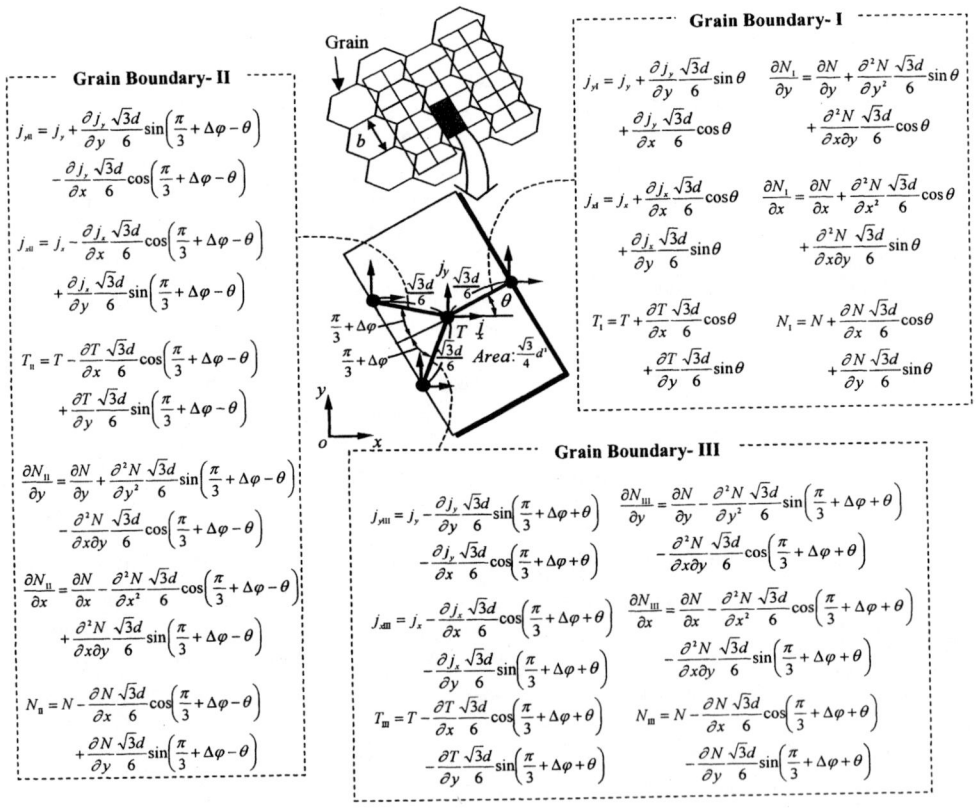

FIGURE 1. A Model of Polycrystalline Line Structure.

Now, a model of microstructure of grains is introduced as shown in Fig. 1, and the atomic balance is calculated in an unit square region. In the model, θ is the angle between Grain Boundary-I and the x axis, $\Delta\varphi$ is a constant related to the relative angle between grain boundaries, and d is the average grain size. The components of current density vector, temperature, atomic density, and the gradient of atomic density at the end of each Grain Boundary-I, -II, and -III are shown in Fig. 1. By substituting the components of current density and atomic density gradient along the grain boundaries, temperature and atomic density into Eq. (1), the atomic flux at the end of each grain boundary is obtained, where the sign of direction from the unit region to outward is defined as positive. After multiplying the effective width of grain boundary δ and unit thickness to the every atomic flux at the ends of Grain Boundary-I, -II and -III, the number of atoms migrating along the grain boundaries per unit time is added each other. The formula of the sum is simplified by neglecting the minute terms and using the law of current conservation, and is divided by the volume of the unit region, $\sqrt{3}d^2/4$. Thus, the number of atoms decreasing per unit volume and unit time in the case that the angle between the unit region and the x-axis is θ, $AFD^*_{gb\theta}$, is given by

$$AFD^*_{gb\theta} = C^*_{gb}N\frac{4}{\sqrt{3}d^2}\frac{1}{T}\exp\left\{-\frac{Q_{gb}+\kappa\Omega(N-N_T)/N_0-\sigma_T\Omega}{kT}\right\}\times$$

$$\left\langle\; \sqrt{3}\Delta\varphi\left\{(j_x\cos\theta+j_y\sin\theta)Z^*e\rho-\frac{\kappa\Omega}{N_0}\left(\frac{\partial N}{\partial x}\cos\theta+\frac{\partial N}{\partial y}\sin\theta\right)\right\}\right.$$

$$-\frac{d}{2}\Delta\varphi\left\{\left(\frac{\partial j_x}{\partial x}-\frac{\partial j_y}{\partial y}\right)Z^*e\rho\cos2\theta-\frac{\kappa\Omega}{N_0}\left(\frac{\partial^2 N}{\partial x^2}-\frac{\partial^2 N}{\partial y^2}\right)\cos2\theta+\left(\frac{\partial j_x}{\partial y}+\frac{\partial j_y}{\partial x}\right)Z^*e\rho\sin2\theta-2\frac{\kappa\Omega}{N_0}\frac{\partial^2 N}{\partial x\partial y}\sin2\theta\right\}$$

$$-\frac{\sqrt{3}}{4}d\frac{\kappa\Omega}{N_0}\left(\frac{\partial^2 N}{\partial x^2}+\frac{\partial^2 N}{\partial y^2}\right)-\frac{\kappa\Omega/N_0}{kT}\left[\frac{\sqrt{3}}{4}d\left\{Z^*e\rho\left(j_x\frac{\partial N}{\partial x}+j_y\frac{\partial N}{\partial y}\right)-\frac{\kappa\Omega}{N_0}\left(\frac{\partial N}{\partial x}\frac{\partial N}{\partial x}+\frac{\partial N}{\partial y}\frac{\partial N}{\partial y}\right)\right\}\right.$$

$$-\frac{d}{2}\Delta\varphi\left\{Z^*e\rho\left(j_x\frac{\partial N}{\partial x}+j_y\frac{\partial N}{\partial y}\right)-2\frac{\kappa\Omega}{N_0}\frac{\partial N}{\partial x}\frac{\partial N}{\partial y}\right\}\sin2\theta\right] \qquad (2)$$

$$\left.+\frac{\sqrt{3}d}{4T}\left\{\frac{Q_{gb}+\kappa\Omega(N-N_T)/N_0-\sigma_T\Omega}{kT}-1\right\}\left\{Z^*e\rho\left(j_x\frac{\partial T}{\partial x}+j_y\frac{\partial T}{\partial y}\right)-\frac{\kappa\Omega}{N_0}\left(\frac{\partial N}{\partial x}\frac{\partial T}{\partial x}+\frac{\partial N}{\partial y}\frac{\partial T}{\partial y}\right)\right\}\;\right\rangle$$

where $C^*_{gb}=\delta D_0/k$, the first term in the angle brackets on the right-hand side is concerned with the atomic flux divergence at the triple point [18], and the other terms are concerned with the flux divergence in the grain boundary itself [19]. If $AFD^*_{gb\theta}$ takes positive value, it means the depletion of atoms. On the other hand, if $AFD^*_{gb\theta}$ takes negative value, it means the accumulation of atoms.

The angle θ takes arbitrary value in actual. It is needed to consider the flux divergence in the whole range of θ, i.e., from 0 to 2π. As taking notice of the atomic flux divergence contributing to the formation of void, the expectation of the only positive value of $AFD^*_{gb\theta}$ within the range of 0 to 2π is obtained, where the negative value of $AFD^*_{gb\theta}$ is treated as zero since it does not contribute to the void formation. The sum of the value of $AFD^*_{gb\theta}$ and its absolute value is divided by two to extract the only positive value of $AFD^*_{gb\theta}$, and the expectation of the extracted value is obtained. Thus, the atomic flux divergence concerning the void formation in polycrystalline line, AFD^*_{gen}, is derived as

$$AFD^{*}_{gen} = \frac{1}{4\pi} \int_{0}^{2\pi} \left(AFD^{*}_{gb\theta} + \left| AFD^{*}_{gb\theta} \right| \right) d\theta \ . \qquad (3)$$

DERIVATION METHOD OF FILM CHARACTERISTICS

The film characteristics included in AFD^{*}_{gen} are d, $\Delta\varphi$, Q_{gb}, Z^{*} and C^{*}_{gb}. The average grain size, d, can be measured using FIB (focused ion beam) equipment. Concerning with $\Delta\varphi$, the value obtained from the uncovered metal line, which is made of the same film as covered metal line, is employed [6]. A method to determine the other characteristics has been derived, treating the center region of the straight-shaped line in which the current density and temperature can be regarded as being constant. First, let N be approximately equal to N_0 because the change in N from N_0 is estimated small percentage at most, considering the magnitude of stress in metal line. According to Blech [13,14], the atomic density gradient depends on passivation existence, and is in inverse proportion to line-length. The atomic density gradient is assumed to be linear within the center region of a line in the initial stage of electromigration damage. The product, $\kappa \partial N / \partial x$, should be determined as the characteristic which depends on passivation existence and on line-length of the straight line used. The film characteristics, $Q^{*}_{gb}[=Q_{gb}-\sigma_T\Omega]$, Z^{*}, C^{*}_{gb} and $\kappa \partial N / \partial x$, are determined using the straight shaped line as follows.

Accelerated tests are performed for a certain period of time. The straight line is subjected to input current density, j_1, under three kinds of substrate temperature, T_{s1}, T_{s2} and T_{s3}. Then, let the temperature in the center region of line be T_1, T_2 and T_3 in the cases that substrate temperature is T_{s1}, T_{s2} and T_{s3}, respectively. Let us call each experimental condition Condition-1: j_1 and T_1, Condition-2: j_1 and T_2, and Condition-3: j_1 and T_3, respectively. In addition to these conditions, the acceleration test is performed under a current density, j_4, different from j_1. In this condition, the substrate temperature is controlled so that the temperature in the center region of line, T_4, is approximately equal to T_2. Let us call this experimental condition Condition-4: j_4 and $T_4(\cong T_2)$. In order to determine the substrate temperature, FEM (finite element method) analysis of electrothermal problem is performed as described later. The void volume is measured within the center region of the line after current stressing for a certain period of time.

On the other hand, considering the center region of the straight line, AFD^{*}_{gen} is simplified by neglecting the terms associated with the temperature gradient, current density gradient and the gradient of atomic density gradient in Eq.(2). The area of the center region, the line thickness, the net current-applying time which is obtained by subtracting incubation period from a current-applying time, and the atomic volume are multiplied to the simplified AFD^{*}_{gen}. The product represents the theoretical volume of void. By regarding the product and the volume of void obtained experimentally as equivalence, the following equation is derived, where the incubation period is defined as the period from start of current-applying to the beginning of increase in electrical resistance of metal line by void formation.

$$V_j = A \times t_j \times thick \times \frac{4}{\sqrt{3}d^2} \times \frac{C^*_{gb}}{T_j} \exp\left(-\frac{Q^*_{gb}}{kT_j}\right)\left(Z^* e\rho_j j_j - \frac{\Omega}{N_0}\kappa\frac{\partial N}{\partial x}\right) \times$$
$$\frac{\sqrt{3}\Delta\varphi}{\pi}\left\{\sqrt{1-\left(\frac{a_j}{b}\kappa\frac{\partial N}{\partial x}\right)^2} - \frac{a_j}{b}\kappa\frac{\partial N}{\partial x}\cos^{-1}\left(\frac{a_j}{b}\kappa\frac{\partial N}{\partial x}\right)\right\} \tag{4}$$

where

$$a_j = \frac{\sqrt{3}}{4}\frac{\Omega/N_0}{kT_j}d \tag{5}$$

and

$$b = \sqrt{3}\Delta\varphi \tag{6}$$

A subscript j represents the number of each condition, V_j the measured void volume, A the area of center region, t_j the net current-applying time, *thick* the line-thickness and ρ_j the resistivity of line at T_j. The unknown film characteristics in AFD^*_{gen} can be obtained using the least-squares method. Namely, the characteristics are determined so that the sum of squares of a residual between the logarithms of both sides in Eq.(4) takes its minimum value. The sum, F, is expressed by

$$F = \sum_j\sum_i\left[\ln V_{ij} - \ln\left\langle A \times t_{ij} \times thick \times \frac{4}{\sqrt{3}d^2} \times \frac{C^*_{gb}}{T_j}\exp\left(-\frac{Q^*_{gb}}{kT_j}\right)\left(Z^* e\rho_j j_j - \frac{\Omega}{N_0}\kappa\frac{\partial N}{\partial x}\right)\times\right.\right.$$
$$\left.\left.\frac{\sqrt{3}\Delta\varphi}{\pi}\left\{\sqrt{1-\left(\frac{a_j}{b}\kappa\frac{\partial N}{\partial x}\right)^2} - \frac{a_j}{b}\kappa\frac{\partial N}{\partial x}\cos^{-1}\left(\frac{a_j}{b}\kappa\frac{\partial N}{\partial x}\right)\right\}\right\rangle\right]^2 \tag{7}$$

where a subscript i represents the number of data measured in each experimental condition. By this method, the film characteristics can be obtained as the optimized parameters, which approximate the all experimental data, and this method makes data-handling easy.

The distributions of current density and temperature, which are required to obtain the film characteristics are obtained by FEM analysis. The fundamental equations in the analysis are expressed as follows;

Governing equation concerning electrical potential ϕ_e:
$$\nabla^2\phi_e = 0 \tag{8}$$

Ohm's law:
$$j = -\frac{1}{\rho_0}\text{grad }\phi_e \tag{9}$$

Equation of steady-state heat conduction [1]:
$$\lambda\nabla^2 T + \rho_0 j \cdot j + (\rho_0\alpha j \cdot j - H)(T - T_s) = 0 \tag{10}$$

where λ is the thermal conductivity, H the constant concerning the heat flow from the line to around the line, and $\nabla^2=\partial^2/\partial x^2+\partial^2/\partial y^2$. The constants ρ_0 and α are obtained by measuring the electrical resistance of the straight metal line under low current density enough to neglect the temperature rising. The constant H is obtained so that the electrical resistance of the metal line, which is calculated based on the temperature distribution from FEM analysis, equals to the measured value. Thus all unknown film characteristics are determined from only simple experiment using the straight shaped line.

EXPERIMENTAL VERIFICATION OF AFD^*_{gen}

Preparation of Sample

The specimens were prepared by following procedure. First, Al thin film was deposited by vacuum evaporation on the silicon substrate covered with a silicon oxide layer. The specimens were patterned by normal photolithography and annealed at 673 K for 70 min. After that, the silicon substrate was divided into half. The surface of one of the halves was coated with polyimide and was annealed in N_2 for 30 min at 408 K, 30 min at 473 K, 30 min at 573 K and 30 min at 673 K, continuously, to cure the polyimide layer. The thickness of the polyimide layer was 2.3 μm. Another half of the substrate, which was not covered with polyimide, was also annealed under the same condition as uncovered one so that both lines were actually fabricated under the same conditions. The dimensions of the Al lines are shown in Fig. 2. The uncovered metal line is called Sample 1 and the covered metal line is called Sample 2. Examples of FIB observation of Al grains are shown in Fig. 3. It is known that the difference in microstructure is one of the key factors of electromigration failure. It was confirmed that the average grain size was the same, approx. 0.7 μm, in both samples, and there was no difference in the microstructure of both lines.

Influence of Passivation on Lifetime of Metal Line

Prior to determination of film characteristics, we made sure whether the passivation by polyimide affected electromigration damage. Both unpassivated metal line and passivated one are subjected to D.C. current until the lines fail. The influence of the passivation on lifetime of the lines is investigated.

The accelerated tests were performed using the experimental set-up shown in Fig. 4. The silicon substrate was put on the hot-stage and the substrate temperature was kept

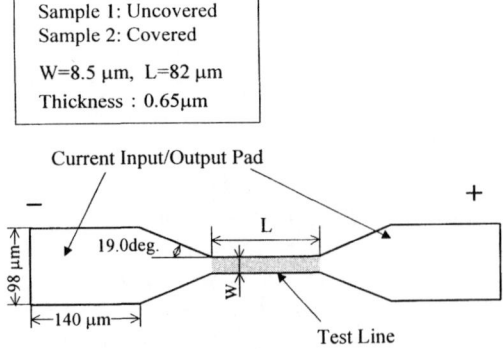

Sample 1: Uncovered
Sample 2: Covered

W=8.5 μm, L=82 μm
Thickness : 0.65μm

Current Input/Output Pad

−   +

98 μm

19.0deg.

L

W

140 μm

Test Line

FIGURE 2. Dimensions of Specimen.

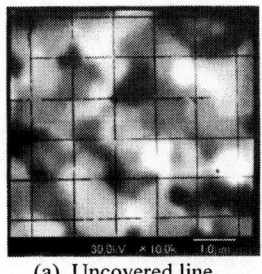
(a) Uncovered line

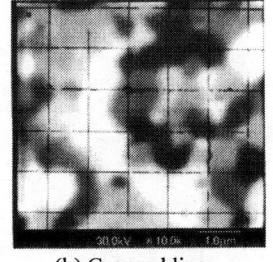
(b) Covered line

FIGURE 3. FIB Observation of Microstructures.

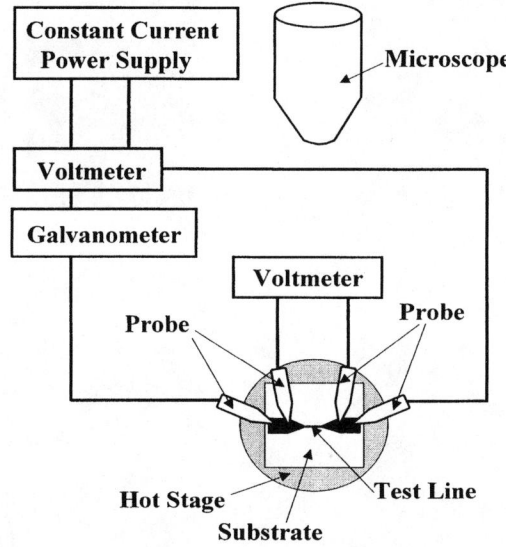

FIGURE 4. Experimental Set-Up.

at 473K. The input current density, 5.5MA/cm^2, was supplied to both the uncovered line and covered one. Ten specimens were used in both cases. After experiment, the passivation film was removed by chemical-etching, and the both lines were observed by FE-SEM (field emission-scanning electron microscope).

The frequency distribution of failure site and mean time to failure measured in uncovered line and covered one are shown in Fig.5. The lifetime of covered line was 1.8 times longer than that of uncovered one. The FE-SEM observation of anode end of the metal line whose passivation was removed is shown in Fig.6. Even in the case of passivated line, the hillock was formed at the time of failure. However, it seemed that the hillock in passivated line was collapsed. Therefore, it was thought that the hillock was hard to be formed because of the passivation film. Figure 7 shows the FE-SEM observation of the line just before failure. It was realized that the voids were formed in a slit-like shape even in the covered metal line. It, therefore, was considered that the dominant diffusion-path was grain boundary in the covered line as well as the uncovered one.

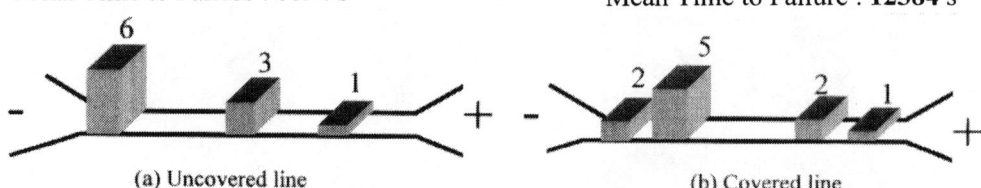

Mean Time to Failure : **6894** s Mean Time to Failure : **12384** s

(a) Uncovered line (b) Covered line

FIGURE 5. Frequency Distribution of Failure Site and Mean Time to Failure.

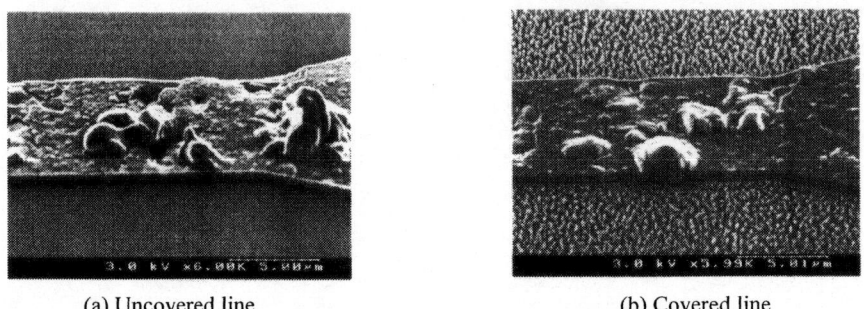

(a) Uncovered line (b) Covered line

FIGURE 6. Comparison of Shape of Hillock between Uncovered Line and Covered One.

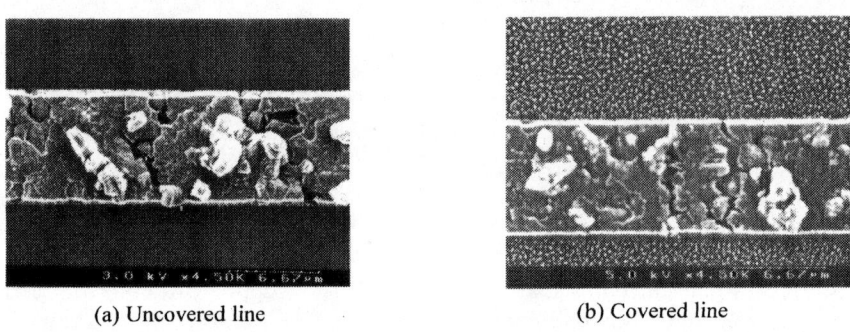

(a) Uncovered line (b) Covered line

FIGURE 7. FE-SEM Observation of Lines Just before Failure.

Derivation of Film Characteristics Using AFD^*_{gen}-Based Method

Experimental Method

In order to apply the AFD^*_{gen}-based method to unpassivated metal lines and passivated ones, the accelerated tests were performed for a certain period of time, using the experimental set-up shown in Fig. 4. To obtain the incubation period during which there was no void nucleation, the change in electrical potential drop in the line was monitored. The three different substrate temperatures were selected as 458, 473 and 488 K. For each temperature, the uncovered (Sample 1) and covered (Sample 2) lines were subjected to a D.C. current with density of 5.5 MA/cm^2 (Condition-1, -2 and -3). In addition, the acceleration test (Condition-4) was performed under a current density of 4.0 MA/cm^2 in the case of Sample 1. In this case, the substrate temperature was kept at 490 K so that the temperature at the center of the line was nearly equal to that in the case of substrate temperature 473 K (Condition-2). In the case of Sample 2, the substrate temperature was maintained at 415 K and the input current density of 8.5 MA/cm^2 was supplied. Twenty-five specimens were used in each testing condition. After D.C. current was applied, the passivation film was removed by chemical etching, and the metal line was observed by FE-SEM to measure void volume. The observed area of the center region was a rectangle of 17 μm length along the longitudinal axis × line width, in both Sample 1 and Sample 2. After the total area of the voids formed within the observed region was obtained by image-processing of the FE-SEM image, the volume of the voids was inferred by multiplying the film thickness to the total area of the voids. The volume of voids was also measured within the cathode end region in the case of Sample 1, and was utilized for determination of $\Delta\varphi$ [6]. The observed area of the cathode end region was a rectangle of 9 μm length along the longitudinal axis × line width. The current-applying time was different for each sample and each experimental condition because it was set within the time range when electrical resistance increases linearly. The current-applying time and incubation period of each sample under each testing condition are listed in Table 1.

Results and Discussion

The film characteristics in Sample 1 and 2 were obtained by the AFD^*_{gen}-based method, as shown in Tables 2 and 3, respectively. The constant Q^*_{gb} for Sample 2 was evaluated to be almost the same as that for Sample 1. Constants Z^* and C^*_{gb} for Sample 1 were not fixed, but the product of these values always took a constant value. Though Z^* and C^*_{gb} for Sample 1 could not be separated, the product of these values obtained for Sample 1 was close to the product of Z^* and C^*_{gb} obtained for Sample 2. And, it was thought that the value of Z^* obtained for Sample 2 was valid, because it was within the range of the values reported previously; -1~-15 [13,14,20,21]. On the other hand, $\kappa \cdot \partial N/\partial x$ for Sample 1 was much smaller than that for Sample 2. Thus, the result supports the previous reports, namely, that the back flow in an uncovered metal line is much less than that in a covered metal line [13,14].

TABLE 1. The Current Applying Time and Incubation Period in each Experimental Condition.

	Condition 1	Condition 2	Condition 3	Condition 4
Sample 1	4800s(576s)	3000s(348s)	1800s(177s)	4800s(555s)
Sample 2	10800s(2040s)	5400s(1040s)	2400s(516s)	1800s(456s)

Averaged incubation time is shown in parentheses.

TABLE 2. The Film Characteristics Included in AFD^*_{gen} in Case of Sample 1.

	Q^*_{gb} [eV]	$C^*_{gb} \cdot Z^*$ [Kμm^3/Js]	$\kappa \cdot \partial N/\partial x$ [J/μm^7]
Sample 1	0.490	-1.25×10^{25}	Approx. -0.005

TABLE 3. The Film Characteristics Included in AFD^*_{gen} in Case of Sample 2.

	Q^*_{gb} [eV]	Z^*	C^*_{gb} [Kμm^3/Js]	$\kappa \cdot \partial N/\partial x$ [J/μm^7]
Sample 2	0.505	-8.7	1.15×10^{24}	-0.366

From comparison of Q^*_{gb} for Sample 1 with that for Sample 2 it was found that the values of Q^*_{gb} for these samples coincided approximately, and the values were close to the value reported for grain boundary diffusion, 0.48eV [4]. This result agrees with the observation that the voids in the covered metal line are formed in a slit-like shape as well as that in the uncovered metal line shown in Fig.7. The formation of a slit-like void means that the main diffusion path of atoms is the grain boundary. From the agreement of Q^*_{gb} in Sample 1 with that in Sample 2, it was concluded that the influences of the thermal stress caused by the covering with a passivation layer and the interface diffusion between the metal line upon Q^*_{gb} is small enough to be neglected in this study. On the other hand, Lloyd and Steagall [22] reported that there was the effect of hydrogen in the polyimide on activation energy. However, the activation energy of Sample 2 was almost the same as that of Sample 1. It was thought that the effect of hydrogen on the activation energy for Al diffusion was negligible in this study.

The values of Q^*_{gb} and $Z^* \cdot C^*_{gb}$ were almost the same in Sample 1 and 2, but there was a difference in $\kappa \cdot \partial N / \partial x$ (see Tables 2 and 3). This result shows that AFD^*_{gen} is able to extract the influence of the back flow due to the atomic density gradient appropriately.

The following function, G^*, is defined by arranging the logarithm of both sides in Eq. (4):

$$G^* =$$

$$\ln \left[\frac{V_y T_j}{\frac{4 \times A \times t_y \times thick \times C^*_{gb} \times \Delta \varphi}{\pi d^2} \left(Z^* e \rho_j j_j - \frac{\Omega}{N_o} \kappa \frac{\partial N}{\partial x} \right) \left\{ \sqrt{1 - \left(\frac{a_j}{b} \kappa \frac{\partial N}{\partial x} \right)^2} - \frac{a_j}{b} \kappa \frac{\partial N}{\partial x} \cos^{-1} \left(\frac{a_j}{b} \kappa \frac{\partial N}{\partial x} \right) \right\}} \right] \quad (11)$$

The film characteristics obtained were substituted into G^* and it was plotted against $1/T$ as shown in Fig. 8. The slope represents $-Q^*_{gb}/k$. It was realized that the least-squares method functioned effectively and approximated well the experimental data.

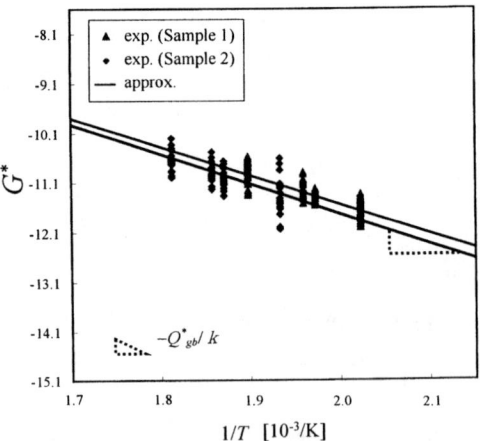

FIGURE 8. Approximation by Least-Squares Method.

84

Furthermore, the correlation coefficients were −0.82 in the case of Sample 1 and −0.74 in the case of Sample 2.

It was shown that the AFD^*_{gen} -based method was able to reflect the effect of passivation on the atomic diffusion mechanism accurately and to determine the film characteristics appropriately. Therefore, the usefulness of the governing parameter AFD^*_{gen} was shown through the discussion on the validity of the film characteristics obtained from the AFD^*_{gen}-based method.

ACKNOWLEDGMENTS

This work was partly supported by Japan Society for the Promotion of Science under Encouragement of Young Scientists 12750063, Grant-in-Aid for Scientific Research (B)(2) 1055024, Research Fellowships of the Japan Society for Promotion of Science for Young Scientists 08427 and the Murata Science Foundation. A part of this work was performed in the Venture Business Laboratory of Tohoku University.

REFERENCES

1. Kirchheim, R., and Kaeber, U., *J. Appl. Phys.*, **70**, 172-181(1991).
2. Nikawa, K., *Proceedings of the IEEE International Reliability Physics Symposium*, 175-181(1981).
3. Marcoux, P. J., Merchant, P. P., Naroditsky, V., and Rehder, W. D., *Hewlett-Packard Journal*, 79-84(1989).
4. Black, J. R., *Proceedings of IEEE*, **57**, 1587-1593 (1969).
5. McPherson, J. W., *Proceedings of the IEEE International Reliability Physics Symposium*, 12-18 (1986).
6. Sasagawa, K., Nakamura, N., Saka, M., and Abé, H., *Trans. ASME, J. Elect. Pack.*, **120**, 360-366(1998).
7. Sasagawa, K., Nakamura, N., Saka, M., and Abé, H., *Trans. Jpn. Soc. Mech. Eng.*, **65**, 469(1999), in Japanese.
8. Sasagawa, K., Naito, K., Hasegawa, M., Saka, M., and Abé, H., *Advances in Electronic Packaging 1999*, ASME **EEP-26-1**, 227-232 (1999).
9. Sasagawa, K., Hasegawa, M., Saka, M., and Abé, H., *Theoretical and Applied Fracture Mechanics*, **33**, 67-72(2000).
10. Sasagawa, K., Naito, K., Saka, M., and Abé, H., *J. Appl. Phys.*, **86**, 6043-6051(1999).
11. Sasagawa, K., Naito, K., Saka, M., and Abé, H., *Advances in Electronic Packaging 1999*, ASME **EEP-26-1**, 233-238 (1999).
12. Sasagawa, K., Naito, K., Kimura, H., Saka, M., and Abé, H., *J. Appl. Phys.*, **87**, 2785-2791(2000).
13. Blech, I. A., *J. Appl. Phys.*, **47**, 1203-1208(1976).
14. Blech, I. A., *Acta Materialia*, **46**, 3717-3723(1998).
15. Lloyd, J.R., and Smith, P. M., *J. Vac. Sci. Technol.*, A **1**, 455-458(1983).
16. Ainslie, N. G., d'Heurle, F. M., and Wells, O.C., *Appl. Phys. Lett.*, **20**, 173-174(1972).
17. Lloyd, J. R., Smith, P. M., and Prokop, G. S., *Thin Solid Films*, **93**, 385-395(1982).
18. Attardo, M. J., and Rosenberg, R., *J. Appl. Phys.*, **41**, 2381-2386(1970).
19. Blech, I. A., and Meieran, E. S., *Appl. Phys. Lett.*, **18**, 263-266(1967).
20. Tu, K. N., *Physical Review*, B**45**, 1409-1413 (1992).
21. Wang, P. -C., Cargill III, G. S., Noyan, I. C., and Hu, C. -K., *Appl. Phys. Lett.*, **72**, 1296-1298 (1998).
22. Lloyd, J. R., and Steagall, R. N., *J. Appl. Phys.*, **60**, 1235-1237(1986).

Influence of Grain Boundary Type on Electromigration in Damascene Copper Lines

Horst Wendrock, S. Menzel, T. G. Koetter, D. Rauser, K. Wetzig

Institute for Solid State and Materials Research (IFW) Dresden
P. O. box 270016, 01171 Dresden, Germany

Abstract. The preceding trend of miniaturization in microelectronics industry leads to drastic increase of current density in the interconnect lines. Additionally, copper as new material and the double damascene technology poses new questions concerning the interconnect damage by electromigration (EM). In the work presented here, the influence of the grain boundary network on the formation of voids and hillocks in Cu interconnects is investigated. For this purpose, electromigration tests of unpassivated Cu lines (width 1 to 4 μm) were carried out in situ in a SEM, coupled with acquisition of their complete microstructural and orientation state by means of Electron Backscatter Diffraction (EBSD). After EM testing the line damages were investigated using a Focused Ion Beam device for studying the microstructural details in normal direction and around the defects. It was found that the strings of large angle grain boundaries in the electron flow direction are decisive for the formation of voids and hillocks. This was found for Cu lines produced by PVD as well as for electroplated lines, though they had rather different microstructure and texture. FIB cuts through hillocks showed that they have grown sometimes epitaxially, sometimes non- epitaxially, and they often consisted of more than one grain. The effect of misorientation state of a grain boundary on its transport properties is discussed and compared with former results on Al interconnects.

MOTIVATION

According to the ongoing miniaturization in the production of ULSI circuits in microelectronics industry, and attended by the transition from aluminum to copper as material for interconnects, research on the mechanisms and details of current induced damage in small and thin interconnects has been intensified during the last years. These processes, summarized under the name electromigration, are understood as diffusion of vacancies modified by momentum transfer from the moving electrons. In the interconnects, different paths are assumed to contribute to the flow of material /1/:

⇒ *Diffusion along surfaces, grain boundaries, interfaces, and through the volume.*

The diffusivity and the activation energy of these processes vary over a wide range, depending strongly on the details of interconnect technology /2, 3/. Compared to the electromigration behavior of Al, in Cu both surface and interface diffusion should be more important because Cu does not form a self-passivating oxide layer like Al. Another difference is that Cu has more anisotropy in the elastic-plastic constants, and thus the mechanisms of stress formation and relaxation in the interconnects are more complicated.

CP612, *Stress-Induced Phenomena in Metallization:* Sixth Int'l. Workshop, edited by S. P. Baker et al.
© 2002 American Institute of Physics 0-7354-0058-X/02/$19.00

The newly introduced "Double Damascene" technology causes further problems with complicated stress states and early failures in the region of vias due to the formation of voids.

In this context, the work presented here will give some results concerning the influence of grain boundaries (GB) on the EM behavior in unpassivated damascene Cu lines produced by sputtering (PVD) and by electroplating (EP). At first we recorded the microstructure by use of the Electron Backscatter Diffraction (EBSD) technique, and then we performed EM tests under observation inside a Scanning Electron Microscope (SEM) to find the initial points for the growing defects, and to correlate them with the initial microstructure /4, 5/. Afterwards we studied the hillocks by means of a Focused Ion Beam device (FIB) to reveal the spatial microstructure of the defects and of their vicinity.

EXPERIMENTAL DETAILS

Two types of unpassivated Damascene Cu interconnect lines were used:
a) PVD lines produced by magnetron sputtering, depth 500 nm, width 1 to 4 μm, line length 200 to 800 μm, barrier of 30nm Ta and 50 nm TaN
b) EP lines with and without via-contacting, depth 400 nm, width 1 to 3.8 μm, 30nm Ta barrier

The PVD lines were reflowed directly after the deposition at 570°C, whereas the EP lines have been tempered at 350°C before CMP. The reticles containing the lines were packaged, contacted by bonding and then mounted onto a flat heating stage custom-made for the SEM. The temperature gradient in the housing has proven to be not more than 20 K when heated to 250°C; with highest current loading ($\leq 10^{11}$ Am^{-2}) the line temperature was about 30 K higher than in the heating stage.

The observations during tests were carried out in a SEM DSM 962 (LEO, Germany) with LaB$_6$ cathode. All EBSD measurements were also performed in this SEM with the system CHANNEL4 (HKL-Technology, Denmark). The pressure was about 10^{-4} Pa in the chamber during the EM-tests. After the tests, at the Cu surfaces no significant oxide layer, but a carbon layer was found by AES measurements, known as contamination layer from the residual gas in the SEM. It seems to be caused from the EBSD mappings where a strong electron beam (1..2 nA probe current in a spot of 30-60 nm) stands for a rather long time (0.2-0.3 sec) at every point of the line.

With the system used, a lateral resolution of EBSD in Cu of 200 nm can be estimated. All maps were done with a stepsize of 100 nm in both directions with 30 kV acceleration voltage and about 1 nA probe current. The EBSD patterns were taken with frame integration time of 0.16 s and an average cycle time of 0.25 s.

RESULTS AND DISCUSSION

Microstructure and texture

The microstructure and texture were studied by EBSD mappings of some parts of the interconnect (see figs. 3 and 5). The mean grain size was between 0.5 and 1 μm in all cases, EP lines showed a wider size distribution than PVD lines (when twin boundaries disregarded). Fig.1 shows that there are distinct differences in texture, PVD lines show a very strong (111)-fibre texture in normal direction whereas in EP lines the (111)-fibre texture is weak and some side textures are found. Furthermore, the statistics of grain boundary misorientation angles should be essential for diffusion properties. In fig. 2 the frequency distributions of GB angles for PVD and EP Cu lines are given. Obviously, in PVD lines the low (< 10°) and high angle (10-55°) grain boundaries dominate, whereas the EP lines have almost only high angle and twin boundaries (55-60° with (111) rotation axis, Σ3 in the CSL nomenclature).

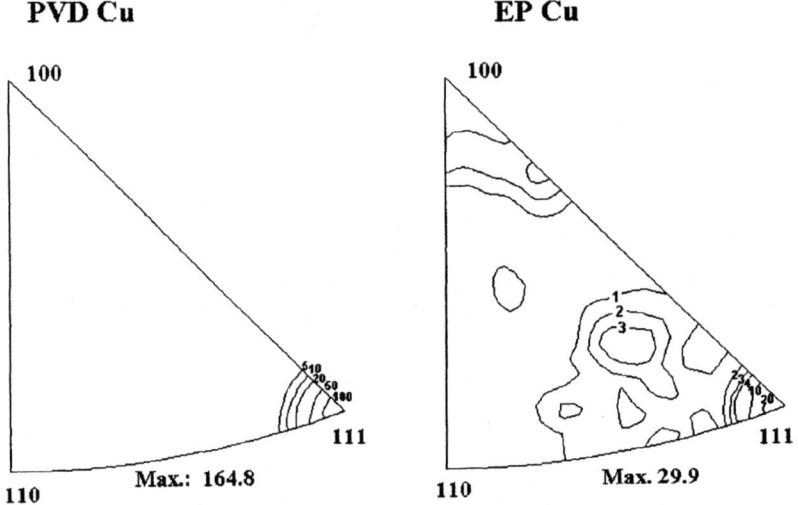

FIGURE 1. Inverse Pole figures of PVD- and EP-Cu lines in surface normal direction, each 2 μm wide, derived from EBSD mappings, numbers are intensities in m.u.d. values.

Additionally, after the tests the lines were studied by Focused Ion Beam (FIB) technique. It was found that the microstructure is two-dimensional at the most places, only in clusters of small grains (<0.5 μm) sometimes a layered structure with burried grain boundaries parallel to the surface was observed. (see figs. 4 and 6).

This fact must be kept in mind when trying to estimate the most probable path of diffusion from the grain boundary images at the line surface.

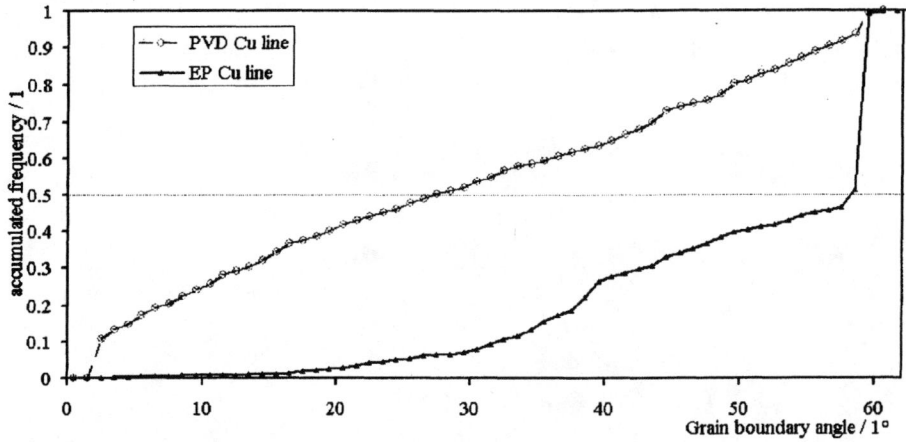

FIGURE 2. Histogram of grain boundary misorientation angles in PVD- and EP-Cu lines, each 2 μm wide, measured by EBSD from a 20 μm part of the line.

EM tests on PVD Cu lines

In all EBSD maps and SEM images shown below the electron flow direction is from top to bottom. Grain areas are drawn in the maps according to their deviation from (111) in normal direction, and additionally the grain boundaries are shown as lines of 3 types. The low angle GB are light grey, the high angle GB black, and the twin GB white.

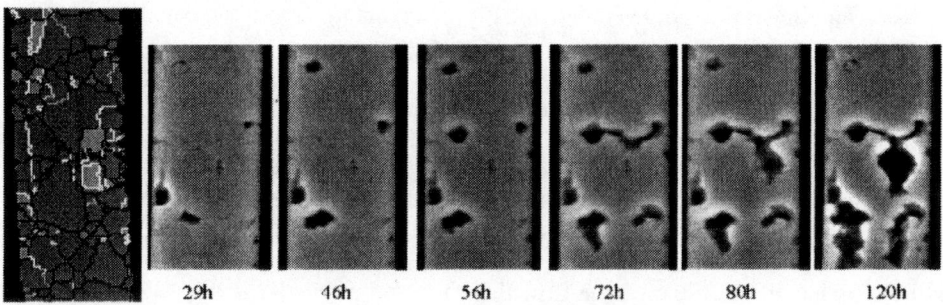

| 29h | 46h | 56h | 72h | 80h | 120h |

FIGURE 3. EBSD mapping and sequence of damage development during an EM test, PVD Cu line 4 μm wide, test at T=310 °C, j=2.5 10^{10} Am^{-2}, current flow from top to bottom, grains are grey value coded with deviation angle from (111) in normal direction, lines show the grain boundary types; black lines = high angle GB, light grey = low angle GB, white= twin GB (Σ3).

Fig. 3 shows the results of an EM test with in situ SEM observation and prior complete EBSD mapping. Because of the many small grains it is difficult to recognize locations of diffusivity divergences in the grain boundary network. However, the sites of void formation can be correlated mostly with a beginning high angle GB in electron flow direction. The formation of hillocks observed at other parts of the line could mostly be correlated with the end of a high angle GB in electron flow direction. The low angle GB, however, seem to have no influence on the damage formation.

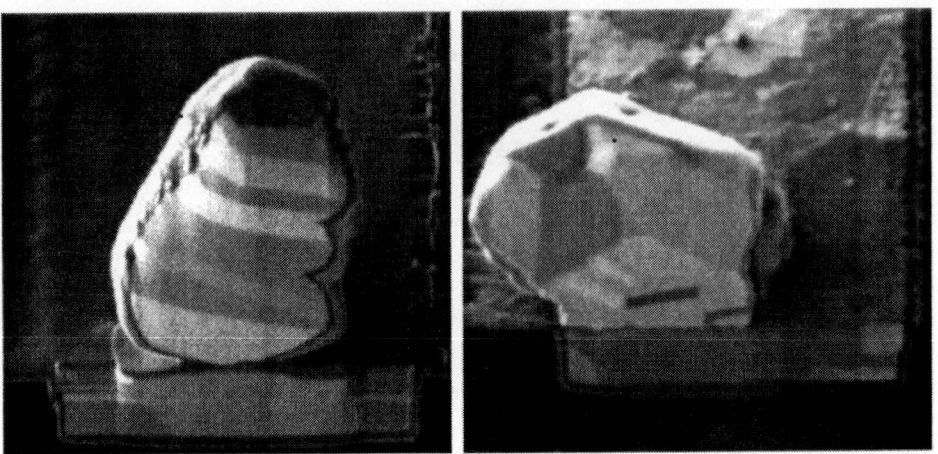

FIGURE 4. FIB cuts through two hillocks after EM test of a 2 μm wide PVD Cu line, tested at T=330 °C, j=7·10^{10} Am^{-2}, cutted across the line and tilted at 45°, electron flow direction from top to bottom; images are taken in the FIB with ion induced secondary electrons and show distinct orientation contrast.

Looking at the microstructure of hillocks formed on PVD Cu lines, we have found that they are clearly non-epitaxially grown and are polycrystalline themselves (see fig. 4). They could have been generated like extrusions at a GB triple point near the surface. In some cases also regular, mostly hexagonal shaped whisker hillocks were found.

EM tests on electroplated Cu lines

The same experiments as for PVD lines have been performed on EP lines. In fig. 5 some results are displayed for a 3.5 μm wide line. Note that the EP lines had a rather short lifetime compared with the PVD lines; but unfortunately we could not carry out enough tests to get statistically significant data. In these lines, connected regions without high angle GB in electron flow direction and only with a twin GB network are observed. They seem to act as blocking zones (shortly BZ, marked in fig. 5), because void formation is found in the downwind of such zones, and hillocks are formed in the upwind of them. In EP lines, again the network of high angle GB (especially the longitudinally directed ones) is decisive for the damage formation. Twin GB do not seem to influence the damaging process, like the low angle GB in the case of PVD Cu lines.

In the tests of EP lines, we did see less hillock formation than in the case of PVD lines. This could be a sign for a somehow increased surface and/or interface diffusion, but could also be explained with the assumption that electroplated copper has a higher density of vacancies than PVD copper. However, in the cases of hillock formation we found in some FIB cuts (see fig. 6) that they have grown epitaxially through the line. So by these findings the hypothesis of increased interface diffusion in EP Cu lines becomes likely.

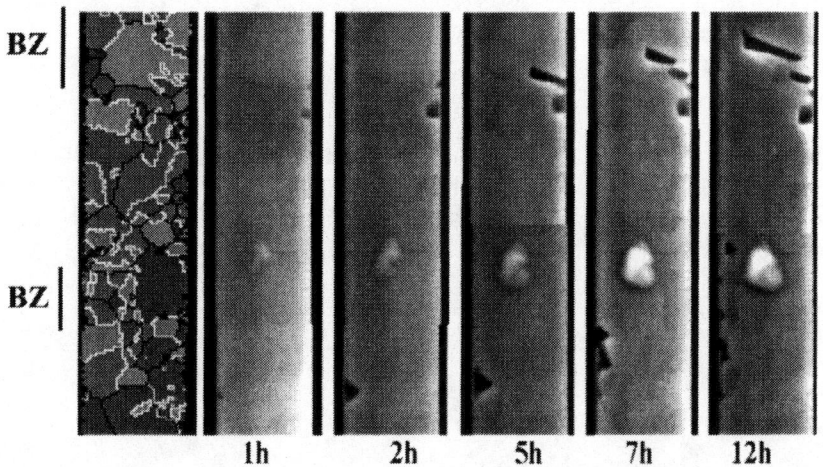

FIGURE 5. EBSD mapping and sequence of damage development during an EM test, EP Cu line 3.5 μm wide, tested at T=280 °C, j=3·10^{10} Am^{-2}, drawing like fig. 3.

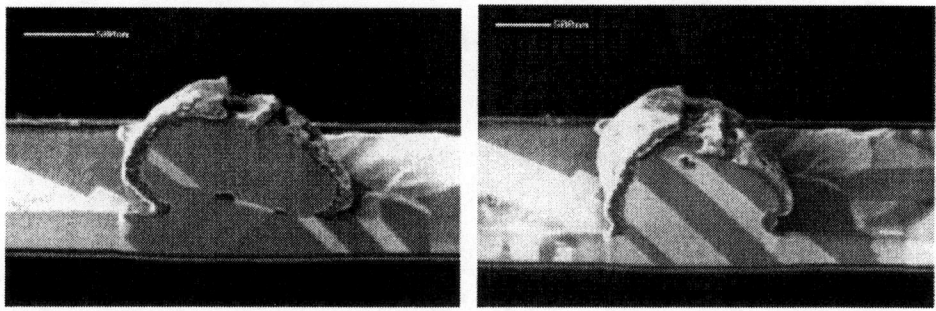

FIGURE 6. FIB cuts through two hillocks after EM test of a 3.5 μm wide EP Cu line, tested at T=260 °C, j=3·10^{10} Am^{-2}, cutted along the line, imaging conditions similar to fig. 4, electron flow direction from left to right.

The surface diffusion seems to be hindered enough to have no influence on the process. A hint for this can be seen in the FIB cuts of hillocks in fig. 6, where a cap layer is visible which has probably been formed by some carbon contamination produced in the SEM during the EBSD maps.

EBSD maps during current loading

With narrower EP Cu lines with a bamboo-like microstructure, some EM tests were performed with repeated EBSD maps of the whole line. The main part of the line remained unchanged over the time, but at some locations grains were rebuilt, as can be seen in fig. 7. We have seen observed moving twin GB, and the formation and disappearing of small grains near the side wall of the line (grain No. 5 and neighbors in fig. 7). Void formation was found only at the sidewalls and at those sites where grain structure had changed before. It is still under question if there is a correlation between these sites of damage and any special grain orientation at that place, because the resolution of our EBSD system is not high enough to see grains smaller than 200 nm. After such an experiment, the carbon contamination layer caused by the EBSD work in the SEM was found to be about 100 nm thick. We removed it with the FIB by a selective carbon mill attachment, and took a final image of the line after test (fig. 7 right).

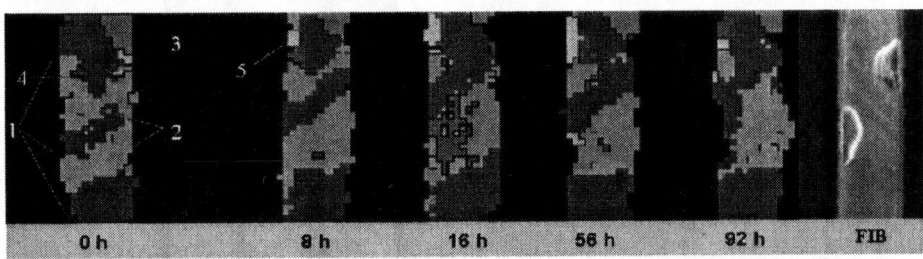

FIGURE 7. Series of EBSD mappings during current loading of a 1 μm wide EP line, tested at T=255°C, j=8·10^{10} Am^{-2}, electron flow direction from top to bottom.

Note that this kind of lines showed a very high lifetime inspite of the rather large current density. This coincides with a similar effect for Al bamboo-lines, so this also indicates that the grain boundary diffusion influences the EM resistance of Cu lines analogously to Al lines.

CONCLUSIONS

The electromigration behavior of Cu lines produced by PVD and by electroplating show some differences but are both controlled by the network of high angle grain boundaries. Interruptions of this network are supposed to be the reason of void- and hillock-formation.

For the growth of hillocks, different mechanisms are to be considered. In the case of PVD Cu lines, non-epitaxial growth has been found. It can be understood as the consequence of a good interface adhesion at the bottom of the line, and of an extrusion mechanism at grain boundary triple junctions near the surface. In electroplated lines, hillocks were found to have epitaxial contact to the line. So one may suppose that the growth starts at the bottom of the line, perhaps due to a less perfect adhesion to the interface.

By repeated EBSD mapping of Cu lines during thermal and current loading, it was shown that in narrow lines with bamboo structure some grains at the sidewall are rebuilt in a complex way, and the places of these changes are clearly correlated with the later formation of voids. These processes are not yet understood well and need further investigation with increased resolution.

ACKNOWLEDGMENTS

The authors wish to thank C. Wenzel and H. Schuehrer for help concerning the PVD Cu interconnect samples. Furthermore they are grateful to AMD Saxony Manufacturing GmbH for intensive supplement and helpful discussions.

This work was funded by the Saxony Ministry of Science and Art.

REFERENCES

1. J. R. Lloyd, J. J. Clement; Thin solid films 262 (1995), 135-141.
2. C.-K. Hu, J. M. E. Harper, Materials Chemistry and Physics; 52 (1998) 5-16.
3. A. Gladkikh, M. Karpovsli, A. Palevski, Y. S. Kaganovski; J. Appl. Phys. 31 (1998) 1626-1629.
4. A. Buerke, H. Wendrock, T. G. Koetter, S. Menzel, K. Wetzig, A. v. Glasow; MRS Symp.Proc Vol. 563, eds. C. A. Volkert, A. H. Verbruggen, D. Brown, San Francisco, 1999, 109-114.
5. K. Wetzig, H. Wendrock, A. Buerke, T. G. Koetter; Proc. 5th Int. Workshop on Stress Phenomena in Metallization, eds O. Kraft et al., Stuttgart, 1999, pp. 89-99.

Evaluation of Temperature Rise Due to Joule Heating and Preliminary Investigation of Its Effect on Electromigration Reliability

S. Shingubara, S. Miyazaki, H. Sakaue, and T. Takahagi

Hiroshima University, Graduate School of ADSM(advanced science and matter),
Kagamiyama 1-3-1, Higashi-Hiroshima 739-8530, Japan
E-mail:shingu@hiroshima-u.ac.jp

Abstract. A demand to increase current density of ULSI interconnections with shrinkage of feature sizes inevitably brings about temperature rise due to Joule heating. We studied how temperature distribution changes with an increase in the current density and further investigated its effect on electromigration-induced failures. We evaluated a temperature rise in a single level Al-alloy interconnection by the use of IR-CCD camera and 3-D fluid dynamical finite element analysis. A build-up of a significant temperature gradient was observed with an increase in the current density. Failure analysis of electromigration accelerating tests revealed that the degradation of MTF at high current density conditions was due to occurrence of a new failure mode of "evaporation mode". The evaporation mode is most likely caused by an annihilation of a void. Further discussions concerning the relationship between the evaporation mode and the Joule heating effect are given.

INTRODUCTIONS

As the miniaturization of ULSI devices proceeds, an operating current density inevitably increases due to the scaling scenario. A number of interconnection levels increases steadily in order to avoid a delay of signals also. These circumstances cause an increase in the temperature of ULSI chips by Joule heating effect. The temperature rise due to Joule heating would affect electromigration (EM) reliability since EM MTF(median time to failure) is in general a function of $\exp(E_a/kT)$, where T is an absolute temperature, and E_a is an activation energy, and k is the Boltzmann constant (1-3). It is apparent that temperature rise is delicately dependent on thermal conductivity of a dielectric film. There is certainly a trade-off between the lowering of operating temperature and the lowering of dielectric constant of insulator films. The aims of the present study are to estimate three dimensional temperature distributions due to Joule heating quantitatively, and further to investigate its effect on electromigration reliability. A special attention was paid to understand a current density dependence of EM MTF at high current density conditions where Joule heating is not negligible.

CP612, *Stress-Induced Phenomena in Metallization:* Sixth Int'l. Workshop, edited by S. P. Baker et al.

EXPERIMENTAL METHODS

In order to elucidate the effect of Joule heating clearly, we designed an interconnect pattern in which an electric current concentrates only at the specified area. The schematic diagram of the interconnect pattern is shown in Fig.1. The width of the lead patterns is 4.0 μm, and in between the lead wires, there is a narrow wire of

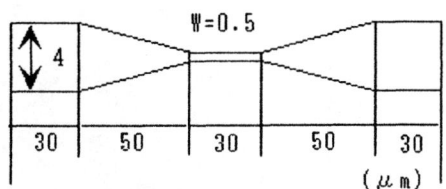

**Fig.1 Schematic diagram
of interconnection pattern**

0.5 μm width and 30 μm length. A film thickness is 0.6 μm. Since the current density is the highest at this narrow part, temperature rise due to Joule heating is observable at high current density conditions. We used aluminum interconnection without dielectric passivation film for simplicity. We measured temperature distribution of the narrow part of interconnection by an infrared microscopy attached with a high-resolution CCD camera (THEMOS-100, Hamamatsu Photonics Corp.) during direct current (DC) current stressing test, at the ambient temperature of 120 °C. A spatial resolution of the temperature measurement is about 2 μm that is determined by the wavelength of the infrared light. However, we could observe temperature increase caused by Joule heating of the 0.5 μm width interconnection. This is because the dielectric film nearby the interconnection was also heated up due to the heat conduction from the interconnection. Infrared light intensity of the one unit of CCD camera comes from a sum of lights from interconnection and those from dielectrics. The observed temperature does not directly correspond to the temperature of 0.5 μm width interconnection, so that we calibrated an absolute value of temperature from the resistance measurement of the stressed interconnection. From a resistance increase caused by the current stress, we estimated average temperature of the narrow part of interconnections, and calibrated the observed temperature. We carried out three-dimensional thermal fluid dynamical finite element (TFDFE) analysis for further estimation of three-dimensional temperature distribution.

RESULTS AND DISCUSSIONS

One example of the observed two-dimensional temperature distribution is shown in Fig.2. In this case, DC current of $23.1 MA/cm^2$ was stressed to Al-1%Si-0.5%Cu interconnection without any dielectric passivation film. There is a significant Joule heating effect at this condition of the current density. The maximum interconnect temperature was 330 °C, which was 210 °C higher than the ambient temperature. The temperature rise existed only in a limited area of about 50 μm in length, which is 1.6 times larger than the current crowded area of 30μm length.

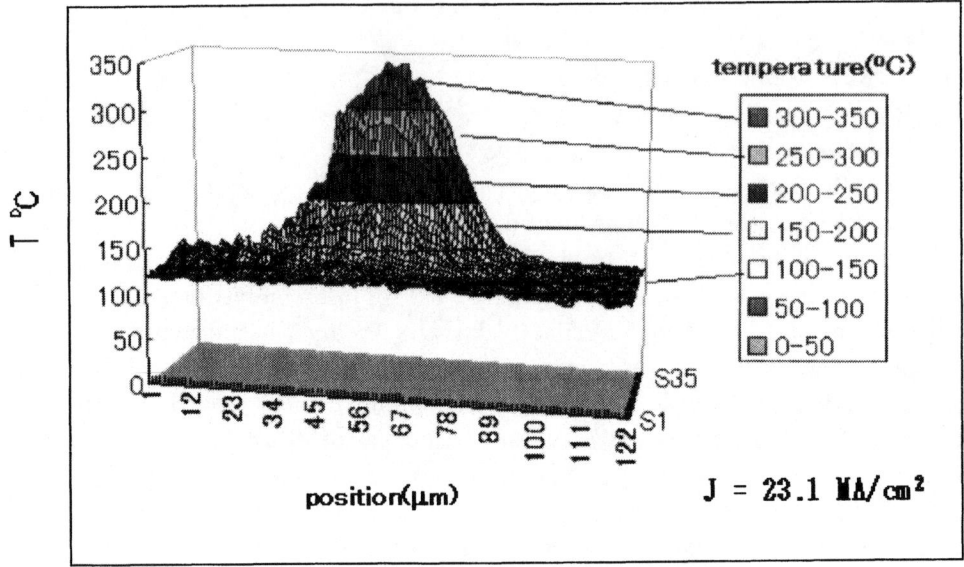

Fig.2   Two-dimensional mapping of the temperature during the DC current stressing test.   The ambient temperature is 120 °C, and the current density is 23.1MA/cm².

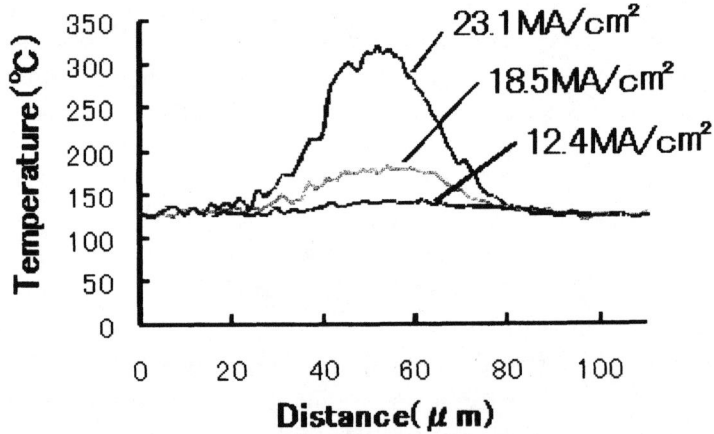

Fig.3   Temperature distribution along the interconnection measured by THEMOS-100.

Ambient temperature is 120 °C.

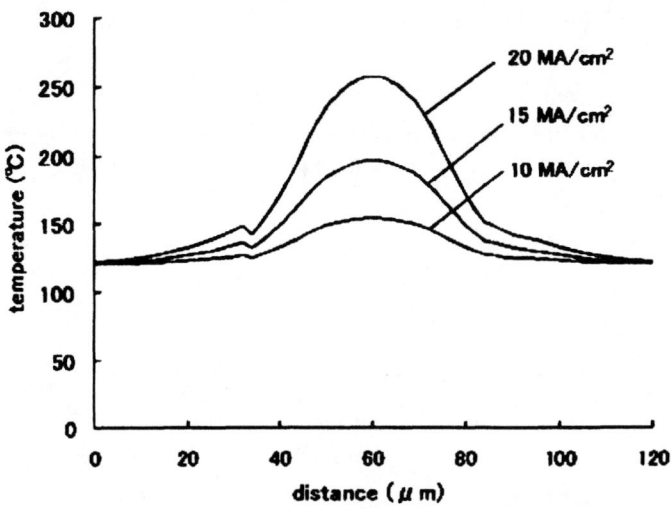

**Fig.4 Temperature distribution along the
interconnect calculated by TFDEF**

Temperature gradient exists even in the narrow part of the interconnection, and it is significantly large at the region where line width changes.   Temperature distributions along the direction that is parallel to the electric current are shown in Fig.3. Temperature rise is significant when the current density increased from 12.4 MA/cm^2 to 23.1 MA/cm^2.   We estimated an average temperature at the narrow part of interconnection for the observed temperature distributions.   There is an nonlinear relationship between the average temperature rise ΔT and the current density J such that $\Delta T = J^n$, where n is 2.1-2.2.

We also estimated temperature distribution using three-dimensional thermal fluid dynamical finite element (TFDFE) computer simulation (Fig.4). TFDFE analysis agrees well with the experimental result quantitatively when adequate boundary conditions were used. In this case, both outside boundaries of Al interconnection in x-direction, that is parallel to the longitudinal direction of the interconnection, were set to the ambient temperature. Backside of the Si substrate was set to the ambient temperature also, and there was an air on the top of the Al interconnection. In order to fit the absolute values of the calculated temperature to the experimental ones, we adjusted the thickness and the thermal conductivity of Si-substrate. Since the spatial resolution of the IR-CCD measurement is not better than 2 μm, there are some fluctuations and inaccuracy in the observed temperature distribution.   We could obtain more accurate information of temperature distribution by the use of TFDFE analysis.

We carried out electromigration accelerating test at various current density conditions at which Joule heating effect is not negligible. Electromigration life-time distributions at various current density conditions are shown in Fig. 5, and the log-log plots of current density dependence of Electromigration median time to failure (EM MTF) are shown in Fig 6. The outstanding feature of EM MTF is that it decreased significantly when current density increased above 10 MA/cm^2. In figure 6, the current density dependence estimated by the empirical Black formula and its corrections that took temperature rise due to Joule heating into account is plotted. Since EM MTF is empirically expressed by J. R.Black as follows (1);

$$MTF = A\, J^{-2}\, exp(Ea\,/\,kT),$$

the current density dependence in log-log plot is linear as shown in dark triangles in Fig.6. When a temperature increase averaged over the interconnection due to Joule heating is considered, EM MTF deviates from the Black relation to shorter values with an increase in the current density, which is shown in dark squares in Fig.6. However, the experimental data (dark circles) became much shorter when the current density increased from 10 to 15 M A/cm². The SEM (scanning electron microscopy) micrographs of the failure sites are shown in Fig.7 (case 1). Most of disconnections of interconnection occurred at the area where the line width changed in the cathode site. Since Al atoms migrate toward anode direction due to electromigration, a negative atomic flux divergence is large at this area. Depletion of atoms causes accumulation of vacancies and eventually a fatal void grew as shown in Fig.7-(a) at the relatively low current density conditions. At high current density conditions, we observed a huge disconnection which seemed to come from an evaporation of Al interconnection, as shown in Fig.7-(b). About 4 μm length Al inter- connection had disappeared completely, and some residues which were formed after melting remained. We could clearly discriminate these two failure modes;

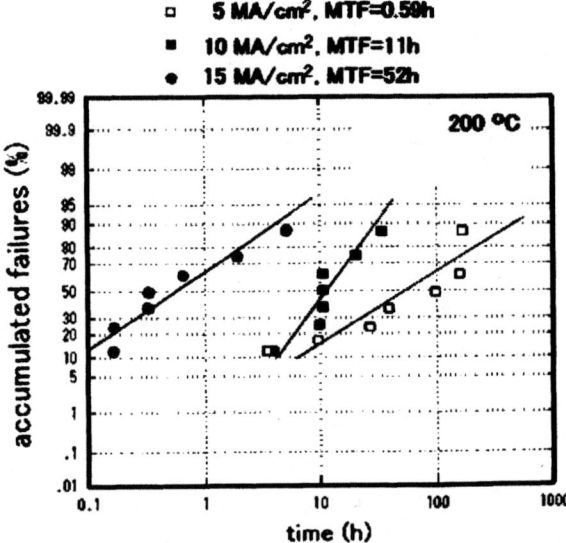

□ **5 MA/cm², MTF=0.59h**
■ **10 MA/cm², MTF=11h**
● **15 MA/cm², MTF=52h**

Fig.5 Electromigration life time distribution at various current densities. Ambient temperature is 200 °C.

the void growth mode, and the evaporation mode, by the SEM observations. We sometimes observed a melted part that was next to the evaporation area. When we carried out EM accelerated test using long (100μm) and wide (2.5μm) Al-alloy interconnections (case 2), we observed these two failure modes too (see Fig.8), however, a change of failure modes from the void growth to the void annihilation occurred at rather low current densities.

These results of failure mode statistics at various current density conditions are summarized in Table 1. In the case 1 which used the Al interconnection pattern described in Fig.1, the all failure modes was the void growth mode at the current density of 5 MA/cm²; and the most of them were the evaporation mode when the current density was 15 MA/cm². These two failure modes were mixed when the current density was 10 MA/cm². In the case 2, where we used the wider and longer interconnections, almost 100 % of the failure mode was the evaporation mode when the current density was 6 MA/cm², while it was the void growth mode when the current density was 2 MA/cm². The common feature of the occurrence of the evaporation mode is that they occurred when the temperature increase due to Joule heating was significant. For example, the average temperature increase was 16.3 °C when the current density was 15 MA/cm² in the Case 1, and it was 29.4 °C when the current density was 6 MA/cm² in the Case 2.

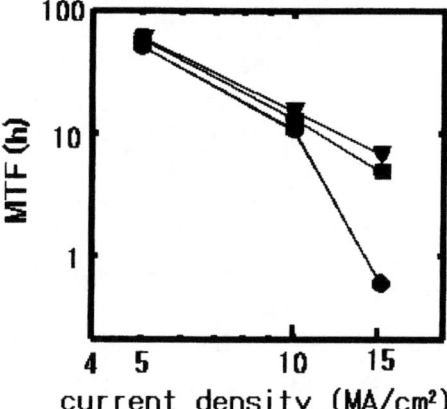

▼ Black`s formula

■ Black`s formula
   (Joule heating considered)

● experimental results

current density (MA/cm²)

Fig.6 Current density dependence of electromigration MTF

(a) Void Growth Mode (J=5MA/cm²)

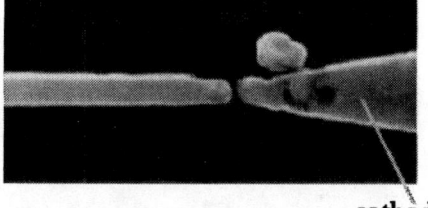

cathode

(b) Evaporation mode caused by void annihilation (J=15MA/cm²)

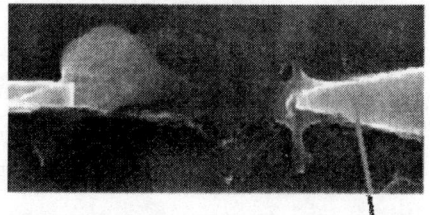

cathode

Fig7. SEM observations of the failure sites after DC stressing test at 200 °C.

99

(a) void growth mode (J=2MA/cm²)

(b) void evaporation mode(J=6MA/cm²)

melt zone ⊢⊣1μm

Fig.8 SEM observation of the two failure modes. L=1000 μm, W=2.5μm

Similar phenomena were reported for the electromigration test of Cu interconnection at high current density conditions   recently (4).

The mechanism of the evaporation mode is most likely to come from an annihilation of a void. Shingubara et al. reported that a huge resistance increase occurred within a second during the DC electromigration test, and further they identified that it was caused by the annihilation of the sub-micron scale void (5).   The resistance increase due to the void annihilation is well

(Case 1) W=0.5 μm. L=30 μm. Al alloy unpassivated. 200 °C

DC current density (MA/cm²)		5	10	15
average temperature increase (°C)		1.8	7.3	16.3
failure mode	evaporation (void annihilation)	0/9	5/8	8/9
	void growth	9/9	3/8	1/9
MTF (h)		52	11	0.59

(Case 2) W=2.5 μm. L=1000 μm. Al alloy unpassivated. 250 °C

DC current density (MA/cm²)		2	4	6
average temperature increase (°C)		3.2	13.0	29.4
failure mode	evaporation (void annihilation)	0/10	4/10	10/10
	void growth	10/10	6/10	0/10
MTF (h)		30	1.3	0.3

Table1. Change in the electromigration induced failure modes at various current density conditions

understood quantitatively if a void is decomposed into a huge number of vacancies.

Instability of a void comes from a change of local stress field from tensile to compressive, and the vacancies created from the void cannot annihilate at the surface since there is a tight interface between aluminum and alumina. They reported a rapid jump of a void, in which a void was decomposed into vacancies during the jump, and later on they confirmed this hypothesis by the molecular dynamic simulations (6,7). The annihilation of a void due to electromigration eventually causes a significant local heating. If we take a vacancy resistivity of 3 $\mu\Omega$cm/atom % (9) into consideration and a void is larger than one-third of the interconnection width, a resistance increase due to a void annihilation can exceed 50 % of the initial resistance in the local area surrounding the void. When the current density is already large to produce significant temperature rise due to Joule heating, the additional increase of the resistance might cause devastating rise of the temperature that exceeds the melting point of Al. The mechanism of the void evaporation mode is illustrated in Fig.9. At relatively low current density conditions, the void growth is the most dominant failure mode. However, with an increase in the current density, dynamic changes of void shape and concomitant change in the stress distribution within an interconnection become more

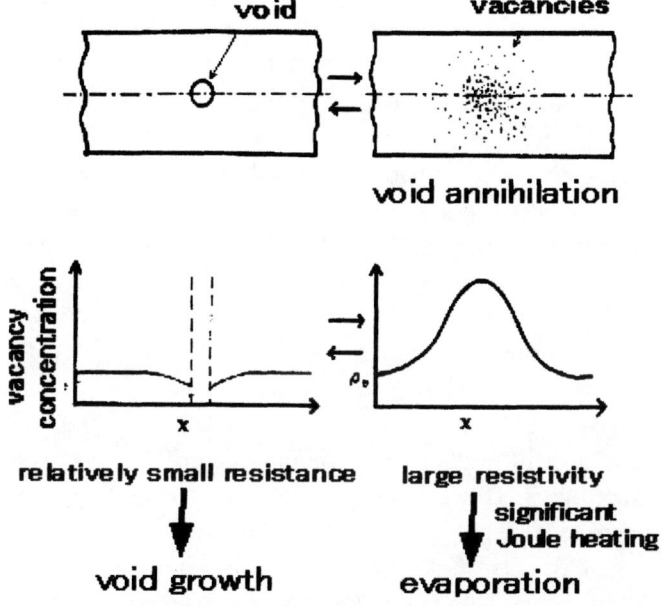

Fig.9 Schematic illustration of electromigration-induced failure modes; the void growth mode, and the void annihilation modes caused by the void anihilation.

important. Movement of voids, their coalescence and gradual diminishment were observed by many researchers (5, 9-12). Non-equilibrium properties of electromigration become more significant when current density is higher, and instability of a void caused by a change in the local stress field from tensile to compressive occurs with a certain probability. When an enough large void is annihilated to form a huge number of vacancies, significant Joule heating arises, and interconnection is evaporated or melted. When a small void is annihilated into vacancies, soon later they are coalesced each other to form another void since the state with too much excess vacancy concentration is thermodynamically unstable, and the observer may recognize this phenomena as a void jump which is concomitant with a sudden change in the resistance (5).

A relationship between the increase in the current density and the void evaporation phenomena is not well understood. A high current density stress accelerates Joule heating as well as electromigration drift velocity. The temperature increase due to Joule heating effect is almost proportional to a square of the current density as discussed before, while the electromigration drift velocity is linearly dependent on the current density. The change of the electromigration induced failure mode may be due to the nonlinear relationship to the current density, so that Joule heating effect is considered to play an essential role against the void evaporation phenomena. From the temperature distributions shown in Figs.3 and 4, it is easily expected that a compressive stress is the largest at a central part of the narrow part of interconnection, since thermal expansion coefficient of Al is much larger than that of dielectrics. The compressive stress of the center of the narrow part at 15 MA/cm^2 is estimated to be 93 MPa if we assume Al interconnection is stress-free at the ambient temperature of 120 °C. Thus there is certainly thermal stress gradient along the Al interconnection that comes from the temperature distribution. If a void drifts toward cathode direction (9) and comes across to the central part of the interconnection where a compressive stress is the largest, a stress field around a void gradually changes to compressive one. Such a change of a local stress around a void may cause instability of a void, then the evaporation failure mode would occur.

There is a possibility of thermo-migration induced failures when a temperature gradient becomes significant. However, under DC stress conditions, the effect of thermo-migration seems to be negligible as compared to that of electromigration. For example, at the current density of 10 MA/cm^2 as shown in Fig.4, the maximum temperature gradient is about 2.0 °C/μm and the corresponding thermo-migration driving force $(Q^*/T)(dT/dx)$ is 1.2×10^{-15} [N] if we use a reported value of Al self-thermo-migration (13). While the electromigration driving force Z^*eE is about 5×10^{-11} [N] if we use Z^* value of -12 (at 120 °C) according to the literature (14).

Hence, thermo-migration cannot be dominant in the case of DC stress conditions, and it might be important in the case of AC stress conditions.

CONCLUSIONS

We evaluated a temperature rise in a single level Al-alloy interconnection due to Joule heating by the use of IR-CCD camera and 3-D fluid dynamical finite element analysis. Average interconnection temperature increase is almost proportional to a square of the current density J, however, a build-up of a significant temperature gradient was observed with an increase in the current density. We also carried out DC electromigration accelerated tests at various high current density stress conditions. The current density dependence of electromigration MTF deviated from the Black law to shorter values when the current density increased, and failure analysis revealed that the degradation of MTF at high current density conditions was due to occurrence of a new failure mode of "evaporation mode". The evaporation mode is most likely caused by the annihilation of a void, which is originated from a change of local stress from tensile to compressive, and further it seems to be related to the thermal stress distribution that derives from the significant temperature gradient. The evaporation mode may become serious when ULSI interconnections are operated at higher current density conditions. In those cases, adequate materials design to reduce temperature rise due to Joule heating would be a key issue to improve electromigration reliability. There is a possibility of thermo-migration induced failures when a temperature gradient becomes significant, and further study related to AC electromigration test at high current density conditions is in progress.

ACKNOWLEDGEMENT

The authors would like to express gratitude to Hamamatsu Photonics Corporation for measurements using IR-CCD camera, Themos-100. This study was partially supported by the grant-in-aid of the Japan ministry of education and science.

REFERENCES

1. Black, J.R,   Proc. IEEE   **57,** 1587 (1968).
2. Hunter, William R.,   IEEE Trans. Electron Devices **44,** 304 (1997).
3. Banerjee, K., et al.,   Tec. Dig. IEDM, 65 (1996).
4. Ramanath, G.,
    Proc. of 6th International Workshop on Stress-Induced Phenomena in Metallization, 2001, to be published.
5. Shingubara, S., Kaneko, H., and Saitoh, M.
        J.Appl.Phys.,   69 , 207 (1991).
6. Shingubara, S. , Utsunomiya, I. ,   and Takahagi, T.
    Applied Surface Science, 91, 220 (1995) .
7. Shingubara, S., Utsunomiya, I., and Fujii, T.
    Electronics and Communication in Japan, Part 2,   78 , 82 (1995) .
8. Simmons, R.O. and Bullufi, R.W.   Phys. Rev. **102,** 62 (1960).
9. Shingubara, S., Nishida, H., Sakaue, H., and Horiike, Y.
        Jpn.J.Appl.Phys. **33,** 3860 (1994).
10. Arzt,E., Kraft,O., Nix, W.D., and Sanchez, J.E.,   J.Appl.Phys. **76** , 1563 (1994).
11 Okabayashi,H., Kitamura,H., Komatsu,M., and Mori,H.,
    Proc. of 3rd International Workshop on Stress-Induced Phenomena in Metallization (AIP Conf. Proc. **373**) , 214 (1996).
12. Lee,S., Doan, J., Bravman, J.C., Flinn, P.A., Marieb, T.N., and Ogawa,.S.
    Proc. of 4th International Workshop on Stress-Induced Phenomena in Metallization (AIP Conf. Proc. **418**) , 101 (1998).
13. Oriani,R.A.,   J.Phys. Chem.Solids   **30,** 339 (1969) .
14. Sorbello,R.S.,   J.Phys. Chem.Solids   **34,** 937 (1973).

First-Principle Theoretical Study on the Dynamical Electronic Characteristics of Electromigration in the Bulk, Surface and Grain Boundary

Akitomo TACHIBANA

Department of Engineering Physics and Mechanics
Kyoto University, Kyoto 606-8501, JAPAN
Phone & Fax : +81-75-753-5184
e-mail : akitomo@scl.kyoto-u.ac.jp.

Abstract. New formula for the driving force of electromigration have been found using new concepts of (1) kinetic energy density, (2) tension density, and (3) effective charge tensor density. The new "dynamic" wind charge tensor density $\overset{\leftrightarrow}{Z}_{a\,\text{dynamic wind}}(\vec{r})$ is revealed over and above the conventional "static" wind charge tensor density $\overset{\leftrightarrow}{Z}_{a\,\text{static wind}}(\vec{r})$. Some numerical analysis will be demonstrated for possible application to electromigration reliability problems of ULSI devices where extremely high current densities should be maintained through ultra thin film interconnects. Key issues in this study are the comparative study of the dynamical electronic characteristics for the electromigrating regions in the bulk, surface, and grain boundary.

1. INTRODUCTION

Electromigration is the motion of lattice defects induced by electronic current under the presence of external electric field $\vec{E}_{\text{ext}}(\vec{r})$ [1-7]. The external electric field $\vec{E}_{\text{ext}}(\vec{r})$ at position $\vec{r}$ gives an impetus to the migrating atom and thereby creates electromigration force. The electromigration force $\vec{F}_a(\vec{r})$ for atom a is composed of a direct force and a wind force. The direct force originates from "the action through medium" and is proportional to the $\vec{E}_{\text{ext}}(\vec{r})$, while the momentum transfer on the migrating atom from the electronic current is the origin of the wind force. The discrimination of the two forces has conveniently been represented as effective charge, as so observed experimentally. The effective charge Z_a^* for atom a is then written as

$$Z_a^* = Z_{a\,\text{direct}} + Z_{a\,\text{wind}}, \tag{1}$$

where $Z_{a\,\text{direct}}$ is the direct charge associated with the direct force and $Z_{a\,\text{wind}}$ is the wind charge associated with the wind force.

Theoretical works have exclusively been concentrated in calculating the effective charge Z_a^* [8-14].

CP612, *Stress-Induced Phenomena in Metallization:* Sixth Int'l. Workshop, edited by S. P. Baker et al.
© 2002 American Institute of Physics 0-7354-0058-X/02/$19.00

The migrating atoms tunnel as well as hop across the energy barrier, forming measurable current. The current of the migrating atoms is characterized by the magnitude and direction it carries, which is represented by the probability flux density $\vec{S}_a(\vec{r})$. The ratio of the change in $\vec{S}_a(\vec{r})$ to that in time, multiplied with the mass m_a defines the electromigration force density $\vec{F}_a(\vec{r})$:

$$m_a \frac{\partial}{\partial t} \vec{S}_a(\vec{r}) = \vec{F}_a^S(\vec{r}). \tag{2}$$

To measure the Z_a^*, the $\vec{E}_{ext}(\vec{r})$ is scanned up and down, when the $\vec{F}_a(\vec{r})$ shifts in step with $\vec{E}_{ext}(\vec{r})$. The ratio of the change in $\vec{F}_a(\vec{r})$ to that in $\vec{E}_{ext}(\vec{r})$ defines the effective charge tensor density $\ddot{Z}_a^*(\vec{r})$:

$$\ddot{Z}_a^*(\vec{r})eN_a(\vec{r}) = \frac{\partial \vec{F}_a^S(\vec{r})}{\partial \vec{E}_{ext}(\vec{r})}, \tag{3}$$

where $N_a(\vec{r})$ is the position probability density designating the position $\vec{r}$ at which the atoms are most likely to be found. If the $\ddot{Z}_a^*(\vec{r})$ is averaged over certain region of space, then it can be the Z_a^* observed experimentally in that region.

This is the second paper of our study on the driving force of electromigration. In the first paper of the series, we have studied static electronic properties of electromigrating systems as a function of surface orientations [15].

Recently we have presented the regional density functional theory of "force" of charged particles [16,17]. In the present paper, we shall apply the theory to the electromigration and the unified treatment of the electromigration force and the effective charge will be shown.

2. CHARACTERISTIC DENSITIES OF ELECTROMIGRATION

2.1. Non-Relativistic Limit of QED

From the picture of "action through medium," the force is exerted through medium. The interaction of electrons and nuclei is hence mediated by the electromagnetic field using the electric field $\vec{E}(\vec{r})$ and magnetic field $\vec{B}(\vec{r})$. The quantum mechanical theory is the quantum electrodynamics (QED). Taking the non-relativistic limit, the QED energy density operator $\hat{H}_{QED}(\vec{r})$ reduces to [16]:

$$\hat{H}_{\text{non-relativistic QED}}(\vec{r}) = \frac{1}{8\pi}\left(\hat{\vec{E}}^2(\vec{r}) + \hat{\vec{B}}^2(\vec{r})\right) + \hat{T}(\vec{r}). \tag{4}$$

It should be noted that the interaction energy of the matter fields with the electromagnetic field disappears due to the variational constraint. Then the explicit contribution of the matter fields to the energy density operator $\hat{H}_{\text{non-relativistic QED}}(\vec{r})$ should only be the kinetic energy density operators:

$$\hat{T}(\vec{r}) = \sum_{\alpha} \hat{T}_{\alpha}(\vec{r}), \tag{5}$$

$\hat{T}_{\alpha}(\vec{r})$ being the kinetic energy density operator of the matter particle α:

$$\hat{T}_{\alpha}(\vec{r}) = -\frac{\hbar^2}{2m_{\alpha}} \cdot \frac{1}{2} \left(\hat{\chi}_{\alpha}^{+}(\vec{r}) \hat{D}_{\alpha k}^{2}(\vec{r}) \hat{\chi}_{\alpha}(\vec{r}) + \hat{D}_{\alpha k}^{+2}(\vec{r}) \hat{\chi}_{\alpha}^{+}(\vec{r}) \cdot \hat{\chi}_{\alpha}(\vec{r}) \right), \tag{6}$$

where $\hat{D}_{\alpha k}(\vec{r})$ denotes the covariant derivative operator. The $\alpha = e$ represents electron with charge $Z_e = -1$, while the $\alpha = a$ represents atom a with charge Z_a. The matter particle field operator is $\hat{\chi}_{\alpha}(\vec{r})$ which satisfies the canonical anticommutation relationships for Fermions and commutation relationships for Bosons.

Maxwell's equation of motion for the electromagnetic field is as follows [16]:

$$\text{rot}\hat{\vec{E}}(\vec{r}) + \frac{1}{c}\frac{\partial \hat{\vec{B}}(\vec{r})}{\partial t} = 0, \tag{7}$$

$$\text{div}\hat{\vec{B}}(\vec{r}) = 0, \tag{8}$$

$$\text{div}\hat{\vec{E}}(\vec{r}) = 4\pi\hat{\rho}(\vec{r}), \tag{9}$$

$$\text{rot}\hat{\vec{B}}(\vec{r}) - \frac{1}{c}\frac{\partial \hat{\vec{E}}(\vec{r})}{\partial t} = \frac{4\pi}{c}\hat{\vec{j}}(\vec{r}). \tag{10}$$

The charge density operator $\hat{\rho}(\vec{r})$ and the charge current density operator $\hat{\vec{j}}(\vec{r})$ are given as follows:

$$\hat{\rho}(\vec{r}) = \sum_{\alpha}\hat{\rho}_{\alpha}(\vec{r}), \quad \hat{\rho}_{\alpha}(\vec{r}) = Z_{\alpha}e\hat{N}_{\alpha}(\vec{r}), \tag{11}$$

$$\hat{\vec{j}}(\vec{r}) = \sum_{\alpha}\hat{\vec{j}}_{\alpha}(\vec{r}), \quad \hat{\vec{j}}_{\alpha}(\vec{r}) = Z_{\alpha}e\hat{\vec{S}}_{\alpha}(\vec{r}). \tag{12}$$

For the matter density operator

$$\hat{N}_{\alpha}(\vec{r}) = \hat{\chi}_{\alpha}^{+}(\vec{r})\hat{\chi}_{\alpha}(\vec{r}), \tag{13}$$

the probability conservation is represented as the continuity equation

$$\frac{\partial}{\partial t}\hat{N}_{\alpha}(\vec{r}) + \text{div}\hat{\vec{S}}_{\alpha}(\vec{r}) = 0, \tag{14}$$

where the probability flux density operator

$$\hat{S}_{\alpha}^{k}(\vec{r}) = \frac{1}{2m_{\alpha}}\left(-i\hbar\hat{\chi}_{\alpha}^{+}(\vec{r})\hat{D}_{\alpha k}(\vec{r})\hat{\chi}_{\alpha}(\vec{r}) + i\hbar\hat{D}_{\alpha k}^{+}(\vec{r})\hat{\chi}_{\alpha}^{+}(\vec{r}) \cdot \hat{\chi}_{\alpha}(\vec{r}) \right) \tag{15}$$

satisfies the equation of motion

$$m_{\alpha}\frac{\partial}{\partial t}\hat{\vec{S}}_{\alpha}(\vec{r}) = \hat{\vec{F}}_{\alpha}^{S}(\vec{r}). \tag{16}$$

107

The right hand side is the force density operator $\hat{\vec{F}}_\alpha^S(\vec{r})$ composed of the tension density operator $\hat{\vec{\tau}}_\alpha^S(\vec{r})$ and the Lorentz force density operator $\hat{\vec{L}}_\alpha^S(\vec{r})$:

$$\hat{\vec{F}}_\alpha^S(\vec{r}) = \hat{\vec{\tau}}_\alpha^S(\vec{r}) + \hat{\vec{L}}_\alpha^S(\vec{r}),\tag{17}$$

$$\hat{\vec{L}}_\alpha^S(\vec{r}) = \hat{\vec{K}}_\alpha^S(\vec{r}) + \frac{Z_\alpha e}{c}\hat{\vec{S}}_\alpha(\vec{r}) \times \hat{\vec{B}}(\vec{r}),\tag{18}$$

$$\hat{\vec{K}}_\alpha^S(\vec{r}) = \hat{\vec{E}}(\vec{r})\hat{\rho}_\alpha(\vec{r}).\tag{19}$$

The electric field density operator has intrinsically two origins:

$$\hat{\vec{E}}(\vec{r}) = \hat{\vec{E}}_{ext}(\vec{r}) + \hat{\vec{E}}_{int}(\vec{r}),\tag{20}$$

where $\hat{\vec{E}}_{ext}(\vec{r})$ and $\hat{\vec{E}}_{int}(\vec{r})$ denotes the external and internal origins, respectively.

It should be noted that the integral of the tension density operator in the whole space should vanish [16]:

$$\int d^3\vec{r}\,\hat{\vec{\tau}}_\alpha^S(\vec{r}) = 0.\tag{21}$$

2.2. Effective Charge

We shall calculate the electromigration force in the absence of magnetic field. For a field operator $\hat{A}$ let the ensemble average $\langle \hat{A}\rangle = \mathrm{Tr}\hat{A}\hat{\rho}$ over density matrix $\hat{\rho}$ be performed in order to get observable quantity.

For atom a we get the equation of motion,

$$m_a \frac{\partial}{\partial t}\langle \hat{\vec{S}}_a(\vec{r})\rangle = \langle \hat{\vec{F}}_a^S(\vec{r})\rangle,\tag{22}$$

where $\langle \hat{\vec{F}}_a^S(\vec{r})\rangle$ is the electromigration force density at $\vec{r}$:

$$\langle \hat{\vec{F}}_a^S(\vec{r})\rangle = \langle \hat{\vec{\tau}}_a^S(\vec{r})\rangle + \langle \hat{\vec{K}}_a^S(\vec{r})\rangle,\tag{23}$$

The effective charge tensor density at $\vec{r}$ is found to be:

$$\ddot{Z}_a^*(\vec{r}) = \frac{1}{\langle e\hat{N}_a(\vec{r})\rangle}\frac{\partial\langle \hat{\vec{F}}_a^S(\vec{r})\rangle}{\partial\langle \hat{\vec{E}}_{ext}(\vec{r})\rangle}.\tag{24}$$

This is further decomposed as follows:

$$\ddot{Z}_a^*(\vec{r}) = Z_a + \ddot{Z}_{a\,\text{wind}}(\vec{r}),\tag{25}$$

$$\ddot{Z}_{a\,\text{wind}}(\vec{r}) = \ddot{Z}_{a\,\text{dynamic wind}}(\vec{r}) + \ddot{Z}_{a\,\text{static wind}}(\vec{r}),\tag{26}$$

where $\ddot{Z}_{a\,\text{dynamic wind}}(\vec{r})$ is the dynamic wind charge tensor density defined as

$$\ddot{Z}_{a\,\text{dynamic wind}}(\vec{r}) = \frac{1}{\langle e\hat{N}_a(\vec{r})\rangle} \frac{\partial\langle \hat{\vec{\tau}}_a^S(\vec{r})\rangle}{\partial\langle \hat{\vec{E}}_{\text{ext}}(\vec{r})\rangle}, \tag{27}$$

and where $\ddot{Z}_{a\,\text{static wind}}(\vec{r})$ is the static charge tensor density defined as

$$Z_a + \ddot{Z}_{a\,\text{static wind}}(\vec{r}) = \frac{1}{\langle e\hat{N}_a(\vec{r})\rangle} \frac{\partial\langle \hat{\vec{K}}_a(\vec{r})\rangle}{\partial\langle \hat{\vec{E}}_{\text{ext}}(\vec{r})\rangle}. \tag{28}$$

The $\ddot{Z}_{a\,\text{dynamic wind}}(\vec{r})$ is completely new concept, while the $\ddot{Z}_{a\,\text{static wind}}(\vec{r})$ will lead to the conventional $Z_{a\,\text{wind}}$. In our formulation, the Z_a could be assigned as the $Z_{a\,\text{direct}}$, but not necessarily.

3. APPLICATION

Quantum mechanical wave packet propagation has been examined in a model of electromigration in Al using first-principle electronic structure calculations with periodic boundary condition [18]. Generalized-gradient approximation by Perdew and Wang with 30 Rydberg (408 eV) cut-off energy is adopted at 302 K using 16-64 k-points with $40\times40\times56$ (for bulk and grain boundary), $40\times40\times84$ (for surface) FFT grids. Uniform $\vec{E}_{\text{ext}}(\vec{r})$ of strength 1.0×10^{-4} au $(5.14\times10^7$ V/m) is applied [19].

Here, a direct product of the density matrix $\hat{\rho}_a$ of atom a and $\hat{\rho}_e$ of electron is used:

$$\hat{\rho} = \hat{\rho}_a \otimes \hat{\rho}_e, \tag{29}$$

where

$$\hat{\rho}_a = |\psi_a\rangle\langle\psi_a|, \quad \hat{\rho}_e = \sum_j v_j |\psi_j\rangle\langle\psi_j|, \quad \sum_j v_j = N. \tag{30}$$

In this expression, $\hat{\rho}_a$ is the wave packet of atom a, v_j is occupation number of Bloch orbital ψ_j where $j = n\vec{k}$ designates band index n and crystal momentum $\vec{k}$, and N is number of conducting electrons. Since the atomic motion is much slower than that of electron, the ψ_j is parametrically dependent on the atomic position $\vec{r}_a = \vec{r}$ and orthonormalized with respect to the electron coordinate $\vec{r}_e$ as

$$\langle \psi_j(\vec{r}_e;\vec{r})\,|\,\psi_k(\vec{r}_e;\vec{r})\rangle = \int \psi_j^*(\vec{r}_e;\vec{r})\psi_k(\vec{r}_e;\vec{r})d^3\vec{r}_e = \delta_{jk}. \tag{31}$$

Then the wave packet,

$$\psi_a(\vec{r}) = \sum_{\vec{G}_a} c_{\vec{G}_a} e^{i(\vec{k}_a+\vec{G}_a)\cdot\vec{r}}, \quad \psi_a(\vec{r}) = e^{i\vec{k}a\cdot\vec{r}}\phi_a(\vec{r}), \quad \phi_a(\vec{r}+2\vec{\ell}) = \phi_a(\vec{r}), \tag{32}$$

propagates in time according as

$$i\hbar\frac{d}{dt}c_{\vec{G}_a}(t) = \sum_{\vec{G}_a} H_{\vec{G}_a\vec{G}_a'}c_{\vec{G}_a'}(t), \quad \left(c_{\vec{G}_a}(t)\right) = \left[e^{-\frac{i}{\hbar}H_{\vec{G}_a\vec{G}_a'}t}\right]\left(c_{\vec{G}_a'}(0)\right), \tag{33}$$

109

$$H_{\vec{G}_a \vec{G}_a'} = \frac{\hbar^2}{2m_a} \left(\vec{k}_a + \vec{G}_a \right)^2 \delta_{\vec{G}_a \vec{G}_a'}$$

$$+ \frac{1}{2} [-i \frac{\hbar^2}{m_a} \left(\vec{k}_a + \vec{G}_a' \right) \bullet \frac{1}{2\ell_x} \int_{-\ell_x}^{\ell_x} dx \frac{1}{2\ell_y} \int_{-\ell_y}^{\ell_y} dy \frac{1}{2\ell_z} \int_{-\ell_z}^{\ell_z} dz e^{i(\vec{G}_a' - \vec{G}_a) \bullet \vec{r}} \sum_j \frac{V_j}{N} < \psi_j (\vec{r}_e; \vec{r}) | \vec{\nabla} \psi_j (\vec{r}_e; \vec{r}) >$$

$$+ i \frac{\hbar^2}{m_a} \left(\vec{k}_a + \vec{G}_a' \right) \bullet \frac{1}{2\ell_x} \int_{-\ell_x}^{\ell_x} dx \frac{1}{2\ell_y} \int_{-\ell_y}^{\ell_y} dy \frac{1}{2\ell_z} \int_{-\ell_z}^{\ell_z} dz e^{i(\vec{G}_a' - \vec{G}_a) \bullet \vec{r}} \sum_j \frac{V_j}{N} \left(< \psi_j (\vec{r}_e; \vec{r}) | \vec{\nabla} \psi_j (\vec{r}_e; \vec{r}) > \right)^*]$$

$$- \frac{\hbar^2}{2m_\alpha} \frac{1}{2\ell_x} \int_{-\ell_x}^{\ell_x} dx \frac{1}{2\ell_y} \int_{-\ell_y}^{\ell_y} dy \frac{1}{2\ell_z} \int_{-\ell_z}^{\ell_z} dz e^{i(\vec{G}_a' - \vec{G}_a) \bullet \vec{r}} \sum_j \frac{V_j}{N} \frac{1}{2} \left[< \psi_j (\vec{r}_e; \vec{r}) | \Delta \psi_j (\vec{r}_e; \vec{r}) > + \text{c.c.} \right]$$

$$+ \frac{1}{2\ell_x} \int_{-\ell_x}^{\ell_x} dx \frac{1}{2\ell_y} \int_{-\ell_y}^{\ell_y} dy \frac{1}{2\ell_z} \int_{-\ell_z}^{\ell_z} dz e^{i(\vec{G}_a' - \vec{G}_a) \bullet \vec{r}} U(\vec{r}). \tag{34}$$

The initial state is set to be a simple Gaussian centered at $\vec{r} = \vec{R}_a$ in a unit cell:

$$\left(\frac{2\zeta}{\pi} \right)^{\frac{3}{4}} e^{i\vec{k}_a \bullet \vec{r}} e^{-\zeta (\vec{r} - \vec{R}_a)^2} , c_{\vec{G}_a} = \left(\frac{2\pi}{\zeta} \right)^{\frac{3}{4}} \frac{1}{2\ell_x \cdot 2\ell_y \cdot 2\ell_z} e^{-i\vec{G}_a \bullet \vec{R}_a} e^{\frac{\vec{G}_a^2}{4\zeta}}. \tag{35}$$

We get $\left\langle \hat{\vec{S}}_a (\vec{r}) \right\rangle$, $\left\langle \hat{\tau}_a^S (\vec{r}) \right\rangle$, and $\left\langle \hat{K}_a^S (\vec{r}) \right\rangle$, respectively, as follows:

$$\left\langle \hat{\vec{S}}_a (\vec{r}) \right\rangle = -\frac{i\hbar}{2m_a} \left[\psi_a^* (\vec{r}) \vec{\nabla} \psi_a (\vec{r}) - \text{c.c.} \right]$$

$$- \frac{i\hbar}{2m_a} \psi_a^* (\vec{r}) \psi_a (\vec{r}) \sum_j \frac{V_j}{N} \left[< \psi_j (\vec{r}_e; \vec{r}) | \vec{\nabla} \psi_j (\vec{r}_e; \vec{r}) > - \text{c.c.} \right], \tag{36}$$

$$\left\langle \hat{\tau}_a^S (\vec{r}) \right\rangle = \frac{\hbar^2}{4m_a} \left[\vec{t}_1 (\vec{r}) + \vec{t}_2 (\vec{r}) + \vec{t}_3 (\vec{r}) + \vec{t}_4 (\vec{r}) + \vec{t}_5 (\vec{r}) + \vec{t}_6 (\vec{r}) \right] + \text{c.c.},$$

$$t_1^k (\vec{r}) = \psi_a^* (\vec{r}_a) \frac{\partial \Delta \psi_a (\vec{r})}{\partial x^k} - \frac{\partial \psi_a^* (\vec{r})}{\partial x^k} \Delta \psi_a (\vec{r}),$$

$$t_2^k (\vec{r}) = \psi_a^* (\vec{r}) \Delta \psi_a (\vec{r}) \sum_j \frac{V_j}{N} \left[< \psi_j (\vec{r}_e; \vec{r}) | \frac{\partial \psi_j (\vec{r}_e; \vec{r})}{\partial x^k} > - \text{c.c.} \right],$$

$$t_3^k (\vec{r}) = 2 \left[\psi_a^* (\vec{r}) \frac{\partial \vec{\nabla} \psi_a (\vec{r})}{\partial x^k} - \frac{\partial \psi_a^* (\vec{r})}{\partial x^k} \vec{\nabla} \psi_a (\vec{r}) \right] \bullet \sum_j \frac{V_j}{N} < \psi_j (\vec{r}_e; \vec{r}) | \vec{\nabla} \psi_j (\vec{r}_e; \vec{r}) >,$$

$$t_4^k (\vec{r}) = 2 \psi_a^* (\vec{r}) \Delta \psi_a (\vec{r})$$

$$\bullet \sum_j \frac{v_j}{N} \left[2\left\langle \psi_j(\vec{r}_e;\vec{r}) \,\middle|\, \frac{\partial \vec{\nabla}\psi_j(\vec{r}_e;\vec{r})}{\partial x^k} \right\rangle - \frac{\partial \left\langle \psi_j(\vec{r}_e;\vec{r}) \,\middle|\, \vec{\nabla}\psi_j(\vec{r}_e;\vec{r}) \right\rangle}{\partial x^k} \right],$$

$$t_5^k(\vec{r}) = \left[\psi_a^*(\vec{r})\frac{\partial \psi_a(\vec{r})}{\partial x^k} - \frac{\partial \psi_a^*(\vec{r})}{\partial x^k}\psi_a(\vec{r}) \right] \sum_j \frac{v_j}{N} < \psi_j(\vec{r}_e;\vec{r}) \,|\, \Delta \, \psi_j(\vec{r}_e;\vec{r}) >,$$

$$t_6^k(\vec{r}) = \psi_a^*(\vec{r})\psi_a(\vec{r})$$

$$\times \sum_j \frac{v_j}{N} \left[2 < \psi_j(\vec{r}_e;\vec{r}) \,|\, \frac{\partial \Delta \psi_j(\vec{r}_e;\vec{r})}{\partial x^k} > - \frac{\partial < \psi_j(\vec{r}_e;\vec{r}) \,|\, \Delta \psi_j(\vec{r}_e;\vec{r}) >}{\partial x^k} \right], \qquad (37)$$

where $\bullet$ denotes inner product of vector, and

$$\left\langle \hat{\vec{K}}_a^S(\vec{r}) \right\rangle = -\psi_a^*(\vec{r})\psi_a(\vec{r}) \left[\vec{\nabla}v_a(\vec{r}) + \int \sum_j v_j \psi_j^*(\vec{r}_e;\vec{r})\psi_j(\vec{r}_e;\vec{r})\vec{\nabla}\left(-\frac{Z_a e^2}{|\vec{r}_e - \vec{r}|} \right)d^3\vec{r}_e \right]. \qquad (38)$$

Finally, $\ddot{\vec{Z}}_{a\,\text{dynamic wind}}(\vec{r})$ and $\ddot{\vec{Z}}_{a\,\text{static wind}}(\vec{r})$ as follows:

$$\ddot{\vec{Z}}_{a\,\text{dynamic wind}}(\vec{r}) = \frac{\hbar^2}{4m_a\psi_a^*(\vec{r})\psi_a(\vec{r})} \left[\vec{z}_1(\vec{r}) + \vec{z}_2(\vec{r}) + \vec{z}_3(\vec{r}) + \vec{z}_4(\vec{r}) + \vec{z}_5(\vec{r}) + \vec{z}_6(\vec{r}) \right] + \text{c.c.},$$

$$z_1^k(\vec{r}) = 0,$$

$$z_2^k(\vec{r}) = \psi_a^*(\vec{r})\Delta \, \psi_a(\vec{r}) \sum_j \frac{1}{Ne}\frac{\partial v_j}{\partial \left\langle \hat{\vec{E}}_{\text{ext}}(\vec{r}) \right\rangle} \left[< \psi_j(\vec{r}_e;\vec{r}) \,|\, \frac{\partial \psi_j(\vec{r}_e;\vec{r})}{\partial x^k} > -\text{c.c.} \right],$$

$$z_3^k(\vec{r}) = 2 \left[\psi_a^*(\vec{r})\frac{\partial \vec{\nabla}\psi_a(\vec{r})}{\partial x^k} - \frac{\partial \psi_a^*(\vec{r})}{\partial x^k}\vec{\nabla}\psi_a(\vec{r}) \right]$$

$$\bullet \sum_j \frac{1}{Ne}\frac{\partial v_j}{\partial \left\langle \hat{\vec{E}}_{\text{ext}}(\vec{r}) \right\rangle} < \psi_j(\vec{r}_e;\vec{r}) \,|\, \vec{\nabla}\psi_j(\vec{r}_e;\vec{r}) >,$$

$$z_4^k(\vec{r}) = 2\psi_a^*(\vec{r})\vec{\nabla}\psi_a(\vec{r})$$

$$\bullet \sum_j \frac{1}{Ne}\frac{\partial v_j}{\partial \left\langle \hat{\vec{E}}_{\text{ext}}(\vec{r}) \right\rangle} \left[2 < \psi_j(\vec{r}_e;\vec{r}) \,|\, \frac{\partial \vec{\nabla}\psi_j(\vec{r}_e;\vec{r})}{\partial x^k} > - \frac{\partial < \psi_j(\vec{r}_e;\vec{r}) \,|\, \vec{\nabla}\psi_j(\vec{r}_e;\vec{r}) >}{\partial x^k} \right],$$

$$z_5^k(\vec{r}) = \left[\psi_a^*(\vec{r})\frac{\partial \psi_a(\vec{r})}{\partial x^k} - \frac{\partial \psi_a^*(\vec{r})}{\partial x^k}\psi_a(\vec{r}) \right] \sum_j \frac{1}{Ne}\frac{\partial v_j}{\partial \left\langle \hat{\vec{E}}_{\text{ext}}(\vec{r}) \right\rangle} < \psi_j(\vec{r}_e;\vec{r}) \,|\, \Delta \, \psi_j(\vec{r}_e;\vec{r}) >,$$

$$z_6^k(\vec{r}) = \psi_a^*(\vec{r})\psi_a(\vec{r})$$

$$\times \sum_j \frac{1}{Ne}\frac{\partial v_j}{\partial \left\langle \hat{\vec{E}}_{\text{ext}}(\vec{r}) \right\rangle} \left[2 < \psi_j(\vec{r}_e;\vec{r}) \,|\, \frac{\partial \Delta \psi_j(\vec{r}_e;\vec{r})}{\partial x^k} > - \frac{\partial < \psi_j(\vec{r}_e;\vec{r}) \,|\, \Delta \psi_j(\vec{r}_e;\vec{r}) >}{\partial x^k} \right], \qquad (39)$$

and by assigning Z_a as the $Z_{a\,\text{direct}}$,

$$\ddot{Z}_{a\,static\,wind}(\vec{r}) = -\int \sum_j \frac{1}{e} \frac{\partial v_j}{\partial \langle \hat{E}_{ext}(\vec{r}) \rangle} \psi_j^*(\vec{r}_e;\vec{r}) \psi_j(\vec{r}_e;\vec{r}) \vec{\nabla} \left(-\frac{Z_\alpha e^2}{|\vec{r}_e - \vec{r}|} \right) d^3\vec{r}_e, \quad (40)$$

where the change in occupation number is found to be

$$\delta v_{n\vec{k}} = e\tau_{n\vec{k}} \langle \hat{E}_{ext}(\vec{r}) \rangle \bullet \vec{V}_{n\vec{k}} \partial f_0(\varepsilon_{n\vec{k}} - \varepsilon_F)/\partial \varepsilon_{n\vec{k}}, \quad (41)$$

with f_0 being the Fermi distribution function.

In Figs. 1-3 are demonstrated the $\ddot{Z}_{Al\,dynamic\,wind}(\vec{r})$ and the ratio $|\ddot{Z}_{Al\,dynamic\,wind}(\vec{r})|/|\ddot{Z}_{Al\,static\,wind}(\vec{r})|$ in models of the bulk, surface and grain boundary of Al, respectively. Only the results using the initial wave packets are shown. The arrow denotes the diagonal element of the $\ddot{Z}_{Al\,dynamic\,wind}(\vec{r})$ with respect the $\vec{E}_{ext} = \langle \hat{E}_{ext}(\vec{r}) \rangle$ in this case. The $\ddot{Z}_{Al\,dynamic\,wind}(\vec{r})$ exceeds 10^3 in magnitude locally. This may predict some unknown new phenomena in real systems. However, the average in space could be negligibly small. This interesting behavior should be originated from the null-sum rule of the tension density operator from which the $\ddot{Z}_{Al\,dynamic\,wind}(\vec{r})$ is derived; as shown in Eqs. (21) and (27). On the other hand, the $\ddot{Z}_{Al\,static\,wind}(\vec{r})$ has the conventional Feynman-Hellmann form; as shown in Eqs. (28) and (40). Accordingly, its average in space should be comparable to the value in theoretical literatures available [8-14]. The ratio $|\ddot{Z}_{Al\,dynamic\,wind}(\vec{r})|/|\ddot{Z}_{Al\,static\,wind}(\vec{r})|$ demonstrates significant figure at some characteristic points.

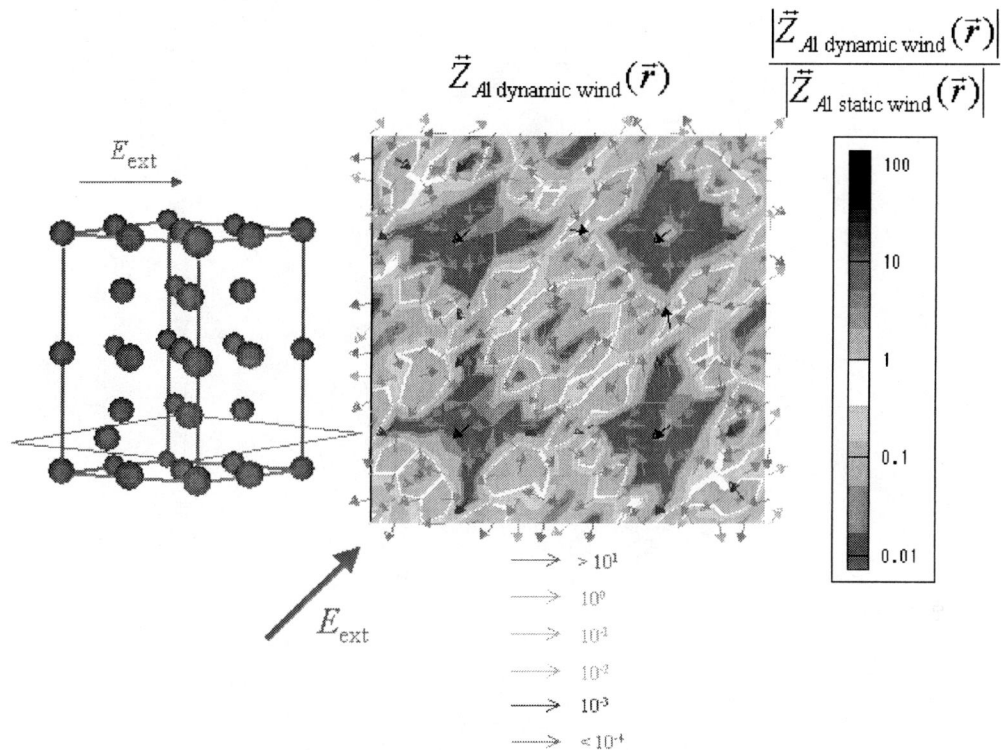

FIGURE 1. $\vec{\vec{Z}}_{Al\,dynamic\,wind}(\vec{r})$ and $\left|\vec{\vec{Z}}_{Al\,dynamic\,wind}(\vec{r})\right|/\left|\vec{\vec{Z}}_{Al\,static\,wind}(\vec{r})\right|$ in the model of a migrating Al atom in the Al bulk.

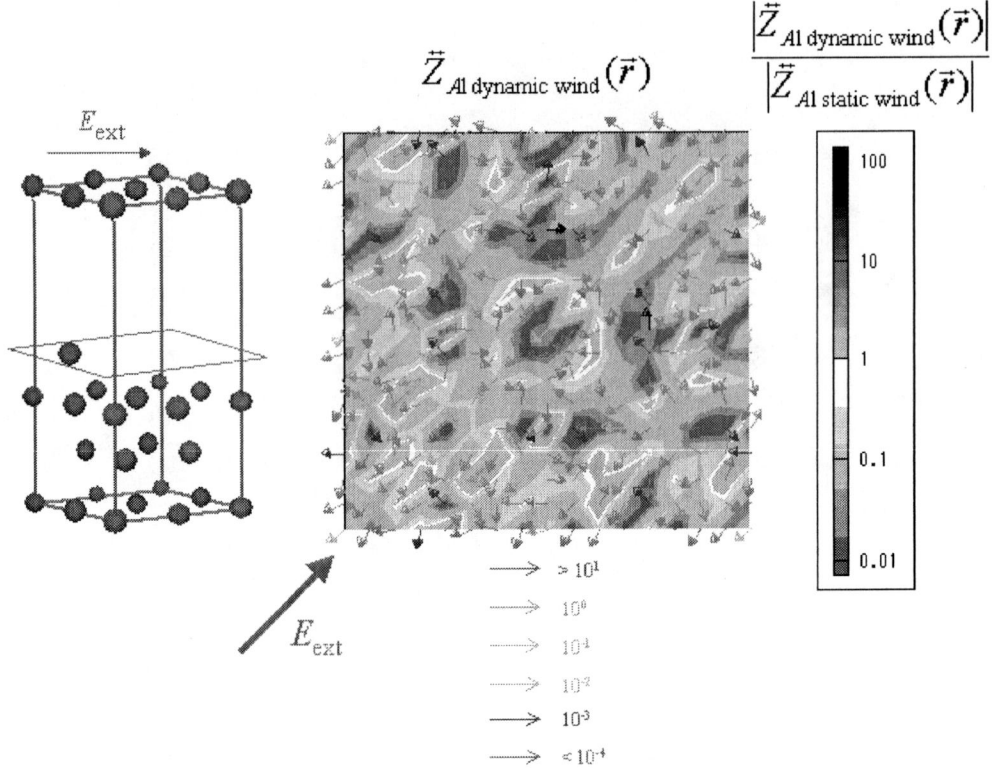

FIGURE 2. $\vec{Z}_{Al\,dynamic\,wind}(\vec{r})$ and $\left|\vec{Z}_{Al\,dynamic\,wind}(\vec{r})\right|/\left|\vec{Z}_{Al\,static\,wind}(\vec{r})\right|$ in the model of a migrating Al atom on the Al (100) surface.

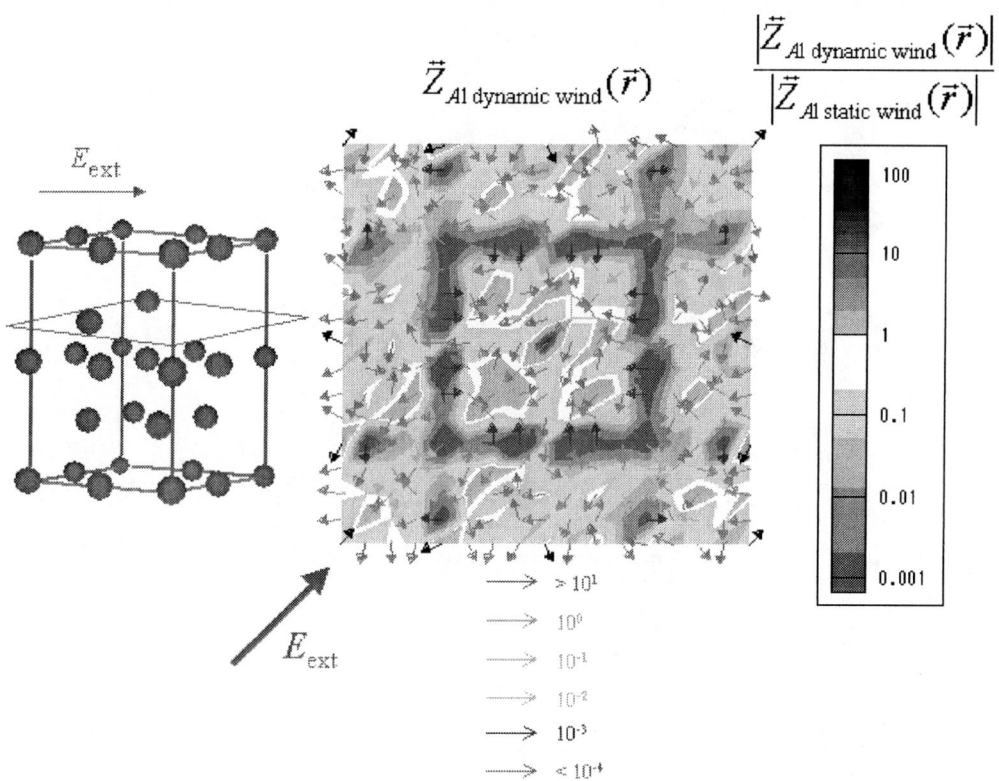

FIGURE 3. $\vec{Z}_{Al\,dynamic\,wind}(\vec{r})$ and $\left|\vec{Z}_{Al\,dynamic\,wind}(\vec{r})\right|/\left|\vec{Z}_{Al\,static\,wind}(\vec{r})\right|$ in the model of a migrating Al atom in the grain boundary: two Al sheets are removed over the Al (100) cross section. A dangling Al atom is put as an obstruction for electromigration.

4. CONCLUSION

(1) *New electromigration force* has been formulated using the non-relativistic limit of quantum electrodynamics (QED). *Kinetic energy density* and *tension density* emerges from the viewpoint of "action through medium" in the quantum field theory rather than "action at a distance."

(2) *New dynamic wind charge* has been revealed over and above the conventional static wind charge and the direct charge. The dynamic wind charge reflects the dynamic interchange of momentums between conducting electrons and migrating atoms.

(3) Quantum mechanical wave packet propagation has been examined in a model of electromigration in Al using first-principle electronic structure calculations with periodic boundary condition. The dynamic wind charge demonstrates significant figure at some characteristic points.

ACKNOWLEDGMENTS

This work has been supported by Grant-in-Aid for Scientific Research from the Ministry of Education, Science and Culture of Japan, for which we express our gratitude.

REFERENCES

1. Huntington, H. B., and Grone, A. R., *J. Phys. Chem. Solids* **20**, 76-87 (1961).
2. Blech, I. A., and Kinsbron, E., *Thin Solid Films* **25**, 327-334 (1975).
3. Blech, I. A., *J. Appl. Phys.* **47**, 1203-1208 (1976).
4. Black, J. R., *IEEE Trans. Electron Devices* **ED-16**,338-347 (1969).
5. Ho, P. S., and Kwok, T., *Rep. Prog. Phys.* **52**, 301-348 (1989).
6. Thompson, C.V., and Lloyd, J. R., *MRS Bulletin*, 19-25 (1993).
7. Kawasaki, H., Gall, M., Jawarani, D., Hernandez,R., and Capasso,C., *Thin Solid Films* **320**, 45-51 (1998).
8. Bosvieux, C., and Friedel, J., *J. Phys. Chem. Solids* **23**, 123-136 (1962).
9. Kumar, P., and Sorbello, R. S., *Thin Solid Films* **25**, 25-35 (1975).
10. Sorbello, R. S., and Dasgupta, B. B., *Phys. Rev. B* **21**, 2196-2200 (1980).
11. Lodder, A., *J. Phys. F: Met. Phys.* **14**, 2943-2953 (1984).
12. Lodder, A., *Solid State Comm.* **71**,259-262 (1989).
13. Sorbello, R. S., *Solid State Phys.* **51**, 159 (1998).
14. Lodder,A., and Dekker, J.P., "The Electromigration Force in Metallic Bulk," in *Stress Induced Phenomena in Metallization,* edited by Okabayashi et al ., American Institute of Physics, 1998, pp.315-328.
15. Iguchi, K., and Tachibana, A., *Surface Science* **159/160**, 167-173 (2000).
16. Tachibana, A., *J. Chem. Phys.* in press.
17. Tachibana, A., *Theor. Chem. Acc.* **102**, 188-195 (1999).
18. Tachibana, A. *et al.* to be published.
19. Periodic Regional DFT Program Package, Tachibana Lab. (2001).

PART II: CYCLIC LOADING EFFECTS

Interconnect Failure due to Cyclic Loading

Robert R. Keller[†], Reiner Mönig, Cynthia A. Volkert, Eduard Arzt, Ruth Schwaiger, and Oliver Kraft

Max-Planck-Institut für Metallforschung, Seestr. 92, D-70174 Stuttgart, Germany
[†]*National Institute of Standards and Technology, Materials Reliability Division, 325 Broadway, Boulder, CO 80305 U.S.A.*

Abstract. The damage generated by AC currents at 100 Hz in interconnects has been studied and compared with mechanical fatigue damage in thin films. The nature of the damage under the two loading conditions is qualitatively similar, supporting the idea that the AC current damage comes from mechanical cycling due to temperature swings on the order of 100 K from Joule heating in the interconnects. In both cases, the damage forms as surface wrinkles within single grains grow in amplitude and extent with time. The possible threat to the reliability of microelectronic and micro-electromechanical systems is further escalated by the observation that soft encapsulation layers do nothing to retard the formation of the damage.

INTRODUCTION

Metallizations in microelectronic devices and micro-electromechanical systems (MEMS) are subjected to a wide variety of time-varying mechanical loading modes, which can potentially induce failures that are not easily predicted through knowledge of monotonic thin film behavior. For instance, power cycling of microelectronic devices due to user operation or energy conservation schemes can lead to significant repetitive thermal strains. In MEMS structures, loading can occur during the repeated action or bending associated with hinged pieces, gears, or cams. In these low frequency applications, mechanical reliability can be compromised because of the consequences of cyclic deformation near surfaces and interfaces. In microelectronic devices, interconnect cross-sections can change, internal stresses can increase, or metal/dielectric delamination can result. In MEMS, cracking or excessive friction may develop.

Direct extrapolation of cyclic mechanical behavior from monotonic testing is generally not feasible, since the damage evolution processes in the two cases are very different. Plastic strain localization becomes more significant in the cyclic case. Complicating matters is the fact that cyclic deformation in small-scale structures is not a well-documented phenomenon. For the case of wide (25 μm) unencapsulated aluminum lines, Philofsky *et al.* [1] showed 30 years ago that AC current testing can

CP612, *Stress-Induced Phenomena in Metallization:* Sixth Int'l. Workshop, edited by S. P. Baker et al.

lead to surface damage associated with slip band formation. The strains in this case are due to temperature cycling, which can easily reach an amplitude of greater than several tens of degrees Kelvin at low frequencies [2]. For the case of blanket metal films, thermal cycling has often been used in conjunction with wafer curvature measurements to study stress evolution, see *e.g.* [3]. Such testing is, however, incapable of separating temperature and stress effects, and is not amenable to large numbers of cycles. Thin film tensile testing has been used with more success to evaluate fatigue behavior in thin films, *e.g.* [4]. Hommel *et al.* [5] showed recently how thin film tensile testing could be used to study cyclic mechanical behavior through the use of compliant polymer substrates.

We consider in this work two types of low frequency cyclic deformation in thin films and interconnects: alternating current (AC)-induced thermal straining of patterned Al interconnects and repeated mechanical loading of Cu and Al films on compliant polymer substrates. The cyclic deformation induced by these forms of loading will be shown to be very similar, despite their origins being considerably different. The focus of this paper is to illustrate surface damage, effects of adjacent materials, and lifetime behavior in cyclically deformed lines and films. Further experimental details are included in [5,6].

ALTERNATING CURRENT-INDUCED DEFORMATION

Several studies in the literature have illustrated that AC currents can lead to severe damage in interconnects [1,6]. The effect is believed to be thermal mechanical fatigue that originates from the cyclic thermal strains due to Joule heating in the interconnects. In this section, the time evolution of damage due to AC currents in Al interconnects, as well as the influence of line width, current density, and various encapsulating layers, is presented.

AC tests were carried out on single level structures that were composed of single Al-1 at. %Si lines, 800 μm long, 0.55 μm thick, and 1.3, 3.3, or 13 μm wide. The lines connected two bond pads and were deposited directly onto oxidized silicon. Tests were carried out on unencapsulated lines, those coated with 2.0 μm of hard-baked photoresist, and those coated with 0.3 μm of reactively sputtered silicon nitride. Each die was attached to a chip carrier using silver paint and the bond pads were connected by ultrasonic wire bonding.

Electrical stressing was performed in a vacuum of approximately 6×10^{-4} Pa (4.5 $\times$ 10^{-6} torr). A signal generator was used to supply the lines with sinusoidal currents at 100 Hz, with DC offsets of less than several mV. Current densities reported in this work refer to the rms current, and are thus a factor $1/\sqrt{2}$ of the maximum current. The sample temperature was monitored using a thermocouple as well as using rms and time-resolved 4 point resistance measurements of the interconnects. The thermocouple, which was attached to the ceramic chip carrier, showed a temperature increase of roughly 100 K during electrical testing of 3.3 μm wide lines at 10 MA/cm^2.

The carrier temperature is a lower bound for the cycling line temperature. Resistance values provided a more localized measurement of line temperatures. During each half cycle, the time-resolved resistance values indicated a temperature amplitude of approximately 100 K (also measured at 10 MA/cm^2 on 3.3 µm lines at 100 Hz), with cycling occurring between approximately 120 and 220°C. Resistance was also used to monitor damage formation during testing; it increased due to heating at the beginning of each test and then remained constant until it rapidly increased just before failure occurred by open circuit.

Fig. 1 shows surface damage in an unpassivated 3.3 µm wide Al-1at.%Si line, after 9 x 10^6 AC current cycles (corresponding to 1.8 x 10^7 temperature cycles during 25 hours of testing at 100 Hz) at 11 MA/cm^2, or after, at which point it failed by open circuit. At a given time during the course of any test, one observes regions that are not at all damaged as well as regions with damage ranging from very slight to very severe. The surface damage shown in Fig. 1(a) is typical of some of the less-severely-damaged regions. Periodic surface wrinkles are observed with a spacing of the order of 0.5 µm. Fig. 1(b) shows an example of more severe damage, where the surface wrinkles are larger in amplitude and extend over a larger area. In this case, the wrinkle spacing is approximately 0.25 µm and large extrusions have formed at the sides of the line. Fig. 1(c) shows the site where the open circuit occurred. The characteristic wrinkles are present on either side of the break in the line and there is evidence (such

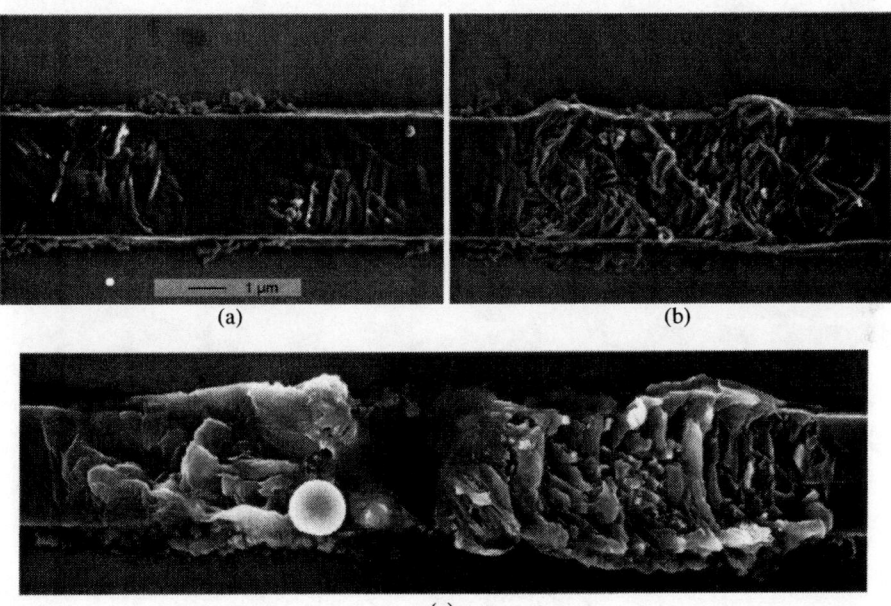

(a) (b)

(c)

FIGURE 1. 3.3 µm wide unpassivated Al-1at.%Si line after 9 x 10^6 AC cycles at 100 Hz and an rms current density of 11 MA/cm^2. Region (a) is only lightly damaged, while region (b) is more severely damaged. Region (c) shows the site where open failure of the line occurred. All images are shown at the same magnification.

as the circular droplet to the left of the break) indicating that the line locally melted just before failure.

Fig. 2 shows FIB images of a 3.3 μm wide Al-1at.%Si line that was tested at 10 MA/cm^2 for approximately 6.5 x 10^6 AC cycles (18 hours at 100 Hz). The channeling contrast in the images shows the grains in the line, which have a diameter of roughly 2 μm. Similar grain sizes were observed by FIB in the contact pads and in untested lines. For lightly damaged regions such as the one shown here, the damage appears to be contained within a single grain. Fig. 2(b) shows a cross-sectional cut through the line. The underlying SiO$_2$ layer and Si wafer are clearly visible, as well as the fact that the Al is very thin (< 0.1 μm) in the troughs of the wrinkles. There are no voids present at the interface between the metal and the underlying substrate. This is in contrast to what is observed in the mechanically cycled films, as will be seen later in the paper.

In order to investigate the effect of encapsulating layers on the formation of surface damage, AC tests were also run on lines encapsulated with hard-baked photoresist or with a silicon nitride film at 100 Hz and an rms current of 11 MA/cm^2. The line encapsulated with 2 μm of photoresist was tested for 4.6 x 10^5 AC current cycles. After the test, the photoresist was dissolved using acetone and the Al surface observed by scanning electron microscopy. The surface morphology on the photoresist-encapsulated sample was indistinguishable from that seen on unpassivated lines, indicating that the photoresist layer did nothing to hinder the formation of damage. The sample encapsulated with 0.3 μm of silicon nitride was tested for 1.3 x 10^5 AC current cycles. The nitride did not adhere well to the Al as can be seen from the cross-sectional FIB images in Fig. 3. The surface wrinkles caused the nitride layer to both bow away and separate from the metal. In addition, FIB cross-sections suggest that the nitride may have reacted with the Al in several local regions along the line. Further studies are underway to elucidate the effect of hard encapsulants, by growing well-adhering overlayers.

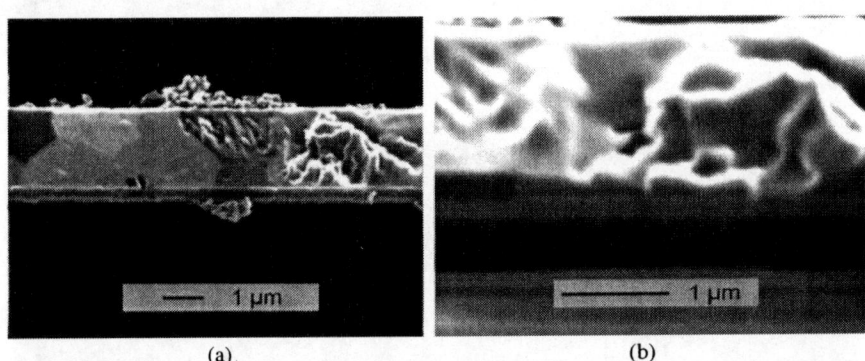

(a)   (b)

FIGURE 2. FIB images of a 3.3 μm wide Al-1at.%Si line after 6.5 x 10^6 AC cycles at 100 Hz and an rms current density of 10 MA/cm^2, (a) at a tilt angle of 30° and (b) at a tilt angle of 45° after cross-sectioning along the line.

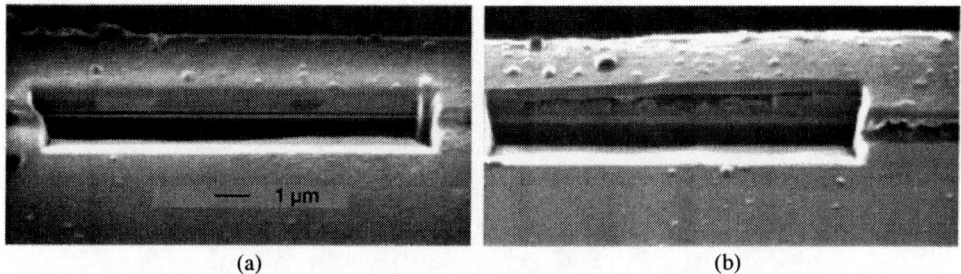

(a)   (b)

FIGURE 3. Silicon nitride-encapsulated, 3.3 μm wide Al-1at.%Si line (a) before and (b) after 1.3×10^5 AC cycles at 100 Hz and an rms current density of 11 MA/cm^2. Sample sectioned and imaged by focussed ion beam microscopy. Note bowing and fracture of overlayer as well as damage on surface of line. Images are shown at same magnification.

In some of the lines, sub-micron wide whiskers were found after testing. They were observed only at sites where surface wrinkles were also present. Fig. 4 shows a typical example: several whiskers, including a long, curved one of approximate diameter 50 nm, have grown from a site where surface wrinkles have formed. In general, it seems that whiskers formed somewhat more often in the wider lines and those in wider lines tended to grow much longer than those in narrower lines. Whether this is due to differences in microstructure or mechanical properties or to the fact that the wider lines were tested at lower current densities is not yet clear.

It is difficult to quantify the overall extent of damage as a function of testing time, linewidth, and current density, because of the varying severity of damage found within a single line. However, some rough trends can be observed. With increasing time under stress, the total surface area that is damaged increases, the amplitude of the wrinkles increases, and the spacing of the wrinkles decreases. Eventually, the damage is severe enough that it leads to an electrical open, presumably by local heating resulting from local thinning.

Some similar trends can be established in regard to the effect of current density. Increasing current density leads to increasing surface area of damaged metal and increasing amplitude of the wrinkles. Whether wrinkle spacing decreases remains uncertain. Increasing current density also leads to shorter times to open circuit. At lower current densities, a smaller fraction of the surface area of a line exhibits damage; this behavior persists until open circuit. However, those regions that do undergo changes can become severely damaged. Such areas sometimes show whisker formation.

The effect of linewidth on damage morphology is less clear although some trends can be observed despite the small number of tests performed in this preliminary work. In both wide and narrow lines, damage first begins in isolated regions of the line. These regions are roughly the size of single grains and, as expected for purely geometrical reasons, span a greater fraction of the linewidth in narrower lines. The transition from a damaged region to an undamaged region is less distinct for the case of narrower lines. Finally, the wrinkles are generally more regularly spaced and their amplitude seems considerably larger in the wider lines. In most cases, lines were tested until they failed by open circuit. In general, the higher the current density, the sooner

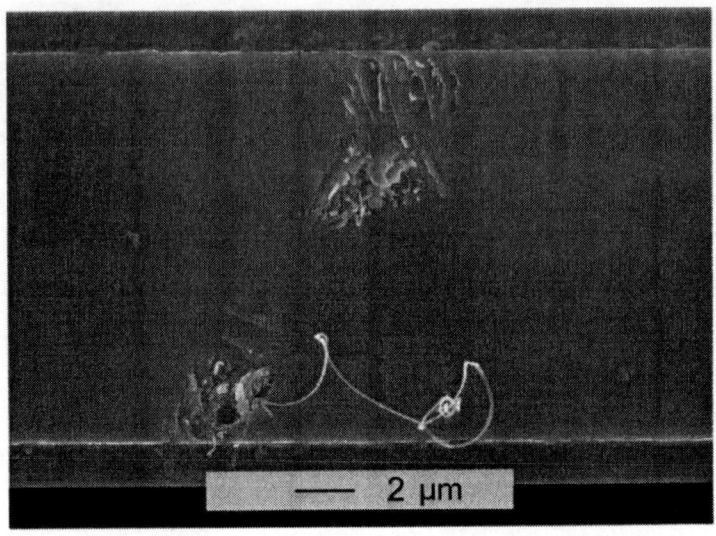

FIGURE 4. Whisker formation in 13 μm wide Al-1at.%Si line after 7.5 x 10[7] AC cycles at 100 Hz and an rms current density of 5.9 MA/cm^2.

the lines failed. Given the small number of tests performed on the effect of linewidth, obvious trends were not clear, but it may be that the narrower lines required a larger current density to reach failure within a given number of cycles than the wider lines. Particularly in comparing tests performed using different current densities or linewidths, the time to reach an electrical open is not necessarily a direct measure of the time required to reach a certain level of damage severity. In defining failure for the purposes of measuring lifetime, it may be more useful to quantify the extent of damage in terms of a fraction of the surface area that has become damaged. Whether this leads to lifetime determinations that reveal clearer trends with current density or linewidth than those seen using time to open circuit remains to be seen.

CYCLIC MECHANICAL-INDUCED DEFORMATION

In the following, the damage morphology that results from cyclic mechanical deformation of continuous films is described. For these tests, Al and Cu films were sputter-deposited onto compliant polymer substrates, and the samples subjected to cyclic tensile loading. During these cycles, the substrate material is deformed elastically while the film undergoes plastic deformation in both tension and compression. The method and details of the sample preparation have been described elsewhere [5]. Owing to the sample structure, the film stress cannot be determined from the externally applied load. Therefore, the stress-strain behavior of the films is analyzed using *in-situ* X-ray diffraction during straining of the samples [5,7]. Figure 5

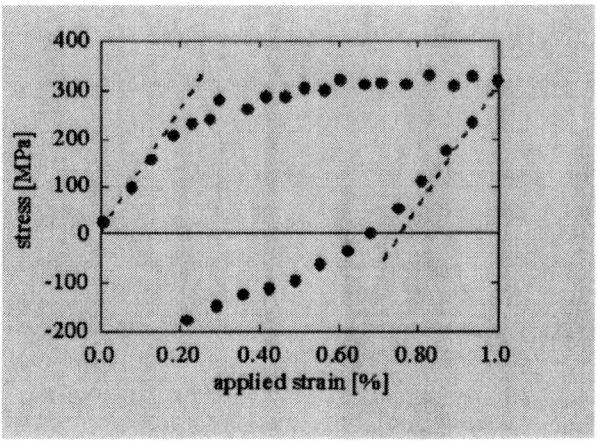

FIGURE 5. Stress-strain behavior of a 1 μm thick Cu film on a 125 μm thick polyimide substrate during a single cycle at room temperature. The film stress was measured by X-ray diffraction.

shows the first cycle from such a tensile test on a 1 μm thick Cu film. On loading, the stress increases linearly with applied strain. The film deforms elastically, as indicated by the dashed line, up to a strain of about 0.15 % and stress of almost 200 MPa. On further loading, the stress-strain behavior is no longer linear and the film is plastically deformed. Due to strain hardening the film stress increases to more than 300 MPa. On unloading of the sample, the substrate contracts and the applied strain is reduced, driving the film into compression. As a result, the film stress first decreases elastically (dashed line) and then makes a transition from elastic to plastic behavior. It has been shown by Kraft *et al.* [8] that this technique can be used for fatigue testing of thin metal films by repeating this cyclic test many times. After testing, damaged regions consisting of cracks and wrinkles were distributed over the entire film. For example, Fig. 6 shows a region of fatigue damage in a 1.0 μm thick Cu film, which was prepared under the same conditions as the one from Fig. 5, but cycled 10,000 times at a frequency of 0.1 Hz rather than just once. One can clearly see the periodic wrinkles formed within single grains, as well as an intergranular crack running from the top to the bottom of the images. The wrinkles have an amplitude of several 100 nm. Careful studies of the wrinkled grains using cross-sectional FIB imaging show that the grains are slightly extruded out of the film and that voids had formed at the interface to the substrate. As the number of cycles increased, the area fraction of damaged regions in the film increased and eventually saturated before the entire surface was damaged. Macroscopic failure or fracture of the samples did not occur, presumably due to the underlying elastic substrate.

In order to compare the damage morphology of the Al interconnects stressed by an AC current to the mechanical fatigue damage in Al films, a 0.5 μm thick Al film on a polyimide substrate was tested. This sample was fatigued with a total strain range of about 0.8% at a frequency of 10 Hz up to 10^6 cycles. The damage in the Al film was very similar to that in the Cu film. Figure 7 (a) shows a FIB micrograph of a damaged

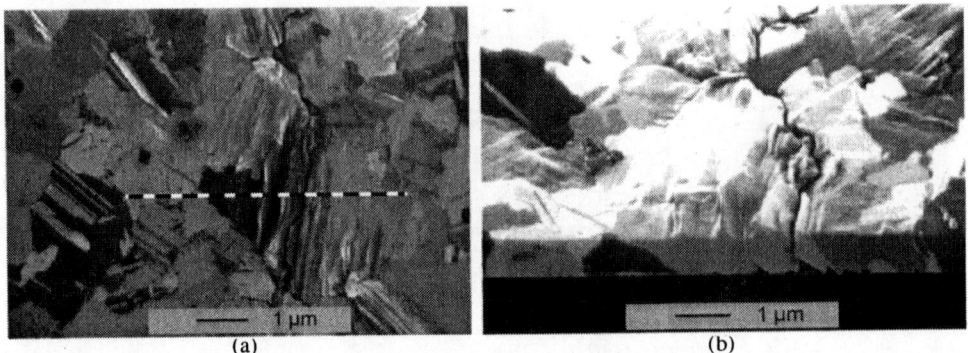

(a)   (b)

FIGURE 6. FIB micrographs showing the fatigue damage in 1 μm thick Cu films after 10^4 cycles at 0.1 Hz with a total strain range of 1.0%: (a) plan-view image and (b) cross-section at an angle of 45°. The cross-section was prepared at the position marked in (a) by the dashed line. The loading direction was horizontal with respect to the micrographs.

region in the Al film. Surface wrinkles and intergranular cracks can be seen. A cross-section of such a damaged region (Fig. 7 (b)) reveals that the wrinkles are confined to a single grain and that large voids have formed at the interface to the substrate, as for the case of Cu. The height of the wrinkles is similar to the ones found in Cu, but the distance between the individual wrinkles is somewhat larger in the Al than in the Cu. This might be related to the presence of the native oxide on the Al surface.

For both the Al and Cu films, the observations are consistent with the following damage evolution during cyclic mechanical deformation: First, extrusions and voids are formed in individual grains, and then, intergranular cracks originate from these voids. This picture is supported by the fact that many damaged single grains were found which contained wrinkles and voids but were not connected to cracks. As the

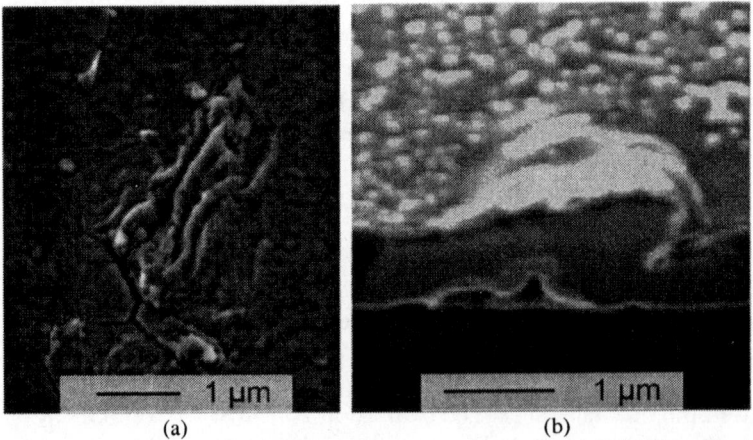

(a)   (b)

FIGURE 7. FIB micrographs showing the fatigue damage in a 0.5 μm thick Al film after 10^6 cycles at 10 Hz with a total strain range of 0.8%: (a) plan-view image and (b) cross-section at an angle of 45°. The loading direction in the fatigue test was horizontal with respect to the micrographs.

number of cycles increases, the number of damaged grains and cracks also increases. Macroscopic failure of the films did not occur because of the supporting, elastic substrate.

CYCLIC DEFORMATION IN FILMS AND INTERCONNECTS

The damage that results from AC current testing of interconnects and from cyclic mechanical testing of films has been presented in the preceding sections and many similarities are apparent. This is not too surprising since, as stated above, the observed damage created by AC currents has been attributed to thermal mechanical cycling from Joule heating in the interconnects. Possible contributions from electromigration are believed to be small [1,6]. Although it is well known that sufficiently large DC currents lead to electromigration damage, the damage generated by AC currents is viewed as negligible in comparison [9]. This is generally attributed to the idea that the early stages of electromigration damage are reversible and no net damage accumulation occurs under pure, symmetrical AC testing conditions [10]. In addition, at the frequencies used here, the time during a single cycle is too short for significant atomic diffusion to occur. Therefore, the severity of the observed damage is expected to be controlled by the amplitude of the thermal stresses generated by the alternating current. Heating within a line undergoing AC cycling scales with the input power and thus with the square of the current. The actual temperature of the line depends on the efficiency of heat dissipation, which is determined by line geometry and the surrounding materials. The resistance measurements revealed that the 3.3 μm lines cycle between roughly 120 and 220°C at an rms current density of 10 MA/cm^2. A 100 K temperature increase represents a 210 MPa decrease in film stress, assuming biaxial, anisotropic elastic behavior in (111)-textured Al. X-ray diffraction measurements of the residual biaxial stress in a blanket film portion of the samples resulted in a room temperature value of 270 MPa. Therefore, assuming elastic behavior, the stress would cycle between approximately +60 and −150 MPa during testing of the 3.3 μm lines at 10 MA/cm^2. Variations in linewidth will change the temperature of the samples, with narrower lines exhibiting both lower average temperatures and smaller temperature amplitudes since they dissipate less power at a given current density. Thus, the exact stresses experienced during current cycling will depend on the linewidth. In addition, variations in current density will change the stress amplitude, with higher current densities driving the stress more compressive and increasing the stress amplitude.

The results from the AC testing can be summarized and compared with literature data for bulk samples using a stress-lifetime (Wöhler, or S-N) plot in Fig. 8. For these samples, the stress amplitude shown was calculated from the cyclic temperature amplitude for each current density as calibrated by resistance measurements. The failure condition was defined as an open circuit. The smooth curve represents data for

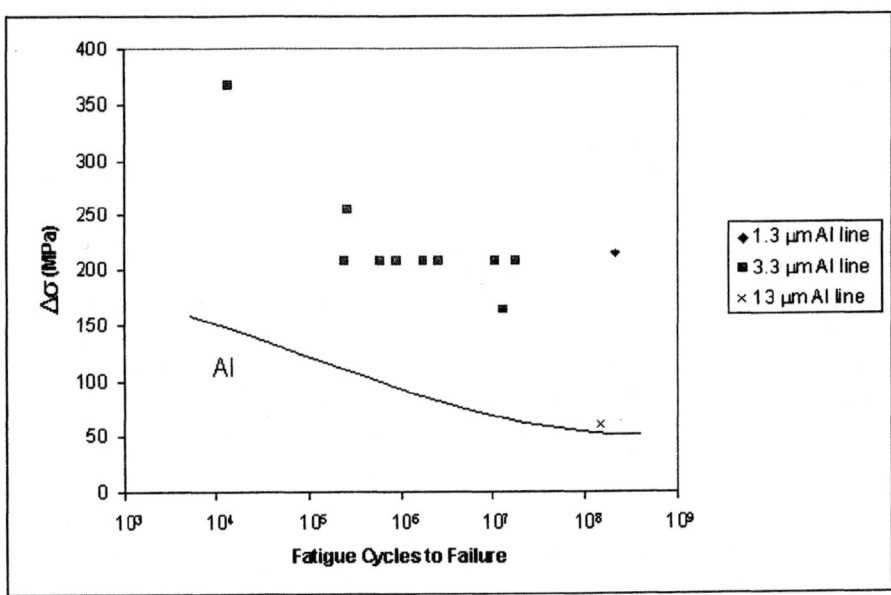

FIGURE 8. Stress amplitude vs. lifetime for AC tested unpassivated Al-1at.%Si lines cycled at 100 Hz. The smooth curve represents fatigue behavior of bulk, fine-grained Al (curve reprinted from reference 11: Acta Metallurgica, vol. 19, A.W. Thompson and W.A. Backofen, "The Effect of Grain Size on Fatigue," pp. 597-606, Copyright (1971), with permission from Elsevier Science).

fatigue of bulk fine-grained Al (20 μm) [11], where failure occurred by macroscopic cracking of the samples. At a given number of cycles, the bulk samples fail at roughly half the stress amplitude required to fail the line samples. However, the general dependence of the number of cycles to failure on stress amplitude is similar for lines and bulk aluminum. There is no evidence of a distinct threshold in stress for inducing fatigue damage in any of the samples. This is consistent with the idea that fatigue can occur even for stresses that are quite a bit lower than the macroscopic yield strength, since significant localized plasticity can occur both in polycrystalline films [12] and in bulk metals (high-cycle fatigue).

Although lifetime data was not collected for fatigued films, it became apparent that the films could also sustain considerably higher stress amplitudes than bulk metals. Several factors may account for the differences in stress amplitude among lines, films, and bulk samples. First, the bulk data was collected for aluminum with a grain diameter of 20 μm, whereas the Al films and lines had an approximate grain diameter of 0.5 μm. While fatigue crack initiation in Al is not strongly grain-size-dependent [11], there is a slight improvement in fatigue life with decreasing grain size. Second, Al films exhibit enhanced yield strengths due to constrained dislocation activity in thin films [12,13]. They can therefore sustain larger stresses at a given applied strain and will experience less plastic strain during cycling. Third, failure defined by cycles to open circuit is not the same as failure defined by cycles to macroscopic cracking. No fatigue cracks were observed in the lines; localized melting was observed instead. Melting presumably occurred when the cross section of the line decreased to some critical value either because of crack formation or because of local plasticity. Cracks were observed in the thin film on polyimide samples but due to the support of the underlying substrate, did not result in immediate failure. In addition, the substrate could act to suppress the strain localization processes that are known to contribute to fatigue fracture by limiting the total amount of displacement in each cycle. In the absence of a Si or polyimide substrate, perhaps more macroscopic displacements could have been developed, thereby making crack formation easier. Fourth, the stress amplitudes in the interconnects were determined assuming biaxial elastic behavior. While this is likely appropriate for the 13 μm lines, it is less applicable to the 3.3 and 1.3 μm lines [14]. Use of uniaxial elastic constants would lower the stress amplitudes considerably and predict a more tensile average stress. Finally, there is some uncertainty in the temperature measurements during AC testing, which translates to uncertainty in stress. Modifications are in progress to improve the accuracy of the temperature measurement.

There are many similarities to be found between the damage evolution of Al lines/films and bulk Al. In early stages of testing of the films and lines, damage takes the form of subtle surface wrinkles that are widely spaced and show little vertical displacement (amplitude). Such damage takes place only in certain grains, and is reminiscent of what is seen in early stages of fatigue of bulk metals, where dislocation intersections with the free surface leave slip traces. Detailed crystallographic analysis would confirm whether the selectivity in initial damage sites correlates to grains of favorable orientation for slip. With further straining, more damage in the lines and films is seen, in the form of a larger surface area fraction covered in wrinkles, greater amplitude, and narrower spacing. This also occurs in bulk metals, where more grains begin to yield, due to either strain hardening or to slip transmission across boundaries between a yielded and an unyielded grain. Greater surface wrinkle amplitude occurs as strains are accumulated when dislocations intersect and leave a grain. An increasing dislocation density leads to narrower slip band spacing, due to operation of additional dislocation sources and the onset of multiple slip system activity.

There are several differences in the damage observed in the AC tested lines and the mechanically tested thin films. For example, after fatigue testing of the thin films, voids were found in the interface between the metal and the substrate, for both Cu and Al. This is in contrast to what is observed in room temperature fatigue cracking of bulk metals, and also in AC-tested lines, neither of which shows evidence of void formation. Schwaiger and Kraft [16] have suggested that void nucleation in thin films may occur by dislocation annihilation processes during cyclic straining, and that such processes play an important role in thin film fatigue. Dislocation annihilation as a mechanism for increasing the concentration of non-equilibrium point defects such as vacancies has been discussed as a possible contributor to damage by Essmann *et al.* [17] for the case of fatigue in bulk Cu. The fact that voids are seen only in the fatigue-tested metal thin film samples may be a consequence of the metal/polyimide interface itself, which may act as a preferred nucleation site for voids. Another reason such voids are not observed in AC-tested lines could be that the lines are strained under considerably higher temperatures, often in excess of 200°C, where the overall mobility of point defects is increased. At elevated temperatures, the point defects may have time to move to grain boundaries or to other interfaces and thus avoid reaching a critical supersaturation for void nucleation.

A second difference between the damage observed in the lines and the thin films is the presence of whiskers: whiskers were observed protruding from the AC tested lines but not from the thin films. It is possible, in contrast to void formation, that elevated temperatures favor whisker growth, which would explain their absence in the room temperature tested thin films. The exact mechanism for the whisker growth is not known, but is likely driven by the presence of compressive stresses during cycling. The fact that whiskers were found more often in wide lines than narrow ones might be because the stresses are more compressive in the wider lines. Larger compressive stresses could result from a more biaxial stress state or from larger temperatures and temperature amplitudes.

As discussed above, the presence of both soft and hard overlayers did nothing to retard the formation of fatigue damage during AC testing. Fatigue crack initiation is highly dependent upon surface roughness, including that which develops during cyclic deformation. It has been shown that crack initiation can be suppressed by coating a metal with a hard overlayer [18,19]. The effect in this case is that topography induced by dislocations reaching the free surface is greatly reduced, as long as the overlayer remains intact. As a result, stress concentrations that lead to crack formation do not develop nearly so quickly. However, the experiment presented here, with a silicon nitride overlayer, did not show the expected suppression of surface damage, probably because the layer was not well adherent. Similarly, the relatively soft photoresist overlayer also did nothing to suppress damage formation. The photoresist did remain adherent during testing, but the surface roughness formation was likely unconstrained by the soft polymer. Since hard-baked photoresist is mechanically similar to some materials now under consideration as low-k dielectrics, AC-induced cyclic deformation may well become a problem in such interconnect systems. In addition, power dissipation will always be present in the interconnects of devices, and so will temperature cycles. When low-k interlevel dielectrics (with corresponding low thermal conductivities) and more metallization layers are introduced into devices, the temperature swings will increase and fatigue may become a very serious reliability threat.

SUMMARY

Observations of damage due to AC currents in Al interconnects have been presented and are attributed to thermal mechanical fatigue from the cyclic temperatures generated by Joule heating. The AC current-induced damage is compared with mechanical fatigue damage in Cu and Al thin films on polyimide substrates, and in most respects looks very similar. In both kinds of samples, the damage is similar to that observed at the surfaces of fatigued bulk specimens. It is therefore concluded that both thin films and interconnects are susceptible to classical fatigue damage, which is generated by the back and forth glide of dislocations. Given the proximity of the surfaces and interfaces in the interconnect and thin film samples, dislocation motion and annihilation is expected to be different from that in bulk material. In particular, sample dimensions may play an important role in determining the nature of the damage that is formed.

The temperature swings generated by the AC currents are typical of those in devices under operating conditions. Based on the fatigue behavior of bulk samples, it is expected that an adherent hard overlayer would inhibit surface fatigue damage formation and a soft overlayer would do little to inhibit damage formation. This may explain the fact that fatigue damage in interconnects encapsulated with well-adhering hard layers such as silicon dioxide or silicon nitride has not been commonly observed. However, if the interconnects are encapsulated with a soft material such as a polymer film, fatigue damage formation may no longer be hindered and failure due to fatigue may occur under normal device operating conditions.

ACKNOWLEDGEMENTS

We thank the NIST Office of Microelectronic Programs and the Max-Planck Visiting Scientist Program for support and Dr. D. Balzar for assistance with X-ray measurements.

REFERENCES

1. Philofsky, E., Ravi, K., Hall, K., and Black, J., "Surface Reconstruction of Aluminum Metallization – a New Potential Wearout Mechanism," in *9th Annual Proceedings of Reliability Physics*, IEEE, New York, 1971, pp. 120-128.
2. Gui, X., Haslett, J.W., Dew, S.K., and Brett, M.J., *IEEE Trans. Electron Dev.* **45**, 380-386 (1998).
3. Sinha, A.K. and Sheng, T.T., *Thin Solid Films* **48**, 117-126 (1978).
4. Read, D.T., *Int. J. Fatigue* **20**, 203-209 (1998).
5. Hommel, M., Kraft, O., and Arzt, E., *J. Mater. Res.* **14**, 2373-2376 (1999).
6. Volkert, C.A., Keller, R.R., Mönig, R., Arzt, E., and Kraft, O., "Fatigue as a Failure Mechanism in Interconnects," to be published (2001).
7. Kretschmann, A., Kuschke, W.-M., Baker, S.P., and Arzt, E., "Plastic Deformation of Thin Copper Films Determined by X-Ray Microtensile Tests," in *Thin Films: Stresses and Mechanical Properties VII*, Mat. Res. Soc. Symp. Proc. Vol. 436, Pittsburgh, 1996, pp. 59-64.
8. Kraft, O., Schwaiger, R., and Wellner, P., to be published in *Mater. Sci. Eng. A*.
9. Rodbell, K.P., Castellano, A.J., and Kaufman, R.I., "AC Electromigration (10 MHz – 1 GHz) in Al Metallization," in *Stress Induced Phenomena in Metallization: Fourth International Workshop*, edited by H.Okabayashi, S.Shingubara, and P.S.Ho, AIP, New York, 1998, pp. 212-223.
10. Shono, K., Kuroki, T., Sekiya, H., and Yamada, H., "Mechanism of AC Electromigration," *Proc. 1990 VMIC Conference*, 1990, pp. 99-105.
11. Thompson, A.W. and Backofen, W.A., *Acta Metall.* **19**, 597-606 (1971).
12. Vinci, R.P., Cornella, G., and Bravman, J.C., "Anelastic Contributions to the Behavior of Freestanding Al Thin Films," in *Stress Induced Phenomena in Metallization: Fifth International Workshop*, edited by O.Kraft, E.Arzt, C.A.Volkert, P.S.Ho and H.Okabayashi, AIP, New York, 1999, pp. 240-248.

13. Nix, W.D., *Metall. Trans.* **20A**, 2217-2245 (1989).
14. Dehm, G., Weiss, D., and Arzt, E., *Mater. Sci. Eng. A.* **309-310**, 468-472 (2001).
15. Sauter, A., *IEEE Trans. Components Hybrids & Manufacturing Technology* **15**, 594-600 (1992).
16. Schwaiger, R. and Kraft, O., *Scripta Mater.* **41**, 823-829 (1999).
17. Essmann, U., Gösele, U., and Mughrabi, H., *Phil. Mag.* **A44**, 405-426 (1981).
18. Alden, T.H. and Backofen, W.A., *Acta Metall.* **9**, 352-366 (1961).
19. Stoudt, M.R., Cammarata, R.C., and Ricker, R.E., *Scripta Mater.* **43**, 491 (2000).

Damaging of Metallization Layers by High Power Surface Acoustic Wave Fields

Siegfried Menzel, Hagen Schmidt, Manfred Weihnacht, and Klaus Wetzig

Institute of Solid State and Materials Research, P.O. Box 270016, D-01171 Dresden, Germany
Email: m.weihnacht@ifw-dresden.de

Abstract. Similar to interconnection lines in integrated circuits patterned metallization structures are used as electrodes and reflecting elements in surface acoustic wave (SAW) devices. As electromigration is known to be a damaging mechanism in interconnects also the metallization of SAW devices tends to be degraded under high power conditions. Corresponding to the source of degradation, namely the elastic waves, the latter is called acoustomigration. Aim of this work is an approach to understand damage mechanisms by corresponding loading conditions and failure occurences in the metallization. Our work is focused on in situ experiments with standing SAWs combining electrical measurements and different microscopic investigations (OM, SEM, FIB), detailed consideration of SAW stress field structure in connection with model experiments, and making use of the route of Cu technology. First results of a comparative study of fully metallized Al and Cu areas supporting high amplitude travelling SAWs are presented showing different damaging features, similar to those known from electromigration.

INTRODUCTION

Surface acoustic wave (SAW) devices are widely used in consumer electronics, in telecommunications, and they are on the way to enter new markets of wireless and sensor applications [1]. Clearly to recognize are trends towards higher input power, higher frequencies or smaller device structures, all together resulting in higher power density levels. It is a well-known fact that finger-shaped interdigital transducers and reflectors appearing as the key elements of SAW devices can be damaged under the influence of the elastic waves, also named acoustomigration, with a growing degree of probability at high power density levels.

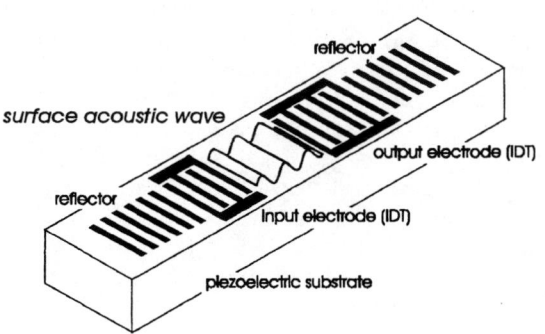

FIGURE 1. Schematic view of a SAW device (resonator).

CP612, *Stress-Induced Phenomena in Metallization:* Sixth Int'l. Workshop, edited by S. P. Baker et al.
© 2002 American Institute of Physics 0-7354-0058-X/02/$19.00

The working principle of SAW devices is based on only few elementary functions simplified shown in Fig. 1. A surface acoustic wave will be launched when applying an ac voltage with appropriate frequency to the electrical input due to the piezoelectric effect of the substrate. This wave has the strongest amplitude for a wavelength coinciding with the period of the interdigital structure of the input electrode and can be received and converted back into electrical energy by a second interdigital transducer (IDT) in an analogous but inverse manner. Beside the shown SAW, the input transducer launches an additional wave in backward direction towards the reflector strips arranged nearby. That wave will be reflected completely, so that it can propagate to the output transducer and interfere with the first SAW. A second reflector is shown beyond the output IDT acting similarly like the first reflector. So, the IDTs together with the reflectors form a cavity for standing acoustic waves named SAW resonator with maximum acoustic energy at a certain frequency which is the resonance frequency. Beside that type of SAW devices, different constructions are also used without reflectors. In that case the SAW travels in only one direction (from input to output IDT). Generally, kind and degree of damaging can expected to depend on the nature of wave, i.e. standing or travelling, and will be studied therefore separately as described below.

When considering mechanical damaging of surface metallizations caused by SAWs one more circumstance is important. Owing to the crystal anisotropy of the piezoelectric substrate material quite different types of surface waves exist: similarly to the 3 polarization types of bulk acoustic waves (2 transverse, 1 longitudinal), SAWs with different polarization type can be found (see Fig. 2). Correspondingly, the ac stress can be of compressive/tensile or shear character, in many cases a combination of both.

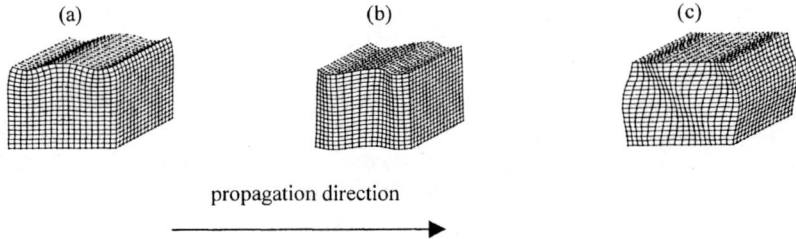

propagation direction

FIGURE 2. SAW types with different polarization of particle displacement (a) sagittal, (b) shear horizontal, (c) longitudinal polarization.

Since many years, it is well-known that in SAW devices a degradation of metallization layers, and in turn of device performance occurs depending on the level of acoustic power and time of operation. Apparently, the finger damaging is caused by the dynamic stress fields of the SAW and, therefore, it is described by the term acoustomigration. In contrast to electromigration, the device performance has a peculiar sensitivity against any degradation of all strip-like elements (both electrodes and reflectors). This is due to the analogue character of working principle, i.e. all device parameters depend directly from the material as well as geometrical parameters.

In connection with the above mentioned trends of SAW applications, efforts have been increased to improve the power durability by investigating the features of damaging and the usefulness of new material systems [2-5]. At the moment, all these attempts still lack a fundamental understanding of the facets of acoustomigration. A few words should be said about the character of acoustic loading. It represents a kind of cyclic mechanical loading and the material degradation is, therefore, a kind of cyclic mechanical fatigue. In contrast to conventional fatigue conditions (low-cycle, high-cycle fatigue) the stress and strain varies locally. Beside that, in case of SAW loading both frequency and number of cycles are far away from values of known fatigue. For that reasons, theory and experiences of "classical" fatigue can hardly be directly compared to acoustomigration. A further point which has to be considered is a possible local heating during acoustic loading.

In this situation, it seems to be worthwhile to make model experiments and SAW field simulations for getting deeper insight in the mechanisms of acoustomigration. In our paper, we present first results of in-situ studies (see also [6,7]) on high power standing waves, of experiments with travelling waves comparing fully metallized Al and Cu areas, and simulation results for representative cases of SAW stress fields.

RESULTS AND DISCUSSION

Experiments with standing waves

Experiments have been carried out in order to get correlations between the visible damaging of SAW structures and the features of electrical characteristics as a function of power level. For these studies we used a one port resonator (i.e. only one IDT for excitation and detection of the SAW) on STX quartz with 430 MHz peak frequency, 60 Al finger pairs for the transducer and 200 Al strips for each reflector. The fingers were free standing and had a width of 1.81 μm and a height of 160 nm. The sample was mounted inside a scanning electron microscope (SEM, Hitachi S3500N) and electrically driven by a network analyzer (HP8753C), combined with a 33 dB amplifier and two −20 dB directional couplers.

The resonator admittance measured as a function of frequency showing the different stages of affecting by input power is depicted in Fig. 3. At low power levels only a reversible frequency shift of the resonance peak due to thermal heating has been observed. Next, the admittance curve exhibits a change of shape, typically appeared an additional shoulder at the high frequency tail (265mW, more pronounced at 650 mW). In this state the peak frequency was reduced after switching off the power. Finally, at an input power of 2W, a further distortion of the admittance curve and shifting down after loading is obvious.

In a direct relationship to the irreversible changes of admittance curves was the formation of hillocks at the strip edges observed simultaneously in the SEM images. As visible in Fig. 4, the hillocking occurs mainly at the strip edges indicating a damaging mechanism which is governed by the standing acoustic wave pattern. That feature will be discussed below. Damaging was observed within the IDT region as well as within the area of reflector strips definitely demonstrating the non-electric effect of hillock formation because the reflector strips are free from electric currents.

The progress of metallization damaging with time has been investigated in a similar manner by comparing electrical measurements with simultaneous SEM observations. The sequence of SEM images depicted in Fig. 5 shows the increasing damage state with time, but it seems that essential alteration takes place at the beginning of loading.

In contrast to Fig. 4 only one strip edge is damaged in Fig. 5. The picture has been made in the outer part of the device and is completely in accordance with the fact that the pattern of damaging coincides with the pattern of acoustic standing wave maxima, because the acoustic wavelength slightly differs from the finger periodicity. That difference can be understood in terms of more detailed resonator simulation including internal SAW reflections at the finger edges. So, one observes more and more asymmetric damaging with respect to the finger center lines when going to the outer parts of the device.

Hillock formation is a typical thin-film phenomenon occuring during film growth [8], thermal cycling [9], and in connection with electromigration [10]. Several mechanisms have been proposed to account for the hillock formation. It is commonly assumed that hillocking takes place under compressive stress. The mass-transfer mechanism involved might be grain-boundary diffusion or grain boundary sliding. In the case of electromigration, hillocking is directly related to void formation as a consequence of diffusion-flux divergences. Here, we found voids only to small extent. The hillock formation under the influence of SAWs might be attributed to an ac stress activation acting in a comparable way as the thermal energy does.

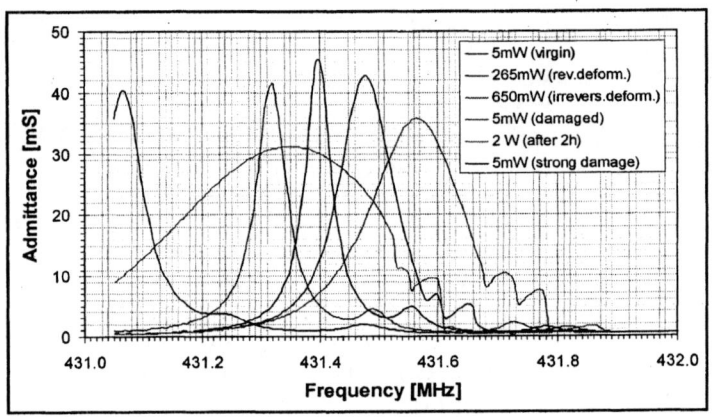

FIGURE 3. Real part of electrical admittance of a 430 MHz one port resonator on STX quartz under different power loading conditions (rev.: reversible, irrevers.: irreversible).

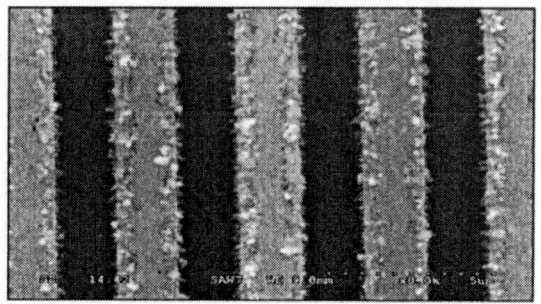

FIGURE 4. Hillock formation at the strip edges in the center of the SAW resonator.

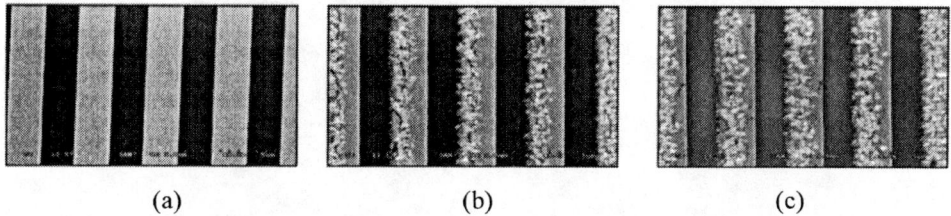

(a)   (b)   (c)

FIGURE 5. Hillock formation in the SAW resonator increasing with loading time ((a) initial state, (b) after 0.5 min, and (c) after 71 min).

Experiments with travelling waves

SAW propagation without interfering backward waves, i.e. travelling waves instead of standing waves, is a common situation in practical devices too. From the point of view of damaging the material system a laterally more homogeneous picture we have to expect. The SAW amplitude distribution radiated from an IDT follows the rules of acoustic diffraction. As a consequence, surface regions with enhanced power density exist outside the IDT which appear to be most intensively damaged. We used test structures with unidirectional IDTs that ensure high SAW amplitudes at given input power because twice the energy will be radiated in one direction compared to bidirectional IDTs. Beside transmitting IDTs receivers have been incorporated into the test structure (see Fig. 6) for monitoring the electric behavior during microscopical studies. The arrangements of Al and Cu metallizations are depicted in Fig. 6 too.

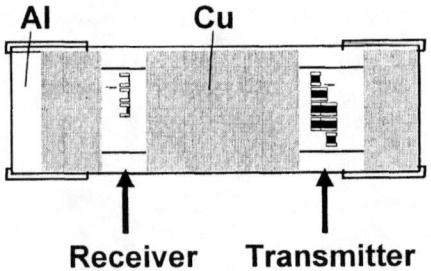

FIGURE 6. Test structure for experiments with travelling waves on LiNbO$_3$ with Al and Cu metallization layers.

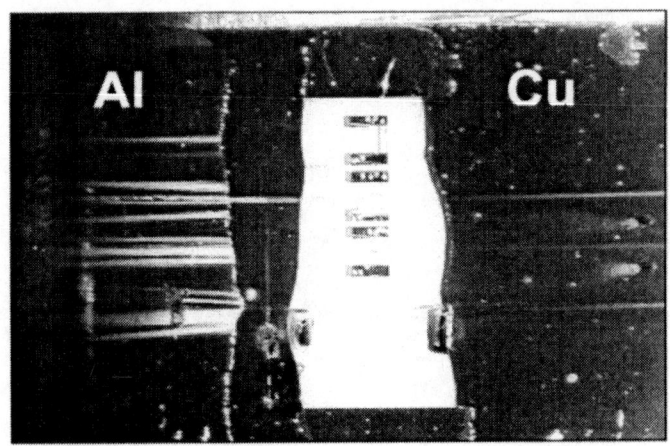

FIGURE 7. Microscopical image of surface roughening in the receiver region in horizontal lines by high amplitude SAWs (the SAW travels from right to left): strongly reduced damaging of Cu layers compared with Al.

The transmitter structure was a set of slightly different IDTs which could be driven simultaneously, so a series of parallel rays have been observed originated by the roughening of metallic layer surfaces due to acoustomigration. In Fig. 7 such rays are visible left from the receiver (the SAWs come from the right edge). The distinct stronger roughening of Al compared with Cu is evident.

More detailed information about the damage behavior can be deduced from SEM and FIB images shown in Fig. 8. Formation of grooves in Cu layers instead of voids and hillocks in Al is recognizable.

Al is known to display an extraordinary high tendency to hillock formation compared to other materials, e.g. Cu. That phenomenon is probably caused by a very low surface diffusivity of Al refered to the grain boundary diffusivity. The observation in Fig. 8 is a hint that acoustic loading causes diffusional processes because they can be made responsible for both the hillock formation in Al and the grooving in Cu rather than e.g. dislocation glide.

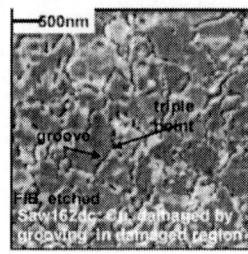 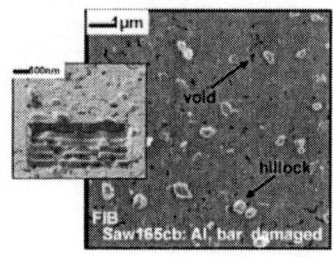 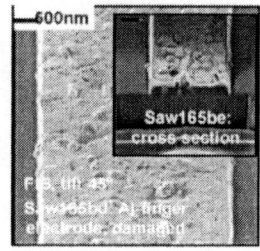

FIGURE 8. SEM and FIB pictures showing the nature of damaging in Cu and Al layers by travelling SAWs.

MODELLING OF SAW FIELDS

Above mentioned diversity of SAW features (see Fig. 2) implies detailed modelling of ac stress fields for understanding of the acoustomigration process. This can be done in the frame of numerical studies what is common since long time in SAW technology. They are based on the equations of motion and the constitutive equations with regard to elastic, dielectric, and piezoelectric material parameters plus all mechanical and electrical boundary conditions at the interfaces and at the free surface. The state of the art is characterized by most comfortable simulation of SAW excitation and propagation in metallization gratings on piezoelectric substrates using FEM/BEM methods (see for example [11]). But, for many purposes an approach considering the wave propagation in a continuously metallized substrate is fully informative. One has to solve the problem using a partial wave analysis [12].

We will do so here in order to understand the damaging patterns of Figs. 4 and 5. Fig. 9 shows the calculated distribution of ac stress components and their local position with respect to the electrodes for the STX quartz-cut using computer programs based on [13]. The T_{11} component (compressive/tensile stress component in wave propagation direction) and T_{22} component (in finger direction) have their maxima between the electrodes and nodes in the finger centers, the shear component T_{12} is 90° phase shifted but has only a small amplitude. Considering T_{11} and T_{22} as responsible for the acoustomigration effect the damaging of finger edges seems reasonable. The asymmetric hillocking on only one finger edge outside the center of resonator (Fig. 5) is the result of the fact that the resonance frequency is slightly different from the synchronous frequency of the grating structure as discussed above. The stress amplitudes correspond to an input power of 1W per 1 mm SAW beam width at 100 MHz. It would be hazardous to estimate a damaging threshold for the ac stress in face of the number of loading cycles in the order of 10^{12} which is beyond the firmed knowledge about high cycle fatigue.

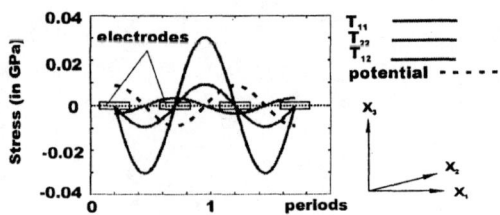

FIGURE 9. Modelling of AC stress components T_{ij}: standing waves on STX quartz, 100 MHz, 1 W/mm (legend see text).

SUMMARY

Surface acoustic wave (SAW) devices suffer from damaging of finger metallization layers under high power loading. The aim of this work is to obtain insight in the underlying mechanisms of degradation features named acoustomigration. In situ experiments have been carried out with standing and travelling SAWs combining electrical measurements with different microscopic investigations. Hillock formation at the edges of Al strip-like layers within a SAW resonator has been observed at high SAW power with a local distribution which is in agreement with calculated ac stress fields. First results of a comparative study of fully metallized Al and Cu areas exhibit different degradation features at SAW loading: hillocking in the case of Al and grooving for Cu, similar to those known from electromigration.

ACKNOWLEDGMENTS

The authors are grateful to R. Kunze for support in electrical measurements, G. Martin for SAW design, R. Wobst for programming work, and V. Weihnacht for helpful discussions.

REFERENCES

1. Hickernell, F.S. "Surface Acoustic Wave Technology – Macrosuccess through Microseism" in: *Physical Acoustics, Vol. XXIV*, ed. by E.P. Papadakis, Acad. Press, San Diego 1999, p. 135.
2. Satoh, Y., Nishihara, T., Ikata, O., Ueda, M., and Ohomori, H., "SAW Duplexer Metallizations for High Power Durability", *IEEE Ultrasonics Symp. Proc.* 1998, pp. 17-26.
3. Takayama, R. et al., "High Power SAW Filter with New Al-Sc-Cu/Ti/Al-Sc-Cu/Ti Electrodes", *IEEE Ultrasonics Symp. Proc.* 1998, pp. 5-8.
4. Kim, Y. et al., "Passivation Layer Effects on Power Durability of SAW Duplexer", *IEEE Ultrasonics Symp. Proc.* 1999, pp. 39-42.

5. Ruile, W., Raml, G., Springer, A., and Weigel, R. "A Novel Test Device to Characterize SAW Acoustomigration", *IEEE Ultrasonics Symp. Proc.* 2000, pp. 275-278.

6. H. Schmidt, K. Franke, F. Höller, G. Martin, M. Roß-Meßemer, and M. Weihnacht, "UV Reflective Modulation Using SAWs with High Amplitude", *IEEE Ultrasonics Symp. Proc.* 2000, pp. 655-658.

7. S. Menzel, H. Schmidt, K. Wetzig, M. Weihnacht, "In situ SEM investigation of stress induced migration in SAW structures", *Proc. of the 12th European Congress on Electron Microscopy (EUREM), Vol. II, Physical Sciences*, July 2000, Brno, Czech, pp. 541-542.

8. Roberts, S. and Dobson, P. J. "The Microstructure of Al Thin Films on Amorphous SiO_2", *Thin Solid Films 135*, 1986, pp. 137-148.

9. Rouay, M. and Aliotta, C. F., "Hillock Formation in Lead Films by Grain-Boundary Sliding", *Philosophical Magazine A*, 42, 1980, pp. 161-184.

10. Kinsbron, E., *Appl.Phys. Lett.* 36, 1980, p. 968.

11. Biryukov, S. and Weihnacht, M., "Elastic Electrode Polarization in a Spatial Harmonic Field and the Natural Boundary Element Method", *IEEE Ultrasonics Symp. Proc.* 2001, P2B-6.

12. Campbell, J. J. and Jones, W. R., *IEEE Trans. SU-15*, 1968, pp. 209.

13. Wobst, R., "The generalized eigenvalue problem and acoustic surface wave computations, *Computing* 39, 1987, pp. 57-69.

PART III: STRESS VOIDING

Cohesive Zone Models of Void Nucleation At Interconnect Interfaces

A.F. Bower

Division of Engineering, Brown University, Providence, RI 02912

Abstract. Numerical and analytical models are used to predict the conditions necessary to nucleate voids by thermal stress in representative interconnect structures. Weak grain boundaries and interfaces in the structure are idealized as cohesive surfaces, which are characterized by their work of separation and the peak stress that the interface can withstand. Analytical expressions are developed to predict qualitatively the influence of line width, barrier layer thickness, bulk material properties, and the cohesive properties interfaces on the conditions necessary to nucleate voids. Detailed three dimensional numerical computations are used to predict the influence of interconnect geometry and microstructural features on void nucleation.

INTRODUCTION

Stress driven nucleation and growth of voids is a leading cause of interconnects failures. Current industry roadmaps require continued reductions in line dimensions, from the sub-micron sizes in use today, to nanometer-scale dimensions within the next decade. In addition, a shift is anticipated from nitride and oxide passivation to polymeric low-k dielectric materials. Furthermore, interfaces in modern interconnect structures have a complex microstructure, which make use of many thin layers of dissimilar materials to achieve desired mechanical, thermal, chemical and electrical properties. Experiments (e.g. Lane *et al*, 2000) show that the mechanical strength of representative interconnect interfaces are strongly sensitive to chemical composition and microstructure, pointing to the need for an improved understanding of the role of atomic scale structure on void nucleation and growth.

Our objective in this paper is to contribute to this effort by conducting parametric studies of stress driven void nucleation in representative interconnect structures. In particular, we will investigate the influence of interconnect geometry, grain structure, elastic mismatch, and the cohesive properties of interfaces and grain boundaries on the conditions necessary to nucleate voids at the interconnect/passivation interface and at grain boundaries within the line.

We will begin by deriving analytical solutions for an idealized interconnect geometry, sketched in Fig. 1. The interconnect is modeled as an infinitely long cylinder, radius a, with elastic constants E_i, v_i and thermal expansion coefficient α_i. The interconnect is surrounded by a barrier layer with outer radius b, and elastic

CP612, *Stress-Induced Phenomena in Metallization:* Sixth Int'l. Workshop, edited by S. P. Baker et al.

constants E_b, v_b and thermal expansion α_b, and is encapsulated within an infinite passivating solid with elastic constants E_p, v_p and thermal expansion coefficient α_p. In this simplified model, we assume that failure occurs by debonding at the interface at $r=a$ between the interconnect and the surrounding barrier layer. This interface is modeled as a cohesive surface, with traction-displacement relation

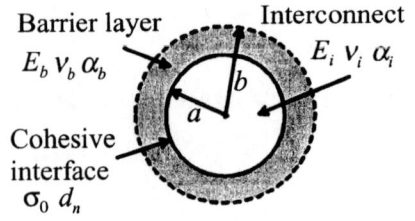

Barrier layer $\quad$ Interconnect
$E_b \, v_b \, \alpha_b$ $\qquad\qquad$ $E_i \, v_i \, \alpha_i$
Cohesive interface
$\sigma_0 \, d_n$
Passivation $\quad E_p \, v_p \, \alpha_p$

FIGURE 1. Simple axisymmetric model of an interconnect surrounded by a barrier layer and encapsulated within passivation.

$$\sigma_n = \sigma_0 \frac{\delta_n}{d_n} \exp(1 - \delta_n / d_n) \qquad (1)$$

where σ_0 and d_n are material constants that govern the interface strength and compliance. Their physical significance is illustrated in Fig 3: σ_0 is the tensile strength of the interface, while d_n specifies the interface separation at which the peak stress occurs. Note also that the work of separation for the interface is related to σ_0 and d_n by

$$\phi_n = e\sigma_0 d_n \qquad (2)$$

where $e=\exp(1)$. The solid is assumed to be free of stress at some initial temperature T_0, and is then cooled to temperature $T = T_0 - \Delta T$. For this simplified interconnect geometry, it is straightforward to derive analytical expressions for the stress state and interface separation as a function of temperature drop ΔT. This in turn allows us to calculate in closed form the critical conditions for interfacial failure as a function of line width, barrier layer thickness, and material and interfacial properties.

While our idealized interconnect structure provides considerable insight into void nucleation at interconnect interfaces, more sophisticated computations are required to investigate in detail the role of interconnect geometry and microstructure. We have therefore conducted numerical simulations of interfacial debonding in three representative interconnect structures, illustrated in Fig. 2. The interconnect structure shown in Fig 2a represents the region of a single crystal line far from any vias. This is the strongest possible interconnect microstructure, and forms a baseline for our parametric studies. The microstructure shown in Fig 2b is used to compute the strength reduction caused by polycrystalline segments within the interconnect. Finally, Fig 2c represents a typical via geometry.

In all computations, we idealize both the line and passivation as isotropic solids, with Young's moduli E_i and E_p, Poisson's ratios v_i and v_p, and thermal expansion coefficients α_i, α_p, respectively. In all numerical simulations, the passivation is modeled as a rectangular block of material with boundaries at least $10a$ from the via,

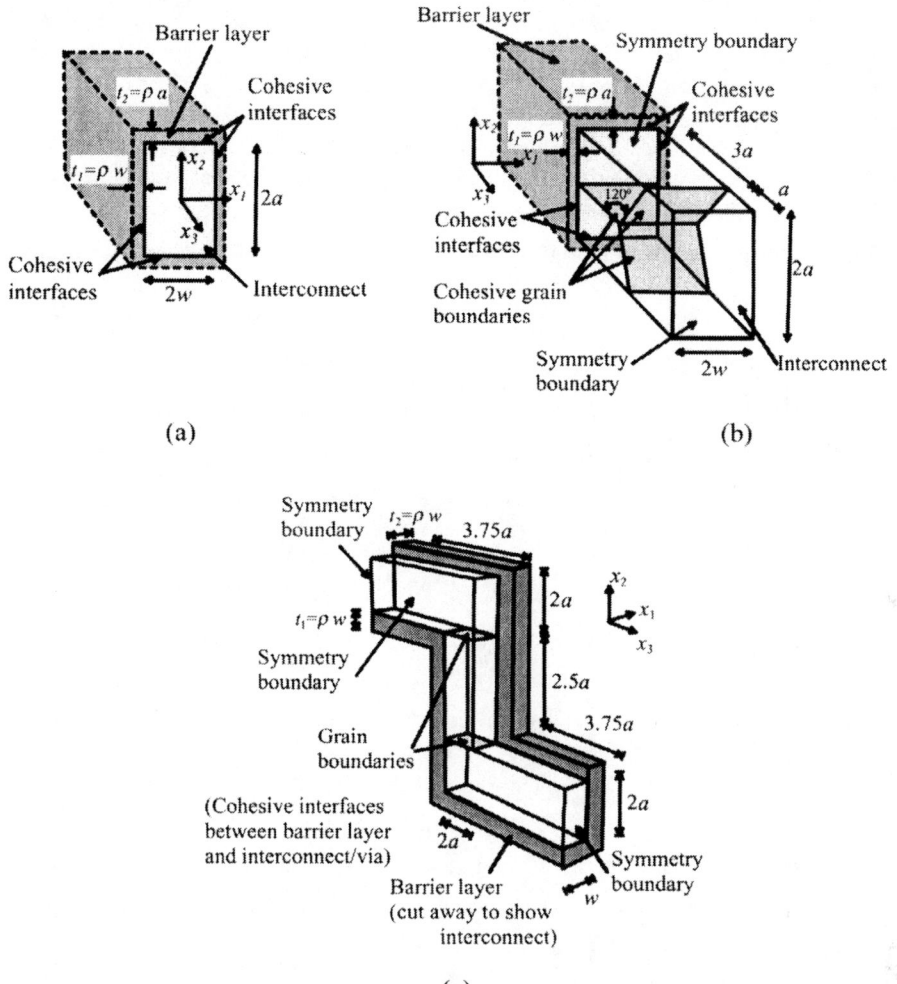

FIGURE 2. Representative interconnect geometries used in numerical simulations. (a) An ideal, single crystal interconnect; (b) A polycrystalline interconnect containing a mixture of bamboo and non-bamboo grains (c) Simple model of a via structure.

to approximate a semi-infinite solid. The barrier layer between the interconnect and passivation is modeled as an elastic solid, with constants E_b, v_b, α_b. The interconnect may contain grain boundaries, and is bonded to the barrier layer through a weak interface. The mechanical behavior of grain boundaries and interfaces is modeled using the cohesive interface formulation developed by Xu and Needleman (1994), described in more detail in the next section. In this approach, the mechanical response

147

of a debonding interface is characterized by an interplanar potential, which specifies the traction acting on the two adjacent material surfaces as a function of their normal and tangential separation. A typical interface response is illustrated in Fig. 3. Under purely normal loading, the constitutive law reduces to the form given in eq. 1. In addition, the model allows for failure induced by tangential tractions under mixed mode loading conditions. Key features of the response are the maximum normal and tangential stress that the interface can withstand σ_0, τ_0; the separation of the interface at maximum normal and tangential stress d_n, d_t and the work done to completely separate the interface Φ_0.

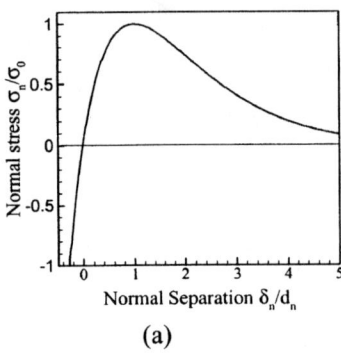 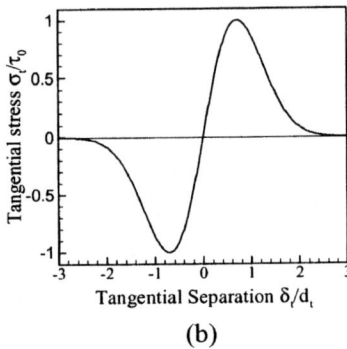

(a) (b)

FIGURE 3: Traction-displacement relations for the cohesive interface model used in analytical and numerical computations. (a) Normal stress-opening displacement behavior; (b) Shear stress-sliding displacement behavior.

Each interconnect structure is assumed to be free of stress at some initial temperature T_0. Subsequently, tensile stress is generated in the line by cooling to operating temperature $T = T_0 - \Delta T$. For each interconnect microstructure, we will compute the critical temperature change ΔT_c at failure as a function of appropriate geometric and material parameters.

THEORY

We begin by summarizing the constitutive relations associated with the Xu-Needleman (1994) cohesive zone interface model. Let $\Re$ denote the volume occupied by the collection of grains, let $\partial\Re$ denote external boundaries, and denote interfaces by Γ. Let n_i denote the outward normal to $\partial\Re$, and let m_i and t_i^α with $\alpha = 1,2$ denote the normal and two mutually perpendicular tangent directions on Γ. Any convenient convention may be used to assign directions to $m_i, t_i^{(\alpha)}$. In addition let $u_i(x_j)$ denote the displacement field in the solid; and denote the stress by σ_{ij}. Assume that the external boundary of the solid is loaded by tractions T_i on part of the boundary

$\partial_2\Re$, while the remainder of the external boundary is subjected to prescribed displacement U_i. We assume infinitesimal deformations, for simplicity. Field quantities within the interconnect and passivation are governed by the usual field equations of linear elasticity.

The normal and tangential separation of the interface then follow as

$$\delta_n = \left(u_i^+ - u_i^-\right)m_i$$
$$\delta_t^{(\alpha)} = \left(u_i^+ - u_i^-\right)t_i^{(\alpha)}$$

(3)

where $u_i^{\pm}$ denote displacements adjacent to the boundary. We define normal and tangential tractions acting on the interface similarly

$$\sigma_n = m_i\sigma_{ij}m_j$$
$$\tau^{(\alpha)} = t_i^{(\alpha)}\sigma_{ij}t_j^{(\alpha)}$$

(4)

The tractions are related to the displacement jumps through an interplanar potential ϕ, as

$$\sigma_n = \frac{\partial\phi}{\partial\delta_n} \qquad \tau_t^{(\alpha)} = \frac{\partial\phi}{\partial\delta_t^{(\alpha)}}$$

(5)

The specific form of the interplanar potential adopted here is taken from Xu and Needleman (1994)

$$\phi(\delta_n,\delta_t) = \phi_n + \phi_n \exp\left(\frac{-\delta_n}{d_n}\right)\left\{\left[1-r+\frac{\delta_n}{d_n}\right]\frac{1-q}{r-1} - \left[q+\left(\frac{r-q}{r-1}\right)\frac{\delta_n}{d_n}\right]\exp\left(\frac{-\delta_t^2}{d_t^2}\right)\right\}$$

(6)

where $\delta_t = \sqrt{\delta_t^{(1)2} + \delta_t^{(2)2}}$, and ϕ_n, d_n, d_t, q and r are material parameters controlling the response of the interface. The physical significance of these parameters is illustrated in Fig 3, which shows the normal and tangential stress-displacement relations for the interface. ϕ_n is the total work of separation under tensile loading; d_n is the relative normal displacement of the interface corresponding to the maximum normal interfacial traction σ_0; d_t similarly controls the shear response so that the peak shear traction τ_0 occurs at a tangential interface displacement $\delta_t = d_t/\sqrt{2}$; q controls the relative strength of the interface under normal and shear loadings, so that

$$q = \frac{\phi_t}{\phi_n}$$

(7)

where

$$\phi_t = \sqrt{\frac{e}{2}}\tau_0 d_t$$

(8)

is the shear work of separation. Recall also that the tensile work of separation is related to the strength of the interface and d_n by

$$\phi_n = e\sigma_0 d_n$$

(9)

The behavior of the solid under a quasi-static cycle of load or temperature is computed using the standard finite element method, based on the principle of virtual work

$$\int_{\Re} \sigma_{ij} \delta\varepsilon_{ij} dV + \int_{\Gamma} \left\{ \sigma_n \left(\delta u_i^+ - \delta u_i^- \right) m_i + \tau^{(\alpha)} \left(\delta u_i^+ - \delta u_i^- \right) t_i^{(\alpha)} \right\} dA = \int_{\partial_2 \Re} T_i \delta u_i dA \qquad (10)$$

where δu_i, $\delta\varepsilon_{ij}$ denote kinematically admissible variations in displacement and strain. To account for interface separation, therefore, a discrete form of the second integral in the preceding equation must be added to the usual finite element equations. Actual computations were performed using the commercial finite element program ABAQUS, with the cohesive zones modeled by an appropriate user element.

RESULTS AND DISCUSSION

The idealized interconnect geometry sketched in Fig 1 provides considerable insight into the role of interconnect size, material properties and interfacial properties on the conditions necessary to nucleate voids at the interface between the line and its surrounding barrier layer. A straightforward calculation shows that the separation δ_n of the interface between interconnect and barrier layer is related to the temperature drop through

$$\frac{\overline{\Delta\alpha}\Lambda}{\sigma_0} \Delta T = \frac{\delta_n}{d_n} [\chi + \exp(1 - \delta_n / d_n)] \qquad (11)$$

Here, Λ and $\overline{\Delta\alpha}$ are an effective stiffness and thermal expansion mismatch, defined by

$$\frac{1}{\Lambda} = \left(\frac{1}{\overline{B}} + \frac{1 - 2v_i}{B_i} \right) \qquad (12)$$

$$\overline{\Delta\alpha} = (\alpha_i - \alpha_p)(1 + v_i) + \frac{(\alpha_b - \alpha_p)(1 + v_b)(B_p - B_b)(\rho^2 - 1)}{(1 - 2v_b)B_p\rho^2 + B_b\rho^2 + B_p - B_b} \qquad (13)$$

where $\rho = b/a$ is a measure of the thickness of the barrier layer compared with the interconnect size; B, with the appropriate subscript, denotes the biaxial modulus $E/(1+v)$ of interconnect (i), barrier (b) or passivation (p), and

$$\overline{B} = \frac{B_b \left[B_p\rho^2(1 - 2v_b) + B_b\rho^2 + B_p - B_b \right]}{(1 - 2v_b)\left(B_p\rho^2 + B_b - B_p \right) + B_b\rho^2} \qquad (14)$$

Finally, the dimensionless parameter

$$\chi = \frac{d_n\Lambda}{a\sigma_0} \qquad (15)$$

quantifies the relative compliance of the interface, d_n / σ_0, compared with the elastic compliance of the interconnect and surrounding materials, a/Λ. The radial stress in the interconnect is related to the interface separation by

$$\frac{\sigma_{rr}}{\sigma_0} = \frac{\delta_n}{d_n} \exp(1 - \delta_n / d_n) \qquad (16)$$

150

Now, suppose that the interconnect is free of stress at some initial temperature T_0, and is then progressively cooled to operating temperature $T = T_0 - \Delta T$. Fig. 4 shows the resulting radial stress in the interconnect as a function of normalized temperature change, for various values of compliance χ. For a stiff interface ($\chi < \exp(-1)$), the stress initially increases with ΔT, until a point of instability is reached. At this point the interface separates catastrophically, with a sudden drop in stress. If ΔT is subsequently reduced, the interface gradually closes, until a second point of instability is reached, where the interface snaps closed. For compliant interfaces ($\chi > e^{-1}$), there is no unstable void nucleation. In this case, the stress initially increases, and then decreases, as the interface smoothly pulls apart. Moreover, the temperature-separation and temperature-stress relations are fully reversible, so that as the temperature change is reduced, the interface progressively closes. For $\chi < e^{-1}$, critical void nucleation and healing temperatures can be calculated as a function of χ: the result is shown in Fig 5.

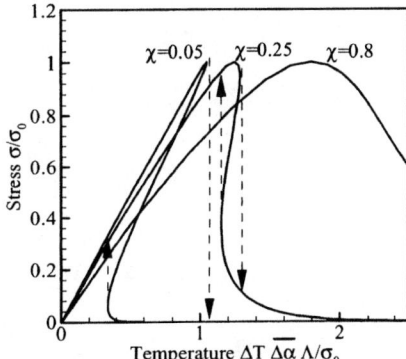

FIGURE 4. Normalized radial stress as a function of normalized temperature for the 1D interconnect model shown in Fig. 1.

For $\chi > e^{-1}$ the failure point cannot be identified unambiguously, so instead we show the normalized temperature change corresponding to the point of maximum stress, which occurs at

$$\frac{\overline{\Delta \alpha \Lambda}}{\sigma_0} \Delta T_f = 1 + \chi \qquad (17)$$

Note that the latter expression provides a safe, and surprisingly accurate, estimate of the void nucleation temperature for all values of χ.

Typical values for the various material properties and geometrical parameters are listed in Table 1. It is difficult to determine values of d_n and σ_0 for the interface with certainty. The work of separation for most practical material systems is of the order of 1Jm^{-2}, and atomic forces typically act over length scales of 0.2-1nm. This gives peak stress values around $\sigma_0 \approx 1 - 10 \text{ GNm}^{-2}$. Atomistic simulations using ab-initio or empirical potentials predict comparable values. Most cohesive-zone based simulations of void nucleation and fracture

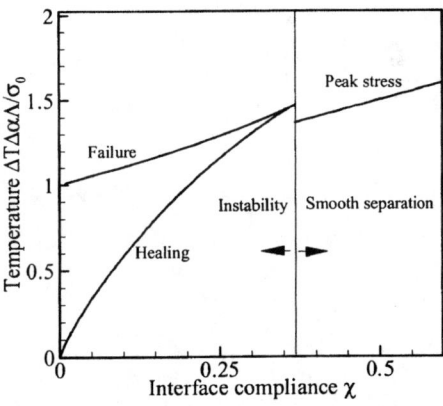

FIGURE 5. Normalized failure temperature as a function of normalized interface compliance for the 1-D interconnect model shown in Fig. 1.

151

TABLE 1. Representative material properties for current and proposed metallization systems.

Line width a	0.07-0.2 μm
Barrier layer thickness b-a	0.02-0.05 μm
Young's Modulus (Cu) E_i	128 GNm^{-2}
Poisson's ratio (Cu) V_i	0.36
Thermal expansion (Cu) α_i	16.0 10^{-6} K^{-1}
Young's modulus (Ta) E_b	160 GNm^{-2}
Poisson's ratio (Ta) V_b	0.3
Thermal expansion (Ta) α_b	6.0 10^{-6} K^{-1}
Young's modulus (SiO$_2$) E_p	71.4 GNm^{-2}
Poisson's ratio (SiO$_2$) V_p	0.16
Thermal expansion (SiO$_2$) α_p	0.51 10^{-6} K^{-1}
Young's modulus (Polyimide) E_p	7.5 GNm^{-2}
Poisson's ratio (Polyimide) V_p	0.35
Thermal expansion (Polyimide) α_p	6.0 10^{-6} – 150.0 10^{-6} K^{-1}
Interface work of separation ϕ_n	1 Jm^{-2}
Interface strength σ_0	0.1-10 GNm^{-2}
Cohesive length d_n	0.1-10 nm

use significantly lower values for the peak stress (and larger values for the cohesive length d_n), however, since the atomic scale values lead to unreasonably high values of stress near defects. This is usually justified by arguing that inelastic processes such as dislocation activity or diffusion increase the effective compliance of the interface, while reducing its strength. These considerations may reduce σ_0 to as low as 0.1GNm^{-2} while increasing d_n to 10nm.

For 0.2 μm copper metallization, with Ta barrier layers and SiO$_2$ passivation, therefore, the normalized interface compliance χ ranges between 0.0036 for an ideal interface ($\sigma_0 = 10GPa$, $d_n = 0.1nm$) to 36 for an interface whose peak strength is significantly reduced by inelastic processes ($\sigma_0 = 0.1GPa$, $d_n = 10nm$). The corresponding range for 0.07 μm interconnects passivated by low-k dielectric is $0.0085 < \chi < 85$.

In view of this uncertainty, it is best to proceed by calculating a safe bound to the void nucleation temperature that is insensitive to the values of σ_0 and d_n. To this end, we re-write the expression for the critical void nucleation temperature in terms of the work of separation for the interface ϕ_n and its peak strength σ_0.

$$\overline{\Delta\alpha}\Delta T_f = \frac{\sigma_0}{\Lambda}\left(1 + \frac{\phi_n\Lambda}{ea\sigma_0^2}\right) \qquad (18)$$

For a fixed work of separation, we observe that an interface with high strength is resistant to failure (for obvious reasons). Surprisingly, an interface with low strength is also resistant to failure, because for fixed ϕ_n a low strength implies a large compliance, which allows the stress in the interconnect to relax. In between these two limits the void nucleation temperature has a minimum value

$$\overline{\Delta\alpha}\Delta T_f^{min} = 2\sqrt{\frac{\phi_n}{ea\Lambda}} \qquad (19)$$

which occurs for an interface strength

$$\sigma_0^{min} = \sqrt{\frac{\phi_n\Lambda}{ea}} \qquad (20)$$

This minimum value provides a safe bound to the conditions necessary to nucleate voids, and is a function of material parameters that can be easily measured or computed using atomistic simulations. With representative values listed in Table 1, we find $\Delta T_f^{min} = 600K$ for current metallization systems, and 1800K for $0.07\,\mu m$ low-k dielectric systems.

Our estimate also predicts the qualitative effects of changes in the properties of the interconnect, barrier layer or passivation, as well as interconnect size and barrier layer thickness. Detailed studies of this nature are left to the reader, but it is easy to see that

1. Resistance to void nucleation increases as interconnect dimensions are reduced;
2. Selecting compliant materials for interconnect, barrier layer or passivation improves resistance to void nucleation;
3. Increasing the Ta barrier layer thickness reduces resistance to void nucleation, although for practical ranges of barrier thickness the void nucleation temperature is altered by less than 15%.

A 1-D model of an interconnect can at best provide only a qualitative estimate of the critical conditions to nucleate voids. We have therefore conducted numerical simulations of void nucleation in several more realistic interconnect microstructures.

In all our numerical computations, properties listed in Table 1 were assigned to passivation, barrier layer and interconnect, and the barrier layer thickness was chosen so that $b/a=1.25$ in each case. The barrier layer/passivation interface was assumed to be perfectly bonded. Grain boundaries and the interconnect/barrier layer interface were all modeled as cohesive surfaces, and were assumed to have identical properties. Specifically, we set the work of separation to 1 Jm^{-2}, and calculated the peak strength under normal loading using equation 20, to give a conservative estimate of the strength of the interconnect. For the remaining constitutive parameters listed in Section 3, we chose $d_t = d_n$, $q=1$ and $r=0$ in all simulations. For numerical purposes, we define the point of failure as the instant when the stress on any interface in the microstructure reaches the cohesive strength σ_0.

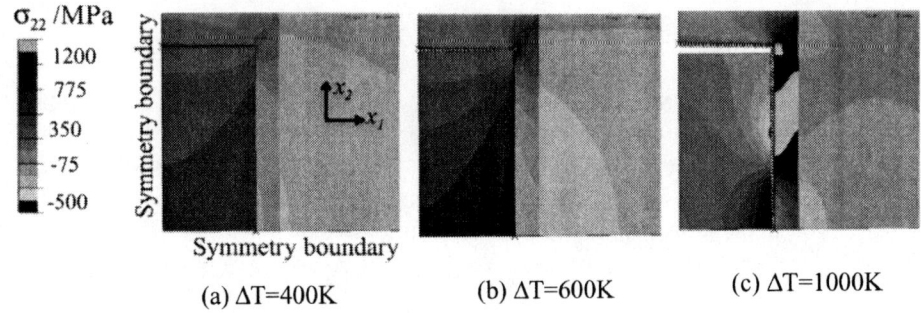

σ_{22} /MPa

1200
775
350
-75
-500

Symmetry boundary

(a) ΔT=400K   (b) ΔT=600K   (c) ΔT=1000K

FIGURE 6. A sequence of contours of vertical stress during void nucleation in the interconnect microstructure shown in Fig. 2(a). Results are shown for $a = 0.2\mu$m, a/w=0.5, $\rho = 0.25$, with SiO_2 passivation (Table 1). Displacements are magnified by a factor of 3.

Consider first the single-crystal line with simple rectangular geometry illustrated in Fig 2a. Our computations show that, for an aspect ratio w/a=1, the interface debonds at a temperature drop $\Delta T = 700K$. The simple axisymmetric model predicts the critical failure temperature for the line to within 15%, suggesting that minor variations in the line geometry do not significantly influence its strength. The failure conditions are weakly sensitive to the line aspect ratio. For w/a<1, failure occurs first at the shorter sides of the interconnect, whereupon a crack propagates down the longer sides of the interconnect, as shown in Fig. 5. Interestingly, there appears to be a line aspect ratio that minimizes the failure temperature for the structure. However, the failure temperature changes by less than 15% in the range 0.125<w/a<1, so this phenomenon has little practical significance.

σ_{33}/MPa

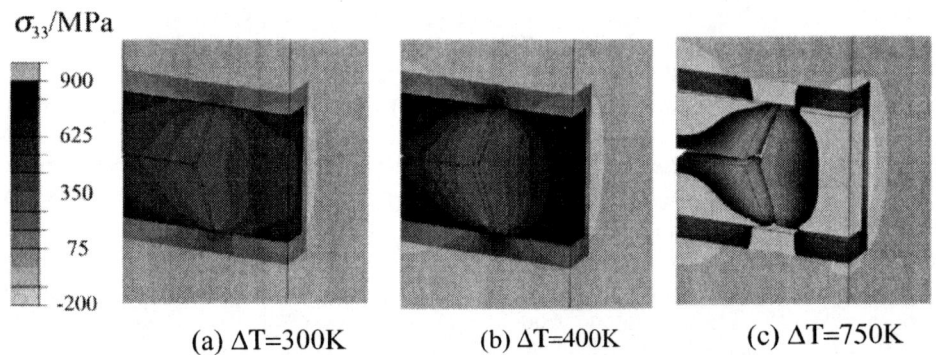

900
625
350
75
-200

(a) ΔT=300K   (b) ΔT=400K   (c) ΔT=750K

FIGURE 7. A sequence of contours of horizontal stress during void nucleation in the interconnect microstructure shown in Fig. 2(b). Results are shown for $a = 0.2\mu$m, a/w=0.5, $\rho = 0.25$, with SiO_2 passivation (Table 1). Displacements are magnified by a factor of 10.

154

Representative results for the microstructure sketched in Fig 2(b) are shown in Fig. 7. This structure fails by debonding of the `bamboo' grain boundaries (these are at 60^0 to the vertical) at a temperature drop of $\Delta T = 350K$. The horizontal grain boundary fails almost immediately thereafter. On further cooling, the interface between interconnect and passivation at the top and bottom of the interconnect fails at $\Delta T = 750K$. Grain boundaries within the interconnect evidently reduce its strength significantly. In practice, this strength reduction is mitigated by the fact that Cu/Cu grain boundaries are likely to have a larger work of adhesion than a Cu/Ta interface. Additional numerical computations suggest that if the work of adhesion for the Cu/Cu grain boundaries exceeds that of the Cu/Ta interface by a factor of 4, the Cu/Ta interface will debond before the grain boundary fails.

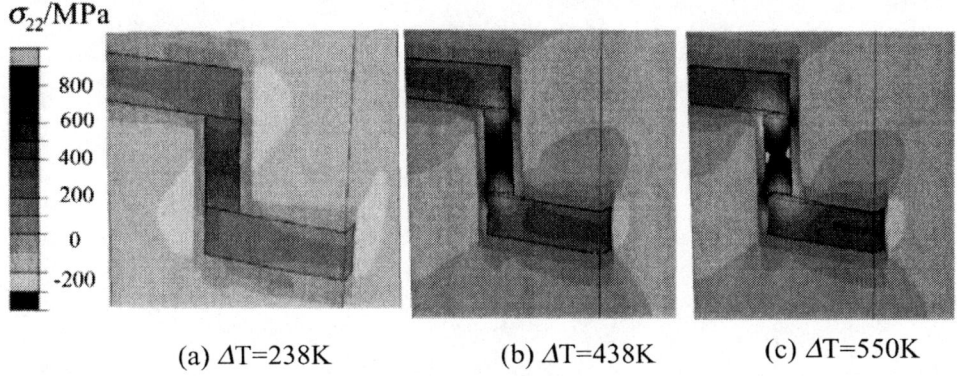

(a) ΔT=238K   (b) ΔT=438K   (c) ΔT=550K

FIGURE 8. A sequence of contours of horizontal stress during void nucleation in the interconnect microstructure shown in Fig. 2(c). Results are shown for $a = 0.2\mu m$, a/w=0.5, $\rho = 0.25$, with SiO_2 passivation (Table 1). Displacements are magnified by a factor of 15.

Finally, Fig. 8 shows results for the representative via structure sketched in Fig 2(c). Failure in this structure occurs by debonding of the Cu/Cu grain boundaries at the top and bottom of the via, at a temperature change $\Delta T = 350K$. On further cooling, the cracks in the grain boundary continue to propagate along the Cu/Ta interfaces adjacent to the interconnect on either side of the via. Full delamination occurs at $\Delta T = 750K$.

CONCLUSIONS

Cohesive zone models offer an appealing technique for evaluating the influence of material properties and line geometry on the failure resistance of interconnect structures. A typical cohesive interface has two key parameters: the work of adhesion, ϕ and the maximum stress the interface can withstand, σ_0 A difficulty in applying these models is that often only the work of adhesion or fracture toughness for the interfaces is known, and detailed stress-separation data are not available. Simple analytical calculations show that for a fixed work of adhesion, there is a critical interface σ_0 that minimizes the resistance to void nucleation. As an engineering approach, this critical value may be used in computations to provide conservative estimates of failure resistance of interconnect structures. On this basis, we have computed critical temperatures required to nucleate voids in several representative interconnect structures. Analytical expressions were developed to predict qualitatively the influence of line width, barrier layer thickness, bulk material properties, and the cohesive properties interfaces on the conditions necessary to nucleate voids. Detailed three dimensional numerical computations were used to predict the influence of interconnect geometry and microstructural features on void nucleation. These show that typical geometrical features associated with grain boundaries and vias generally reduce the strength of the structure by a factor of two. Void nucleation temperatures are only weakly sensitive to details of the interconnect geometry such as line aspect ratio, and the width and height of vias.

ACKNOWLEDGEMENT

This was supported by the Semiconductor Research Corporation Center for Advanced Interconnect Science and Technology (CAIST).

REFERENCES

Lane, M., Dauskardt R.H., Krishna, N. and Hashim, I., *J. Mater.Res.* **15** (1) 203-211 (2000).
Xu, X.P. and Needleman A.N. , *J. Mech. Phys. Solids*, **42** (9), 1397-1434 (1994).

Stress-Induced Voiding in Aluminum and Copper Interconnects

M. Hommel, A. H. Fischer, A. v. Glasow and A. E. Zitzelsberger

Infineon Technologies AG, Reliability Methodology
Otto-Hahn-Ring 6, D-81739 Munich, Germany
martina.hommel@infineon.com

Abstract. Stress-induced voiding (SIV) is a serious reliability problem in metal interconnects. For aluminum a phenomenological model was developed which allows the extrapolation of metallization life times from stress conditions to operation conditions of the integrated circuit. Resistance drift measurements during high-temperature storage (HTS) on wafer-level have been performed and the experimental data could be fitted with that model. The influences of different parameters such as line width, metal level, thermal anneals of certain metal levels during processing and the deposition temperature of the interlevel dielectric material on the SIV behavior are discussed. The SIV behavior of copper dual damascene metallizations has been investigated on via line structures. A linear resistance drift during high-temperature storage has been observed. This is in contrast to aluminum, where a non-linear behavior was found. Failure analysis showed voids inside the via and not in the metal line as it has been observed in aluminum. Stress simulations have been performed in order to explain this behavior. Due to the complex stress state in a copper dual damascene via the temperature dependence of SIV in copper is different from that of aluminum.

INTRODUCTION

During the processing of a multilayered interconnect system, different sources for mechanical stresses in the metallization can occur. One important origin for stresses in metal lines is the deposition of the interlevel dielectric and the subsequent cooling. Due to thermal mismatch between the dielectric material and the metal, high mechanical stresses in the metallization develop. All following thermal treatments may have an influence on the stress state of the metal. Under use conditions stress can be induced by high temperature storage or applying current which can cause electromigration. These effects can be accelerated in reliability tests.

The main driving force for stress-induced voiding is a stress gradient. The subsequent development of stress-induced voids leads to a stress relaxation and to an equalization of the stress gradient. There are two main processes which determine the stress-induced voiding. The first is the driving force by the stress and the second is the diffusion [1]. The temperature dependence of the diffusivity can be described by an Arrhenius law. For RIE-processed structures in aluminum the temperature dependence of the stresses in the lines is well understood. Normally SiO_2 is used as dielectric material for aluminum metallization. Due to the smaller coefficient of thermal

CP612, *Stress-Induced Phenomena in Metallization:* Sixth Int'l. Workshop, edited by S. P. Baker et al.
© 2002 American Institute of Physics 0-7354-0058-X/02/$19.00

expansion of SiO_2 the metal lines are in a state of high tensile stress, when the interconnect is cooled down from a higher deposition temperature of the encapsulating oxide to a lower storage temperature. With such an initial stress state the two described acceleration mechanisms work in opposite direction when the temperature is increased: the tensile stress decreases linearly while the diffusion increases exponentially. As a consequence, a temperature of maximal drift T_{max} exists (s. Fig. 1) [2]. Therefore, reliability tests should be performed around T_{max} to ensure the highest acceleration.

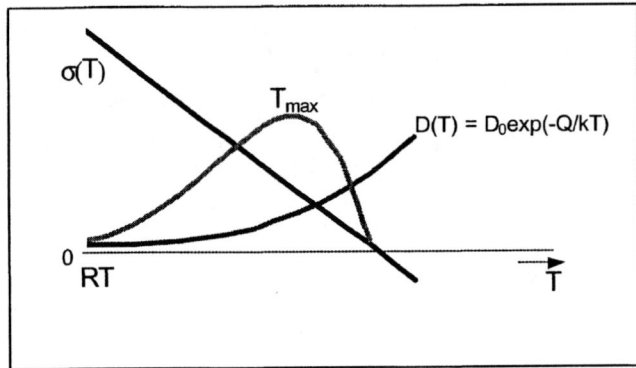

FIGURE 1.   Temperature dependence of the acceleration factors for stress-induced voiding in aluminum.

For an estimation of the reliability properties under usage the SIV behavior at higher test temperatures has to be extrapolated to use conditions. For the stress relaxation mechanism various models have been developed. Several phenomenological models exist [e.g. 2, 3.] which are mostly based on power-law creep. Some physical models have been published, where numerical [4, 5] as well as analytical models [6, 7] are described.

Phenomenological Model For Stress-induced Voiding In Aluminum

Median Time To Failure

The experimental approach to the problem is a high temperature storage (HTS) test, where the samples are kept in an oven at constant temperature. The resistance of test structures is measured after a certain stress intervals. In the phenomenological model described in the following, diffusional creep is assumed to be the major stress relaxation mechanism during the HTS test [8, 9]. For this approach Gibb's equation was used, which describes the relation between the plastic strain rate and the external stress at a certain stress temperature T_{str} [10]:

$$\dot{\varepsilon} = a \frac{D}{k T_{str}} \sigma \; . \tag{1}$$

D is the diffusivity, a is a constant which contains microstructural and geometrical parameters and k is the Boltzmann constant. In a (quasi) steady state at a constant storage temperature, the total strain rate which is the sum of plastic and elastic strain rate is zero. The elastic strain can be replaced by Hooke´s law using an effective elastic modulus E_{eff} for the multilayer system:

$$\dot{\varepsilon}_{pl} = -\dot{\varepsilon}_{el} = -\dot{\sigma}/E_{eff} \; . \tag{2}$$

Gibb´s equation (1) combined with (2) leads to a differential equation for the stress:

$$\dot{\sigma}(t) + a E_{eff} \frac{D}{k T_{str}} \sigma(t) = 0 \; . \tag{3}$$

For the solution of (3) the following approach for σ(t) was used:

$$\sigma(t) = \sigma_0 \exp\left(- t/\tau\right) \tag{4}$$

with a time constant

$$\tau = k T_{str}/a E_{eff} \, D \; . \tag{5}$$

The change in the plastic strain leads to an increase of the void volume which is proportional to the increase of the line resistance. The failure of the line is defined by a certain amount of relative resistance increase and, at a given stress temperature, this resistance increase is obtained at a median time to failure t_{50}. The change in plastic strain which is necessary to produce the failure can be expressed by an integration over time until the median time to failure t_{50}:

$$\Delta \varepsilon_{pl} = \int_0^{t_{50}} \dot{\varepsilon}_{pl}(t) dt \; . \tag{6}$$

With (2) and (4) one gets

$$\Delta \varepsilon_{pl} = \frac{\sigma_0}{E_{eff} \tau} \int_0^{t_{50}} e^{- t/\tau} dt \; . \tag{7}$$

With the assumption that $t_{50} \ll \tau$ which is supported by experimental results, it follows:

$$t_{50} \approx \frac{E_{eff}\,\tau}{\sigma_0(T_{str})}\Delta\varepsilon_{pl} \approx \frac{kT_{str}}{D\sigma_0(T_{str})}\frac{\Delta\varepsilon_{pl}}{a}. \tag{8}$$

The initial thermally-induced stress σ_0 can be calculated

$$\sigma_0(T_{str}) = E_{eff}\left(\alpha_{Al} - \alpha_{ILD}\right)\left(T_{dep}^{*} - T_{str}\right), \tag{9}$$

where α_{Al} and α_{ILD} are the coefficients of thermal expansion of the aluminum metallization and the interlevel dielectric, respectively. A certain amount of the initial thermal stress can be already relaxed in the period between the deposition process and the start of the HTS test. Therefore a reduced initial stress is assumed. This is taken into account by introducing a reduced or "effective" deposition temperature $T_{dep}^{*} < T_{dep}$ of the ILD. From eq. (8) and (9) one gets the following equation for the median time to failure of an interconnect due to the growth of stress-induced voids:

$$t_{50} = C\frac{T_{str}}{T_{dep}^{*} - T_{str}}\exp\left(\frac{Q}{kT_{str}}\right). \tag{10}$$

Q is the activation energy of the appropriate relaxation mechanism. For a given set of identical test structures, C represents a specific constant (C contains parameters which can be assumed to be constant within the considered temperature range).

Application Of The Model To Experimental Data

Stress-induced voiding was investigated on AlCu(0.5%) metallizations with a Ti/AlCu/TiN stack and the described phenomenological model was applied to the experimental data. The resistance was monitored on wafer-level during the high-temperature storage. The failure times were determined at 5 different storage temperatures for various metal levels and line widths. Meander-shaped metal lines without vias were used as test structures. A two-level metallization (A) with a line width $w = 0.3$ μm and a four-level metallization (B) with two different line widths $w = 0.32$ and 0.40 μm were investigated.

A failure criterion of 5% resistance increase was applied. The obtained failure times are well log-normal distributed (s. Fig. 2). A deviation occurs only for higher temperatures $T > 250°C$. The shape factor σ of this particular lot is in a range between 0.2 and 0.3. From these log-normal plots the median time to failure t_{50} can be determined at CDF = 50%.

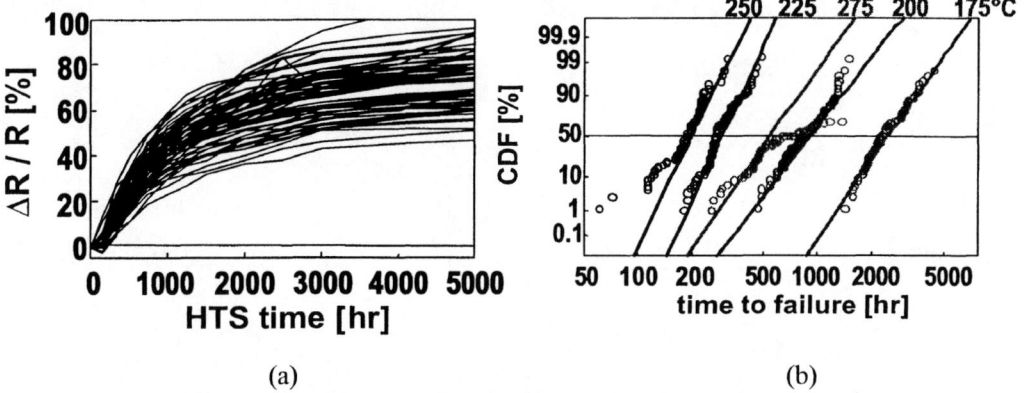

FIGURE 2. (a) Relative resistance drift for HTS temperature = 250°C; (b) Lognormal plots of failure times of M2-structures of metallization B at various storage temperatures for 0.32 μm wide lines.

Fig. 3 shows the failure distribution for two different line widths at two different storage temperatures. The failure time increases with the line width which can be ascribed to a higher stress in narrower lines.

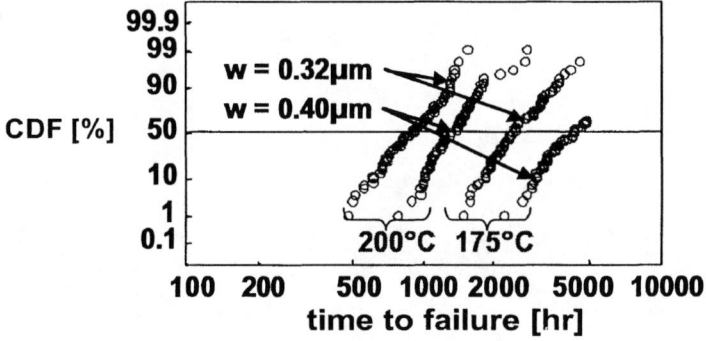

FIGURE 3. Lognormal plots of failure times of M2-structures of metallization B at two different storage temperatures for different line widths.

A nonlinear fit based on the presented phenomenological model using eq. (10) was applied to the t_{50}-data (Fig. 4). For metallization B the lines in M3 have a smaller lifetime than in M2. The M1 level of metallization A with a slightly smaller line width has a much shorter median time to failure compared to metallization B. From the fit the effective deposition temperature T_{dep}^{*} was determined: For metallization B a $T_{dep}^{*} \cong 280°C$ was found. According to the smaller t_{50} of metallization A, the effective deposition temperature determined from the fit was higher: $T_{dep}^{*} \cong 310°C$. Due to the fact that the oxide deposition temperature was 400°C for both metallizations, it can be concluded, that the thermally-induced stress in metallization B has relaxed more than in metallization A.

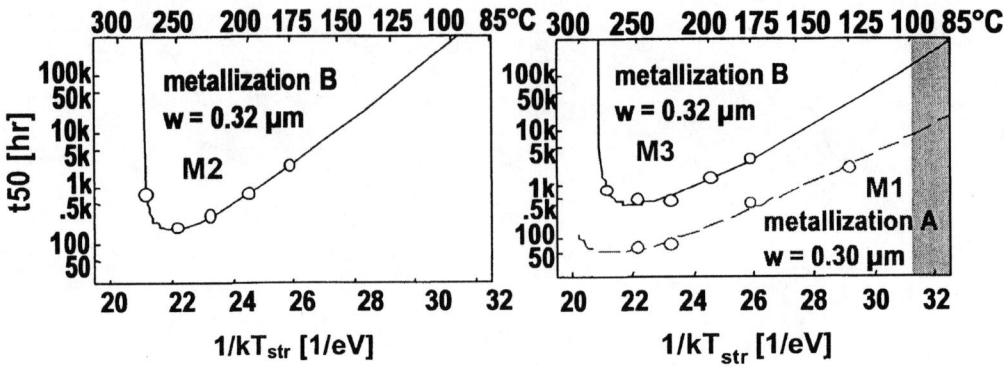

FIGURE 4. Median time to failure t_{50} measured on structures with different line widths in metal levels M2 and M3 at various storage temperatures T_{str}. The fit is based on equation (10). The grey areas indicate the typical operation temperature range between 85 and 100°C.

Physical failure analysis showed different kind of stress voids in the two metallizations. Metallization A had many small voids which were wedge-shaped. Metallization B showed few large voids which extended throughout the whole line width. According to this observation for metallization A an activation energy of about 0.7 eV was found which can be contributed to grain boundary diffusion. In metallization B an activation energy of 1.0 eV was determined, which indicates that there are additonal diffusion mechanisms operating besides grain boundary diffusion, (e.g. volume diffusion), to create the large voids.

Influence of HDP-deposition temperature on SIV

Two different deposition temperatures in a high-density plasma (HDP) process (350°C and 400°C) have been applied to one lot of metallization B with two different line widths in M2:

TABLE 1. Metallizations in M2 with two different line widths and HDP-deposition temperature.

M2 (w = 0.32 μm)	M2 (w = 0.40 μm)	T_{dep} [°C]
◆	■	350
◊	☐	400

For both line widths the median time to failure is much higher for the lower deposition temperature (s. Fig. 5). The HDP-process at the lower temperature induced smaller thermal stresses. With eq. (8) this results in a higher median time to failure t_{50}. A wafermap showed that there is a higher resistance drift in the centre of the wafer. This inhomogeneity can explain the higher shape factors ($\sigma > 0.4$).

162

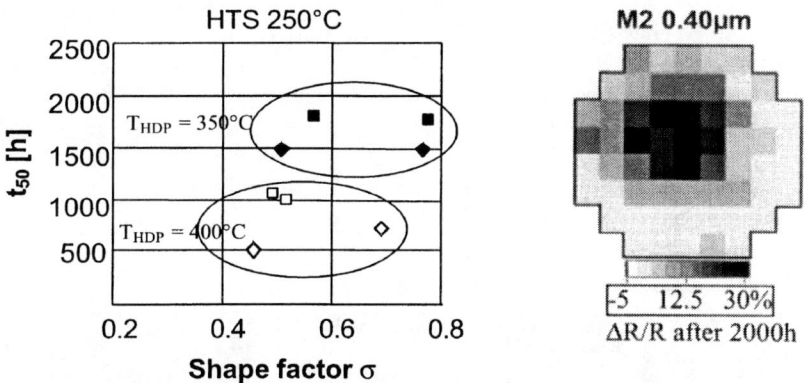

FIGURE 5. Comparison of the t_{50} for two different oxide deposition temperatures and line widths (left); wafer map of the resistance drift for w = 0.4 μm after 2000h storage time (right).

Influence of anneal of certain metal levels on SIV

In a 5-level AlCu-metallization only the first and the fifth level (M1 and M5) have been annealed after deposition. M2, M3 and M4 have not been annealed. Fig. 6 shows the resistance drift of every metal level after a high temperature storage at 250°C for 2000h. Obviously the annealed metal levels M1 and M5 showed no significant drift. The not annealed levels M2 and M4 exhibited remarkable resistance drifts whereas M3 showed a significantly smaller drift. A possible explanation could be the symmetrical position of M3 in the 5-level metal stack. The stress in this level can be assumed to be smaller compared to M2 and M4. Due to the smaller driving force this could explain the smaller resistance drift in M3.

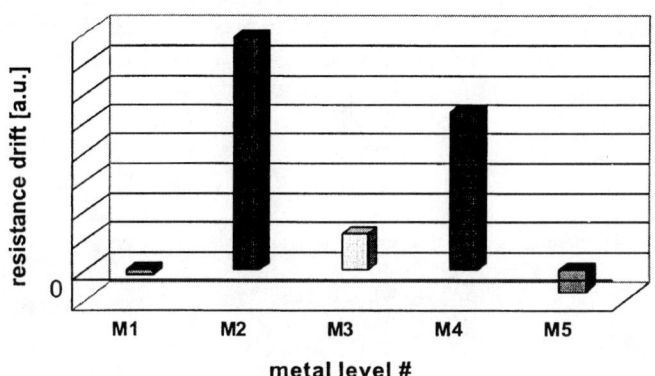

FIGURE 6. Resistance drift in 5 metal levels (M1 and M5 have been annealed after metal deposition. M2, M3 and M4 have not been annealed).

163

Stress-Induced Voiding In Copper

The most important difference of the stress-induced voiding behavior of copper in comparison with aluminum is the dual damascene architecture of the copper metallization. In aluminum metallizations usually tungsten-vias are used. In such metallizations only the metal line is sensitive for HTS failure by voiding. In copper dual damascene structures, where the copper via is susceptible for voiding, stress-induced voids are found at the via side walls or the via bottom (s. Fig.7) and not in the metal line.

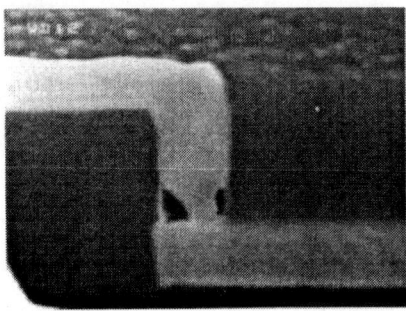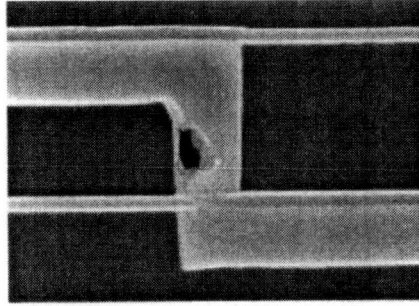

FIGURE 7. Stress-induced voids in the via of a copper dual damascene structure after 2000h HTS at 275°C.

Stress Simulation Of Cu/SiO$_2$

So far very little is known about diffusion and stress relaxation processes in copper dual damascene structures [11]. In order to explain the development of the voids inside the via a stress simulation was performed for a dual damascene line encapsulated by SiO$_2$. The metal stack consists of a TaN/Ta-liner, a Cu seed-layer and electroplated Cu, capped with a Si$_3$N$_4$-film. A viscoplastic model was used in which the subsequent process steps are included in the initial stress state at 25°C (s. Fig. 8). The stress state for temperatures up to 400°C was calculated. In Fig. 8 the component σ_{yy} is shown which gets more compressive with increasing temperature.

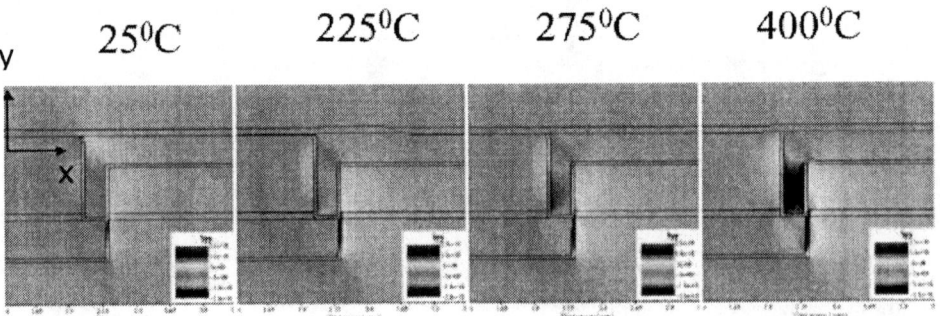

FIGURE 8. Stress simulation in a copper dual damascene structure for the temperature range from 25°C up to 400°C. The stress component σ_{yy} is shown, which gets more compressive with increasing temperature.

Different stress components in the centre of a copper via have been analyzed. The temperature dependence of the stress components σ_{xx}, σ_{yy} and the hydrostatic stress σ_{hyd} in the centre of the copper via are shown in Fig. 9.

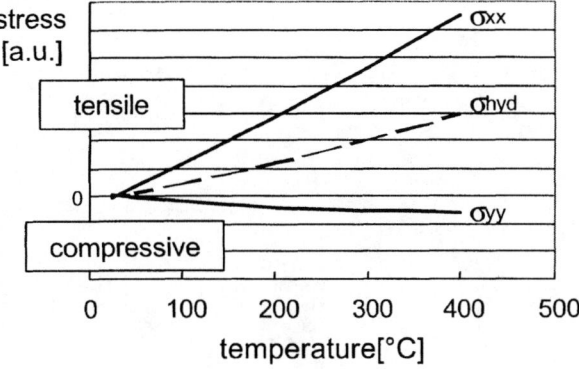

FIGURE 9. Stress components σ_{xx}, σ_{yy} and σ_{hyd} in the centre of the via as a function of temperature.

σ_{yy} gets more compressive with temperature whereas σ_{xx} and σ_{hyd} are tensile and increase with temperature. When the copper metallization is heated up from room temperature to a higher storage temperature the stress gradient between the via and the metal line above the via increases. This can lead to vacancy migration. These vacancies move to grain boundaries which are under compressive stresses (s. Fig. 10a) to relieve the stress and to decrease the stress gradient [12].

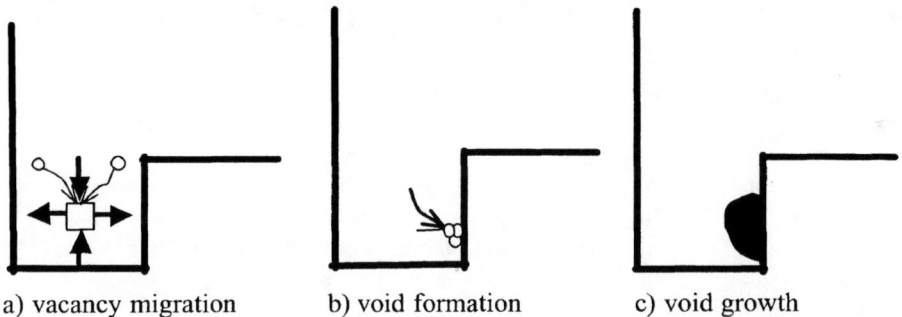

a) vacancy migration   b) void formation   c) void growth

FIGURE 10. Sequence of proposed voiding process in a copper dual damascene via.

Sites of preferential void nucleation are inhomogenities which appear frequently at via side walls or the via bottom (Fig. 10b). At these sites, voids will nucleate easily, preferentially at the intersection of a grain boundary with the interface. The tensile hydrostatic stress will lead to the growth of a nucleated void (Fig. 10c).

For copper both, the diffusivity and the stress gradient as driving forces for SIV, increase with temperature. As a result the acceleration of SIV increases with temperature which is in contrast to the behavior of aluminum, where the resulting driving force has a maximum at a certain temperature. In order to analyze this behavior a high temperature storage experiment with a via-line test structure was performed in which the storage temperature was increased gradually. The resistance drift for the three tested temperatures 225°C, 275°C, and 300°C is shown in Fig. 11. A linear resistance increase was observed. Even after more than 2000h storage at a constant temperature no saturation was observed.

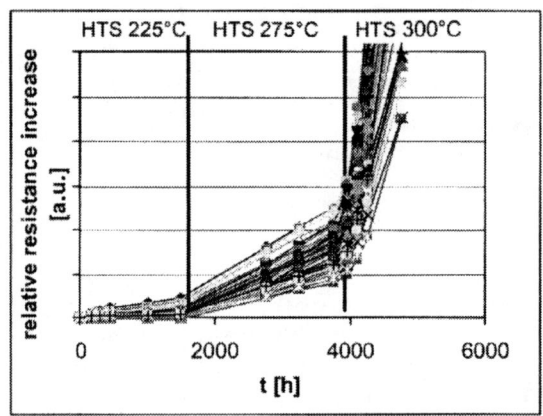

FIGURE 11. Linear temperature dependence of stress-induced voiding in copper.

Influence Of Microstructure On Stress-Induced Voiding in Copper

The microstructure of the metallization plays an important role for stress-induced voiding [13]. After electroplating the copper metallization usually undergoes an annealing step which should stabilize the microstructure. The resistance drift behavior was investigated for two different annealing temperatures. The metallization with the lower annealing temperature shows smaller grains prior to high temperature storage (s. Fig. 12a). Samples with smaller initial grain size showed a significant resistance drift during high-temperature storage whereas the samples with the larger grains did not show any drift. The correlation of the large resistance drift of sample set 1 with microstructural changes was proven by the measurement of the grain size before and after the high temperature storage. The grain size of sample set 1 increased from 215 nm to 316 nm. With increasing grain size the grain boundary volume decreases leading to free vacancies [14] which can generate voids.

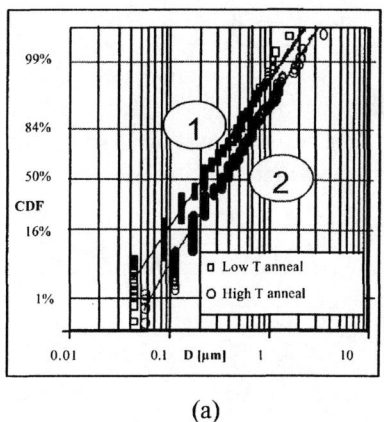 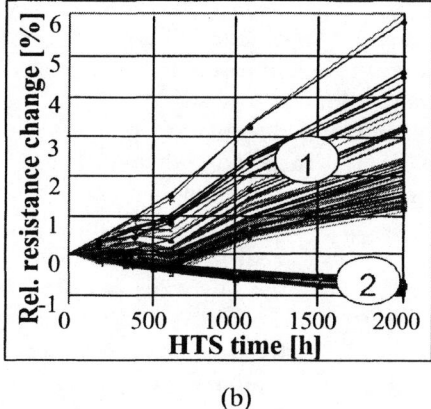

(a)   (b)

FIGURE 12. Influence of microstructure on stress-induced voiding in copper: a) grain size distribution of test structures with two different annealing temperatures prior to the test, b) resistance drift at storage temperature 225°C.

SUMMARY

The stress-induced voiding behavior of aluminum and copper metallization was compared. High temperature storage tests were performed on wafer-level. The resistance was measured during the stress. By defining a failure criterion for the relative resistance increase the median time to failure could be determined experimentally. For aluminum a phenomenological model for the extrapolation of the median times to failure from stress conditions to operation conditions was developed. Diffusional creep was supposed to be the main stress relaxation mechanism. An equation was obtained which is similar to Black's equation for electromigration failure. It describes the relationship between the median time to failure caused by stress-induced voiding and the stress temperature. Only the deposition temperature of the interlevel dielectric and the activation energy of the appropriate stress relaxation mechanism is needed for the extrapolation. Instead of the real deposition temperature in many cases a reduced or effective temperature has been found which is lower than the real deposition temperature. The stress can be partially relaxed between the processing and the start of the high temperature storage test. This is equivalent to a lower deposition temperature. The model was applied to experimental data obtained for aluminum metallization. From the fit of the median times to failure the effective deposition temperature and the activation energy could be determined. Activation energies of 0.7 eV and 1.0 eV were found. This was in correlation with two different kinds of voide shapes: small wedge-shaped voids and large expanded voids, respectively. This indicates that two different relaxation mechanisms like grain boundary diffusion and volume diffusion were acting.

The influence of the high density plasma (HDP) deposition temperature on stress-induced voiding was investigated. The lowering of the HDP temperature from 400°C to 350°C led to an increase of the median time to failure by a factor of two to three. Particular metal levels of a 5-level metallizations were annealed and the drift behavior at 250°C was measured. The annealed metal levels showed no drift. One of the not annealed metal levels (M3) showed a significant smaller resistance drift after 2000h, which could be explained by the symmetrical position of M3 inside the whole metal stack. This metal level should have a minimal stress and therefore the smallest driving force for stress-induced voiding.

In comparison to aluminum the stresss-induced voiding behavior of a copper dual damascene metallization was analyzed. Stress-induced voids were found inside the copper via at the side walls or the via bottom. The stress development in the centre of a copper dual damascene via in the temperature range between 25°C and 400°C was analyzed by stress simulations. The simulations showed that the component σ_{yy} (y is the direction normal to the surface) gets more compressive with increasing temperature and the radial component σ_{xx} gets more tensile. This can lead to a vacancy migration to horizontally oriented grain boundaries. The nucleation sites for voids are preferably the intersections of grain boundaries with the interface of the copper metallization and the liner. At these sites voids can be nucleated and grow under a tensile hydrostatic stress. The temperature dependence of the stress-induced voiding behavior in copper was investigated. In contrast to the saturation behavior found in aluminum, a linear resistance increase in the temperature range between 225°C and 300°C was found. In copper metallization a stable microstructure is important to avoid stress-induced voiding. Copper with two different annealing temperatures was stressed. The metallization with the lower post-plating annealing temperature showed a strong resistance increase due to grain growth, whereas the higher annealing temperature led to a stable behavior with no significant drift.

ACKNOWLEDGMENTS

The authors would like to thank M. Hierlemann from SIM, Infineon Technologies AG, Munich for the stress simulations.

REFERENCES

1. Hu, C. K. et. al, *IBM J. Res. Dev.*, **39** (4) 465-486 (1995)
2. McPherson, J. W., and Dunn, C. F., *J. Vac. Sci. Technol.*, **B5** No. 5, 1321-1325 (1987)
3. Tezaki, A. et. al, *1990 Proc. Int. Rel. Phys. Symp.*, 221-229 (1990)
4. Yost, F. G., *Script. Metall.*, **23**, 1323-1328 (1989)
5. Sauter, A. I. and Nix, W. D., *J. Mater. Res.*, **7** (5), 1133-1143 (1992)
6. Yost, F. G., Amos, D. E., and Romig, A. D., *Proc. 1989 Int. Rel. Phys. Symp.*, 193-201 (1989)
7. Okabayashi, H., *IEEE Transactions on Electron. Devices*, **40** (4), 782-788 (1993)
8. Fischer, A. H., et.al., *Proc. Mat. Res. Soc.*, **612**, D2.6.1-D2.6.6 (2000)
9. Fischer, A. H. and Zitzelsberger, A.E., *IEEE 39th Ann. IRPS Proceedings*, 334-340 (2001)
10. Gibbs, B. G., *Phil. Mag.*, **13**, 589-593 (1966)
11. Kobrinsky, M. J., Thompson, C. V., and Gross, M. E., *J. of Appl. Physics*, **89** (1), 91-98 (2001)
12. Murakami, M., *CRC Critical Reviews in Solid State and Mat. Sciences*, **11** (4) 317-355 (1986)
13. Glasow v., A., and Fischer, A. H., *Proc. of the Advcanced Met. Conference 2000*, in press
14. Sullivan, T.D., *Electron Device Letters*, **14**, (1993)

A Relationship between Film Texture and Stress-Voiding Tendency in Copper Thin Films

J. Koike, A. Sekiguchi, M. Wada, and K. Maruyama

Dept. of Materials Science, Tohoku University
Sendai 980-8579, Japan

Abstract. The origin of stress voiding in heat-treated Cu thin films was investigated in relation to microstructure. Voids were observed at the intersections of twins with grain boudaries or with other twins. Twin interfaces were accompanied by stress concentration due to the elastic anisotropy. Stress concentration was found to act as a driving force for stress voiding. Twin formation and associated void formation could be avoided by controlling the film texture. Texture transition from (111) to (100) was observed in heat-treated films with increasing the film thickness from 200 nm to 300 nm. The (100) oriented films did not show any voids or hillocks. The excellent stress-migration resistance in the (100) oriented films could be attributed to the absence of twins and by small thermal stresses.

INTRODUCTION

For interconnect application of Cu thin films, stress migration is a major reliability problem during processing at elevated temperatures. The most prominent stress-migration failure for Cu is void formation [1-5]. In the case of traditional interconnect lines of Al alloys, the stress migration problem can be alleviated by choosing proper alloying elements and sharpening the (111) texture [6, 7]. However, the use of Cu is motivated by its lower electrical resistivity than Al to reduce resistance-capacitance delay. The resistivity issue becomes increasingly important with decreasing the line width. Then, alloying of Cu increases resistivity and may not be practical for device application. On the other hand, the importance of the texture has been widely recognized in Al interconnects. The effects of film texture on stress voiding, however, has been investigated only briefly [1-4] and are open questions for the reliability of Cu interconnect.

In this paper, we intended to provide new insights on stress voiding in relation to microstructual features and to crystallographic texture. The work is focused on characteristic properties of Cu: the large elastic anisotropy ($2C_{44}/C_{11}-C_{12}= 3.2$ for Cu and 1.2 for Al) and the easy tendency of twinning. The first part of the paper shows that voids are formed at the corners of incoherent twins driven by thermal stress concentration. The second part shows that (100) oriented films are significantly resistant to stress voiding while (111) oriented films are vulnerable.

CP612, *Stress-Induced Phenomena in Metallization:* Sixth Int'l. Workshop, edited by S. P. Baker et al.
© 2002 American Institute of Physics 0-7354-0058-X/02/$19.00

EXPERIMENTAL PROCEDURE

Samples presented for the former part of the result section are electroplated Cu thin films having a thickness of 900 nm. Substrates were Ta 20 nm / SiO$_2$ 1700 nm / Si. A Cu seed layer for electroplating was formed by sputter deposited to a thickness of 50 to 100 nm, following a procedure described in Ref [8]. The samples were kept at room temperature for more than a month, so that, the microstructure at room temperature was stabilized by so-called "self annealing". The samples were subject to a heating and cooling cycle between room temperature and 723 K at a rate of 3.3 K/min in an evacuated chamber. The X ray diffractometer scans showed only a (111) peak both before and after hear treatment, indicating a strong (111) texture.

Samples presented in the latter part of the result section are sputter-deposited Cu thin films on Ta 20nm / Si 500 μm substrates. Sputtered films were employed to better control the film thickness in a range of 50 nm to 900 nm. The as-deposited films of all examined thicknesses had a strong (111) texture. The as-deposited films were heated to 723 K, then cooled to room temperature, using an infrared lamp heater in the sputtering chamber kept at a pressure of 7 x 10^{-7} Pa. Heating and cooling rate was 3.3 K/min. The thickness dependence of film texture after heat treatment was investigated and the effects of the texture on stress voiding was observed.

RESULTS AND DISCUSSION

Stress Voiding Associated with Incoherent Twins

Figure 1 shows images of the heat-treated samples taken by a transmission electron microscope (TEM). Some voids are indicated by arrows in the figure. Voids are formed at the intersections of twins with a grain boundary in (a), and even in the grain interior at the intersections of twins with other twins in (b). The void-related twins are characteristics for Cu and have never been observed in Al. The paper, thus, pays special attention to the twin-related voiding. A possible origin of the twin-related voiding can be understood in terms of stress concentration caused by the anisotropy of elastic moduli. During heat treatment, thermal strain arises in an isotropic manner in an

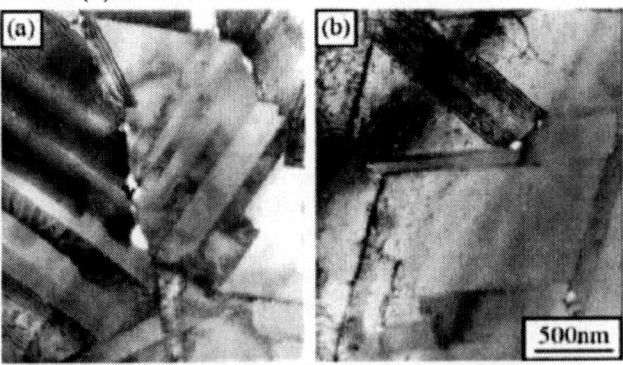

FIGURE 1 TEM images of voids formed at the intersections of (a) twins with a grain boundary and (b) twins with other twins.

170

entire film volume. Because of the orientation differences between twins and their corresponding matrix, elastic moduli for a given strain are also different. This gives rise to variation in thermal stress. In order to satisfy boundary conditions, stress concentration arises at the interface between twins and matrices, acting as a driving force for void formation. This argument is confirmed by calculating stress distribution using a finite element method in combination with TEM orientation analysis of actual twins and matrices. Details for calculation and TEM analysis can be found in Ref. [5].

Figure 2 shows representative examples of more than 10 each cases of voided and unvoided regions including twins and grain boundaries. The left column is for an unvoided region and the right column is for a voided region. TEM images are placed on the top; shear stress (σ_{12}) distribution in the middle; and tensile stress (σ_{22}) distribution is on the bottom. In the figures, rectangles are twins and vertical lines are grain boundaries. Clear relationship can be found between the extent of stress concentration and the void formation tendency. This relationship is valid for all the analyzed regions of the (111) oriented films.

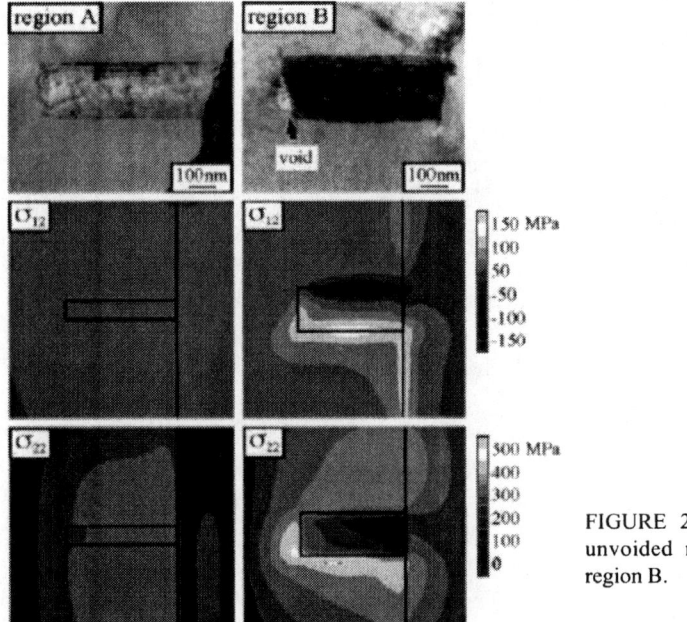

FIGURE 2   Stress distribution of the unvoided region A and the unvoided region B.

Figure 3 shows a typical type of twin interfaces in the (111) oriented films. The orientation of the cubic principal axes of twins and matrices are analyzed by TEM Kikuchi diffraction technique and a typical result is shown in a stereographic projection. A twin plane should be a mirror plane for the twin and the matrix, as indicated by a dotted line with some

171

experimental error. The index of the twin plane is found to be a {322} plane. This index seems odd for a twin plane index of fcc crystals. However, it has been shown in calculation and bicrystal experiment that the {322} twins can exit in a metastable condition at 8 degrees away from the {211} plane, i. e., symmetric incoherent twin interfaces [9, 10]. The atomic arrangement of the {322} twin interface consists of a narrow slab of the 9R structure (BCACABABC stacking instead of ABC stacking of fcc) with a small angle boundary and a large-angle boundary on each side of the 9R structure [9, 10]. Because of the complicated interface structure associated with a large-angle boundary, the properties of the {322} twin interfaces are expected to be similar to those of general large-angle boundaries.

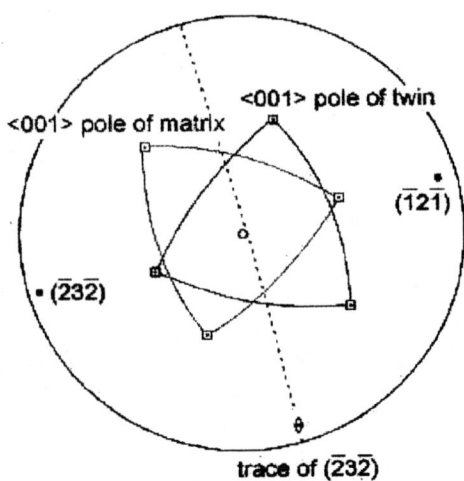

FIGURE 3   Stereographic projection of the twin-plane trace and poles of cubic principal axes for a twin and a matrix.

A good agreement is found between the voiding tendency and stress concentration associated with incoherent twins in the (111) oriented films. However, there is a case in which no voids are formed, yet twins are observed and stress concentration is calculated [11]. This case is found in Cu thin films having a mixed texture. A clear difference from the (111) oriented films is that the twins in the mixed texture are ordinary {111} twins having a coherent interface structure. In this case, the surface orientation of twins are not (111) but mainly (511) for the first order twins of (111) grains and (221) for the first order twins of (100) grains. Second order twins are also observed. Therefore, the distinction of twin type, whether coherent or incoherent, should be made clearly to find a correlation among voiding tendency, stress concentration and twins. Possible reasons for the different voiding tendency between the two films can be attributed to the structural difference of twins. Since the interface energy and the diffusivity is higher in the {322} twins than in the {111} twins, pit-like voids may extend easily from the film surface toward the film/substrate interface.

Annealed Texture and Stress-Voiding Tendency

The macroscopic texture of the heat-treated films was examined by XRD as a function of film thickness. The volume fraction of the (100) oriented grains was calculated from the integrated intensity of each peak that was normalized by structure factor, multiplicity factor, and Lorenz factor. The result is shown in Figure 4 indicating the volume fraction of the (100) grains versus the film thickness. With increasing the thickness to 200 nm, the film texture after heat treatment remains the same (111) texture as in the as-deposited films. With further increasing the thickness to more than 300 nm, the film texture turns to the (100) texture. The texture change by heat treatment occurs abruptly at a thickness between 200 and 300 nm. The formation of the (100) texture by heat treatment has been reported by other researchers [12-15].

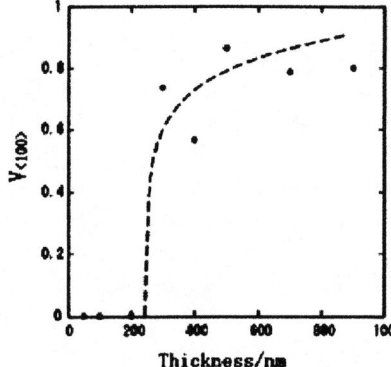

FIGURE 4   The volume fraction of (100) oriented grains in heat-treated Cu thin films for various film thickness.

The 100-nm and 500-nm films are taken as representative cases for the (111) and (100) oriented films, respectively. Local orientation distribution was investigated by analyzing electron back scattering patterns (EBSP). A maximum misorientation angle of 15 degrees was allowed to assign the surface orientation of each grain. Figure 5 shows the distribution of grain orientations of heat-treated films of (a) 100 nm and (b) 500 nm in thickness. Originally color-coded orientation images are converted to gray scale images. Grain boundaries are also indicated by various gray-scale lines according to the type of boundaries. In the figure, (111) and (100) grains are indicated, respectively, by dark-gray and light-gray contrast. As seen in the figure, the 100-nm film is dominated by (111) grains. No (100) grains are found in this film. On the other hand, the 500-nm film is dominated by the (100) grains. Other minority grains are mainly (111) grains. It can be seen that the grain size distribution of the (100) oriented film is bimodal, with larger grains having the (100) orientation while smaller grains having other orientations. The diameter of the (100) grains ranges from 10 to 50 μm.

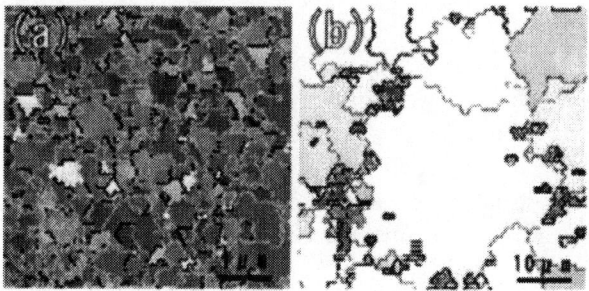

FIGURE 5 EBSP images of (a) 100-nm thick film and (b) 500-nm thick film after heat treatment.

In Figure 5, grain-boundary type is also indicated, based on the coincidence-site-lattice concept. In the (111) oriented film in (a), light-gray lines correspond to $\Sigma 3$ boundaries, while black lines correspond to all other boundary types, mainly general boundaries. It is reminded that the $\Sigma 3$ boundaries that divide two adjacent grains having the same (111) orientation are incoherent twin boundaries. These boundaries are liable to stress voiding [5]. On the other hand, in the (100) oriented film in Fig. 2 (b), light-gray lines correspond to small-angle boundaries with misorientation angles of up to 15 degrees. Black lines correspond to other boundary types, including both special and general boundaries. The $\Sigma 3$ boundaries are rarely found in this film, indicating that the (100) grains are free from twins. The majority of the grain boundaries of (100) grains are either small-angle or special boundaries with Σ values of less than 27.

The surface microstructure after heat treatment is shown in Figure 6 (a) for the (111) film of 100 nm thickness and in (b) for the (100) film of 500 nm thickness. Severe voiding and some hillocks are observed in the (111) film. On the contrary, only a few voids, hillocks and shallow grain-boundary grooves are observed in the (100) oriented film. Although not shown here for a limited space, high-magnification SEM images of the (100) grains show flat surface without any flaws. The location of the voids and hillocks in the (100) film coincides with the area where small (111) grains are found. This indicates that the (100) grains are highly resistant to stress-migration failure. It is noted that the stress migration resistance of the (100) film remains the same by further thermal cycling up to the maximum examined number of three times.

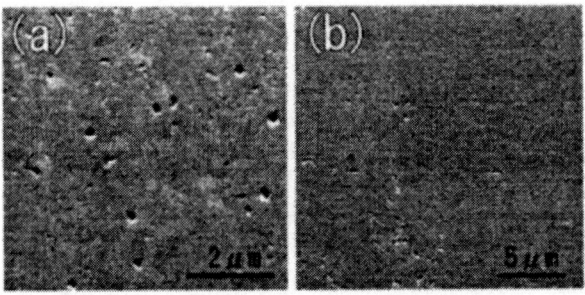

FIGURE 6 SEM images of (a) 100-nm thick film and (b) 500-nm thick film after heat treatment.

The present work shows clear transition of annealed texture from (111) to (100) at a thickness of 200 to 300 nm. This is in agreement with theoretical prediction based on a strain-energy minimization model [16-19]. One may argue that the electroplated film maintained its (111) orientation during heat treatment, despite its thickness of 900 nm. The discrepancy may be caused by the smaller strain-energy driving force in the electroplated film than in the sputtered film. In the first, the electroplated film was kept at room temperature for more than a month, leading to stress relief by self annealing. In the second, the constraint by the film/substrate interface may be weak in the electroplated film because of the segregation of Cl and S at the interface [20]. It would be interesting to sort these out and examine the thickness dependence in the electroplated films.

As for stress migration resistance, a striking difference is observed between the (100) and the (111) oriented films. The (100) films exhibit an excellent stress-migration resistance, while the (111) films show a very poor resistance. The excellent stress-migration resistance of the (100) film can be attributed to the absence of twins, the presence of a large number of special (Σ<27) grain boundaries and small thermal stresses. It is not clear in heat-treated Cu films whether the twins are formed as annealing twins or deformation twins. Whichever is the twin type, the absence of twins in the (100) film may be related to the lack of the twinning driving force by a large stress drop upon the formation of the (100) texture [21] and by the following stress excursion at a low stress level [22]. Conversely, the poor stress-migration resistance in the (111) film is attributed to the presence of a large number of the incoherent Σ3 twin interfaces, as shown in Fig. 5 (a). This is consistent with the electroplated (111) film. These results suggest that a strong (111) texture should be avoided to prevent the formation of incoherent twins and voids. This finding is in a remarkable contrast with a generally accepted view of Al thin films in which a strong (111) texture is preferred for stress migration resistance as well as electromigration resistance [7]. The absence of twins and insignificant elastic anisotropy make Al thin films exempt from excessive voiding as observed in the Cu thin films.

SUMMARY

When Cu thin films maintain the (111) orientation during heat treatment, {322} incoherent twins are formed. Because of a large elastic anisotropy of Cu, the incoherent twins give rise to stress concentration and void formation at twin corners and intersections. With increasing the films thickness, under enough biaxial stressed conditions, the texture of heat-treated films can be changed from (111) to (100). The (100) oriented grains do not contain twins and show an excellent stress-migration resistance. A giant grain size and a good grain-boundary character of the (100) grains may also be advantageous for electromigration resistance.

ACKNOWLEDGEMENT

This work was supported by the Ministry of Education, Science, Sports and Culture of Japan, Grant-In-Aid for Scientific Research (Grant # 13450281).

REFERENCES

1. P. Borgesen, J. K. Lee, R. Gleixner and C.-Y. Li, Appl. Phys. Lett. **60**, 1706 (1992)
2. J. A. Nucci, Y. Shacham-Diamond and J. E. Sanchez, Jr., Appl. Phys. Lett. **66**, 3585 (1995).
3. J. A. Nucci, R. R. Keller, J. E. Sanchez, Jr. and Y. Shacham-Diamond, Appl. Phys. Lett. **69**, 4017 (1996).
4. R. R. Keller, J. A. Nucci and D. P. Field, J. Electro. Mater. **26**, 996 (1997).
5. A. Sekiguchi, J. Koike, Kamiya, M. Saka, and K. Maruyama, Appl. Phys. Lett.79 (2001) in press.
6. F. M. d'Heurle and P. S. Ho, Thin Films-Interdiffusion and Reactions, Ed. by J. M. Poate, K. N. Tu, and J. W. Mayer, John Wiley & Sons, (1978), p. 243.
7. H. Toyoda, T. Kawanoue, S. Ito, M. Hasunuma and H. Kaneko, AIP Conf. Proc. (1996), p. 169.
8. V. M. Dubin, G. Morales, C. Ryu, and S. S. Wong, Mater. Res. Soc. Conf. Proc. **505**, 137 (1998).
9. F. Ernst, M. W. Finnis, A. Koch, C. Schmidt, B. Straumal, and W. Gust, Z. Metallkd. **87**, 911 (1996).
10. D. L. Medlin, G. H. Campbell, and C. B. Carter, Acta metall. **46**, 5135 (1998)
11. J. Koike, A. Sekiguchi, and K. Maruyama, to be published.
12. J. W. Pattern, E. D. McClanahan and J. W. Johnson, J. Appl. Phys. **42**, 4371 (1971).
13. S. D. Dalgren, J. Vac. Sci. Technol. **11**, 832 (1974).
14. T. Ohmi, T. Saito, M. Otsuki and T. Shibata, J. Electrochem. Soc. **138**, 1089 (1991).
15. T. Nitta, T. Ohmi, M. Otsuki, T. Takewaki and T. Shibata, J. Electrochem. Soc. **139**, 663 (1992).
16. E. M. Zielinski, R. P. Vinci and J. C. Bravman, J. Appl. Phys. **76**, 4516 (1994).
17. J. E. Sanchez, Jr. and E. Arzt, Scripta Metall. **27**, 285 (1992)
18. C. V. Thompson, Scripta Metall. **28**, 167 (1993).
19. R. Carel, C. V. Thompson and H. J. Frost, Acta metall. **44**, 2479 (1996)
20. A. Sekiguchi, unpublished results.
21. E. M. Zielinski, R. P. Vinci and J. C. Bravman, Appl. Phys. Lett. **67**, 1078 (1995)
22. S. P. Baker, A. Kretschmann and E. Arzt, Acta mater. **49**, 2145 (2001).

Stress Voiding In Wide Copper Lines

T.M. Shaw, L . Gignac, X-H. Liu, R.R. Rosenberg

IBM Research Division, T.J. Watson Research Center, Yorktown Heights, New York 10598,

E. Levine, P. Mclaughlin. P-C. Wang, S. Greco, G. Biery

IBM Microelectronics Division, Hopewell Junction, New York 12533

Abstract. Hot Stage optical microscopy was used to make in-situ observations of stress void formation in wide (10 –30μm) passivated copper lines. It was found that void nucleation and growth occurred readily where grain boundaries intersected the top copper/dielectric interface. Void closure was observed to occur on heating to temperatures above 300°C indicating that the lines are in compression above this temperature. Isothermal holds at temperatures below 300°C caused void growth to occur with the highest density of voids being observed at 200°C. Cycling to temperatures above the maximum process temperature the structure received (400°C) was found to cause an increase in the density of void nucleation sites activated at the lower void growth temperature. Based on the observations we suggest that localized regions of triaxial stress arise at grain boundaries in wide copper lines causing enhanced void nucleation at the copper/dielectric interface. Possible causes of localized stresses at grain boundaries are discussed including grain boundary sliding and the mismatch of the elastic properties across grain boundaries.

INTRODUCTION

Stress-induced void growth is a common stress relaxation mechanism in metal interconnects that have been passivated with a rigid dielectric. The mismatch in thermal expansion coefficient between the metal lines and the dielectric and substrate can result in a large hydrostatic stress component in narrow lines. The hydrostatic stress provides a driving force for the nucleation and growth of voids by diffusion controlled processes. Detailed experimental and theoretical work on aluminum interconnects have shown that this mechanism can produce significant shifts in the resistance of interconnects and is therefore a reliability concern [1,2]. Studies of copper interconnects have been more limited but show that stress-induced voiding readily occurs and that the density of voids observed depends strongly on thermal treatment before and after passivation, grain boundary misorientation and line width [3,4]. An interesting observation that was made was that a higher density of stress voids occurred in the wider lines. It was suggested that this effect could be caused by an increase in the density of triple point nucleation sites in the wider lines[3]. Thus, even though the hydrostatic component of the stress was reduced, enhanced nucleation resulted in an increase in void density as the line width increased. In this paper we present the results of a in-situ study of stress void formation in wide (10 - 30μm) passivated copper lines. We find that even under conditions where overall stress in the lines is predominantly biaxial void nucleation and growth still occurs.

CP612, *Stress-Induced Phenomena in Metallization:* Sixth Int'l. Workshop, edited by S. P. Baker et al.
© 2002 American Institute of Physics 0-7354-0058-X/02/$19.00

EXPERIMENTAL

For the study, wide copper lines with a range of widths from 10-30μm were prepared by a dual damascene process that has been described in detail elsewhere [5]. Lines were embedded in SiO_2 dielectric and were capped with a thin layer of silicon nitride followed by 0.6 μm of SiO_2. The copper lines were isolated from the dielectric by a thin barrier layer metal. The final passivation was deposited at a temperature of 400°C. Heat treatment of the lines was carried out using a hot-stage optical microscope. This allows direct observations of void closure and formation to be made during a temperature cycle. Micrographs were recorded at frequent intervals during the temperature cycle for later inspection. Isothermal holds were controlled to within 1°C and temperature overshoots on heating were held to less than 2°C. All anneals were conducted in air. Optical micrographs were recorded from five separate regions for accurate counting of the density of voids formed during heat treatment. SEM observations of the voids were made by stripping the passivation using a buffered HF solution.

EXPERIMENTAL OBSERVATIONS

Void formation was readily observed in as-prepared lines using optical and SEM microscopy. In optical micrographs the voids appeared as small dots on the top surface of wide lines (see for example figure (1)). Observations made at temperature using an optical microscope and a hot stage showed that the void contrast was reduced on heating to above 300°C and that complete disappearance of the voids occurred on heating to 350°C. Voids could be easily distinguished from hillocks or particles by their disappearance on heating. On delayering the voids could be seen at the top copper dielectric interface of the lines as shown in figure (2). Optical and Plan view TEM micrographs of the voids indicate that void formation occurred preferentially at grain

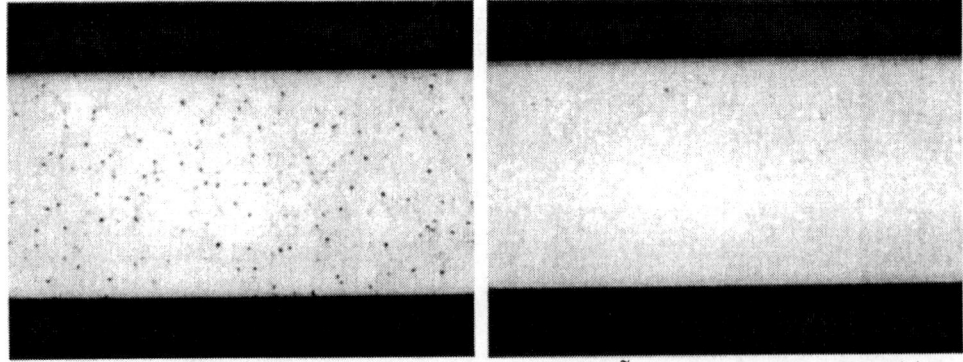

Figure(1) Optical micrographs of stress voids in a 30μm wide copper line (a) Room temperature (b) The same region on heating to 400°C (image taken at temperature)

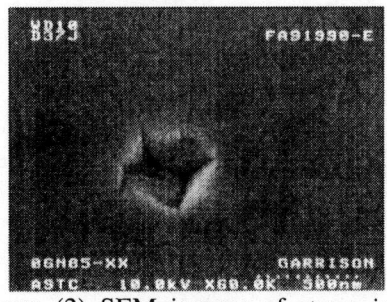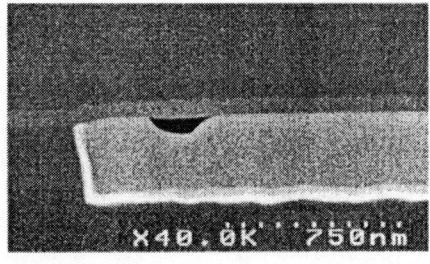

Figure (2) SEM images of stress induced voids formed at the copper nitride interface(a) Plan view (b) cross-section

boundaries and triple grain junctions in the lines. From optical micrographs the initial void density was estimated at .025 voids /μm^2 in a 30 μm wide line. This corresponds to an average void spacing of about 6 μm. The average grain size of copper in the lines was estimated to be 3.4μm.

In order to investigate the conditions under which void formation occurred, lines were first heated to 400°C to close all the voids and then cooled rapidly to a lower temperature for observation during an isothermal void growth anneal. A semi-quantitative assessment of the rate of void growth at different temperatures was made by continuously monitoring a region of a 30 μm wide line during the isothermal hold. The time at which voids first became visible and the time at which no further growth of the voids could be detected were noted and are plotted as a function of temperature in figure(3). No void growth could be detected in holds at temperatures above 300°C. Below 300°C it is apparent that the void growth rate steadily decreases. In figure (4) the void density measured in the same line after isothermal void growth anneals at different temperatures is plotted as a function of the void growth temperature. The highest void density was observed after void growth at 200°C. It is interesting to note that if the sample was first cooled to room temperature after a 400°C void closure anneal, a substantially lower density of voids was observed to form on annealing at the void growth temperature. Based on these observations a standard void growth anneal of 2 hours at 200°C was adopted to investigate the effect of the higher temperature void closure anneal on void formation.

First the effect of the dwell time at 400°C was investigated. The void density after annealing times from 5 to 90 mins at 400°C is shown in figure (5). As can be seen, prolonged annealing at the higher temperature caused only a slight reduction in the density of voids that grow at 200°C. In contrast, increasing the temperature of the high temperature anneal had strong effect on the void density, figure (6). In these experiments a previously unannealed sample was first heated to 350°C for 5 mins and then cooled to 200°C to grow the voids. After measuring the void density at room temperature the same sample was heated to 400°C for 5 mins and the void density after a 200°C anneal measured again. Only a small increase in void density was observed after this heat treatment and a further cycle to 400°C produced no further increase in the void density. However, on heating to 425°C an increase in void density was observed and a further cycle to 440°C produced an additional increase in void density. The increase in void density produced by   anneals at temperatures above 400°C appears to be permanent as annealing at   400°C even for prolonged times did not reduce the density of voids that grew at the void growth temperature back to its original level.   These observations indicate that heating to temperatures above the final passivation temperature has a pronounced effect on void nucleation at the lower temperatures.

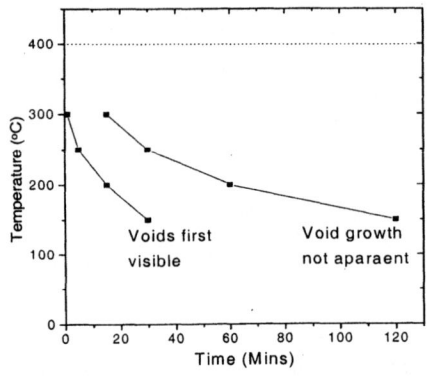

Figure 3. The effect of void growth temperature on void formation.

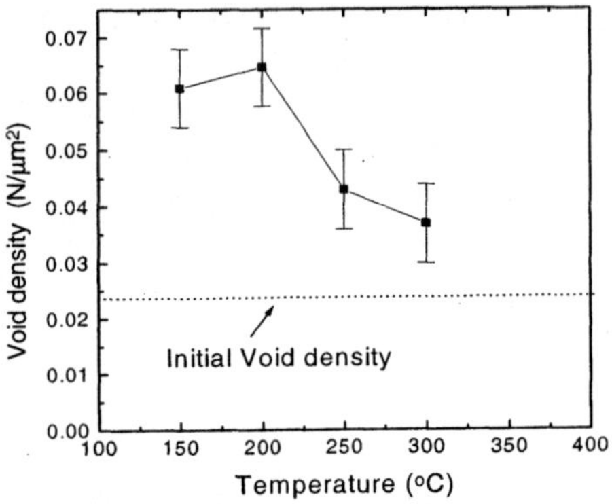

Figure 4. The effect of void growth temperature on the density of voids formed in a **30μm** wide line.

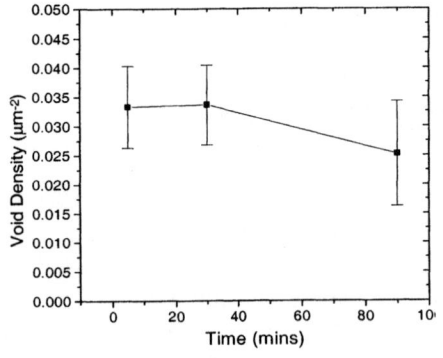

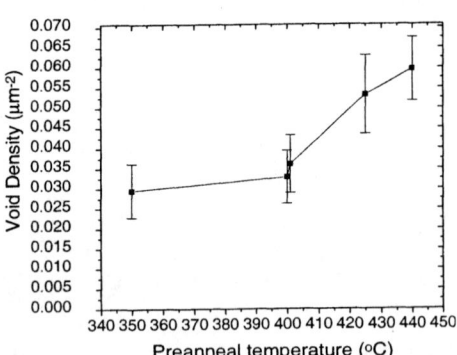

Figure 5. The effect of the annealing time 400°C on the density of voids formed.

Figure 6. The effect of annealing temperature void density.

DISCUSSION

Most of the experimental observations can be rationalized on the basis of the effect that different heat treatments have on the stress state of the lines. Stress analysis of passivated lines shows that for wide lines capped with a thin layer of passivation the passivation provides insufficient constraint normal to the line for a triaxial stress state to develop [6,7]. The line used for most of the current experiments were 30μm wide and were capped with only 0.6 μm of silicon dioxide. It is therefore expected that the lines are predominantly in a biaxial stress state. Previous experiments on blanket copper films have shown that on thermal cycling the biaxial stress in passivated films exhibits a pronounced hysterisis [8,9]. On heating, the thermal expansion mismatch of the copper with the silicon substrate drives the film into compression in the temperature range 200-300°C. In passivated films it is observed that stress relaxation is restricted even at higher temperatures and as a result the compressive stresses in a film increase continually on further heating [8,9]. On cooling the stress in the film quickly converts to tension and by 200°C biaxial stresses in the range 200-300 MPa can occur in the film on cooling from 400°C. Repeated cycles to the same temperature result in a repeatable stress-temperature hysterisis loop as shown schematically in figure (7a).

The optical hot stage observations support the idea that void closure occurs on heating due to the development of biaxial compressive stresses in the lines and that void growth occurs on cooling due to biaxial tension. The observation that void closure starts occurring above 300°C indicates that on heating the lines are stress free at about this temperature. Significant tensile stresses are therefore expected to develop in the lines on cooling below this temperature. It can be expected that both void nucleation and growth follow classic C-curve behavior where driving force for void formation increases on cooling but the diffusion rates rapidly decreases due to reduced thermal activation. The data in figure(3) indicates that the void growth rate decreases over the entire temperature range where void growth is observable. This suggests that void growth is predominately diffusion limited. The maximum in void density shown in figure (4), however, indicates that the void nucleation rate is still controlled by stress at temperatures above 200°C. Thus, processes that alter the stress in the lines can be expected to have strong effect on the density of voids formed at this temperature.

The difference in stress levels between the heating and cooling cycles accounts for the reduced nucleation density observed when a line is heated from room temperature to 200°C as opposed to cooling from a higher temperature. Figure (7a) clearly shows that a higher biaxial stress level is expected on cooling to 200°C and thus void nucleation is expected to be enhanced by this process as observed. The enhanced void density observed on heating to higher temperatures may also be partially caused by an enhancement of the stress level at the void growth temperature as shown schematically in figure (7b). However, this effect cannot account for the observation that the void density remains permanently enhanced after a high temperature anneal. This observation suggests that high temperature cycles create "damage" in the lines that can act as preferential nucleation sites for void growth at the void growth temperature.

While many of the experimental observations can be understood in terms of the effect of thermal cycling on the driving force for void formation the details of the void nucleation and growth process are not well understood. In conventional stress voiding a void grows in response to the hydrostatic stress in a line. In narrow lines with a square cross-section a near pure hydrostatic stress state exists throughout the line providing a large driving force for void nucleation and growth in the entire line. In the wide lines used in the present experiments the stress state is expected to be predominantly one of biaxial tension. As such there is a reduced driving force for the nucleation and growth of voids. Void nucleation is not usually observed in similar wide lines in passivated aluminum structures. These observations suggest that void nucleation at the copper/dielectric interface is enhanced in passivated copper lines.

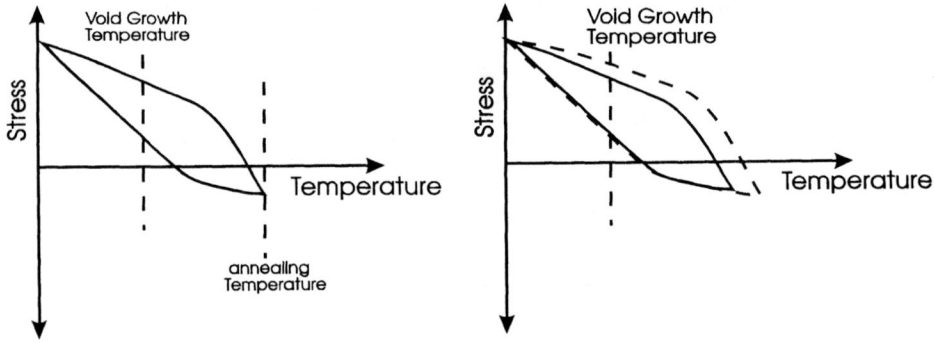

Figure 7. Temperature stress cycles in copper films (a) shows that the stress level in the films higher on cooling to the void growth temperature than on heating (b) shows how increasing the annealing temperature increases the stress level at the void growth temperature

It is significant that void nucleation occurs mainly at grain boundaries. There are at least two mechanisms by which local hydrostatic stresses can develop at grain junctions. Firstly, for inclined grain boundaries, the lower shear resistance of the grain boundary can result in grain boundary sliding. In the absence of a confining dielectric a shear step would form on the surface. In the presence of a dielectric, localized hydrostatic stresses can develop where the grain boundary intersects the dielectric. Such a localized stress field may provide a preferential site for void nucleation at the dielectric/copper interface. A reduction in the adhesive strength of this interface could further enhance void nucleation at this site. It is interesting to note that a region of hydrostatic tension can develop at the grain boundary if either biaxial compression or tension is applied to the line. In order to generate significant levels of hydrostatic stress in the film by this mechanism the grain boundaries must be inclined in the film.

A second factor that could lead to high localized stresses in a biaxial stressed line is the elastic anisotropy of copper. In single crystal copper the Young's modulus along <111> is more than twice that along <100>. The discontinuity in the elastic properties on crossing grain boundaries in copper can be expected to cause significant localized stresses [10,11]. Again, such localized stresses may be sufficient to cause void nucleation. In this case, it would be expected the stress levels generated would be sensitive to the degree of texture in the film. Strong texture which aligned the directions with similar elastic properties in the film would be expected to minimize the local stresses that arise at grain boundaries by this mechanism. A detailed theoretical stress analysis of both these situations is in progress. At the present time it is not possible to determine which, if either of these mechanisms is the dominant cause of void formation; however, it is expected that the amount of voiding that occurs in wide lines is strongly dependent on the grain structure of the line.

CONCLUSIONS

We have observed the formation of stress voids at the copper dielectric interface in wide copper lines. Experimental observations made using a hot stage optical microscope show that void growth occurs under conditions where the line is predominately in a state of biaxial tension. As such the void growth mechanism differs significantly from that observed in conventional stress voiding where void growth is driven by the hydrostatic stresses that develop in narrow lines. Void nucleation was also found to be enhanced by heating to higher temperatures. The observations are consistent with a mechanism in which grain boundary sliding or elastic mismatch across grain boundaries produce local regions of hydrostatic stress that nucleate voids were grain boundaries intersect the surrounding dielectric.

REFERENCES

1. M.A. Korhonen, P. Borgesen, K.N. Tu and C-Y. Li, J. Appl. Phys.73,3790, (1993).
2. T.D. Sullivan, D. P. Bouldin, D. H. Yao in 3rd. Workshop on Stress-Induced Phenomena in Metalization Edited P.S. Ho, J, Bravman, C-Y. Li and J. Sanchez (Published Am. Inst. Phys. Proc. 373 New York 1996) pp. 67-80.
3. J.A. Nucci, Y. Shacham-Diamand, and J. E. Sanchez, Jr., Appl. Phys. Lett. 66(5), 3585 (1995).
4. J.A. Nucci, Y. Shacham-Diamand, and J. E. Sanchez, Jr., Appl. Phys. Lett. 69(26), 4017 (1996).
5. R. Rosenberg, D.C. Edelstein, C.-K Hu and K.P. Rodbell. Annu. Rev. Mater. Sci. 30,229 (2000)
6. P. A. Flinn and C. Chiang, J. Appl. Phys., 67, 2927 (1990).'
7. B. Greenbaum, A. I. Sauter, P. A. Flinn and W. D. Nix, Appl. Phys. Lett., 58, 1845 (1991).
8. M.D. Thouless, K.P. Rodbell and C. Cabral, Jr., J. Vac. Sci. Technol. A 14(4), 2454-2461 (1996)
9. R-M., Keller, S.P. Baker and E. Arzt, J. Mater. Res. (USA) Vol.13, No.5 May 1998 P1307-17
10. D. Chidambarrao, Y,C. Song, and I.C. Noyan, Met. Trans. A 28A, 2515-2515, (1997).
11. P. Su, S. Rzepka, M. A. Korhonen, and C-Y. Li in 5th. Workshop on Stress-Induced Phenomena in Metalization Edited O. Kraft, E. Arzt C.A. Volkert, P.S. Ho, H. Okabayashi (Published Am. Inst. Phys. Proc. 491 New York 1999) pp. 298-303.

Evaluation of Interface Strength between Thin Films in an LSI Based on Fracture Mechanics Concept

Tadahiro Shibutani[†], Qiang Yu[†], Masaki Shiratori[†]
and Takayuki Kitamura[‡]

[†]Department of Mechanical Engineering, Yokohama National University, Yokohama, Japan
E-mail: shibu@swan.me.ynu.ac.jp
[‡]Department of Engineering Science, Kyoto University, Kyoto, Japan

Abstract. Since an electric device is made of multi-layered sub-micron films, the delamination along the interface is one of the major failure mechanisms. Especially, the delamination initiates at the edge of the interface because the stress singularity is generated due to the mismatch of deformation. This paper aims to evaluate the criterion of interface fracture initiation. The stress singularity at the edge of the interface is similar with that at a crack tip in a material. The stress intensity factor characterizes the stress singular field ahead of the crack and provides the criterion of the crack propagation as the fracture toughness on the basis of the fracture mechanics concept. As an analogy to the stress intensity factor, the criterion of the crack initiation from the edge of interface might be evaluated by the stress intensity governing the stress singularity at the edge. An experiment on Si_3N_4/Cu films on a silicon substrate is conducted. The stress singularity appears in the vicinity of the edge where the crack initiates and the fracture toughness of the Si_3N_4/Cu interface is evaluated.

INTRODUCTION

Sub-micron components used in an LSI consist of multi-layered films made of ceramics, intermetallic compounds and metals. The delamination at interfaces between films becomes one of major failure modes. Since the device size shrinks, the film's thickness became several nanometers in the recent years. Therefore, it is important in terms of reliability of LSI to evaluate the interface strength between thin films. Especially, the interface crack usually initiates at the edge of films because of the stress singular field generated by the mismatch between deformations on each side of the interface. On the other hand, the fracture toughness in a cracked material has been evaluated on the basis of the fracture mechanics concept. The stress singularity in the vicinity of the crack tip brings about the crack propagation and the toughness can be evaluated by the stress intensity factor, which characterizes the stress singular field. There is an analogy of the stress singular field of the crack tip and the edge of the interface. The criterion of interface crack initiation (the interface fracture toughness) might be evaluated by the intensity of the stress singularity in the vicinity of the edge.

CP612, *Stress-Induced Phenomena in Metallization:* Sixth Int'l. Workshop, edited by S. P. Baker et al.
© 2002 American Institute of Physics 0-7354-0058-X/02/$19.00

In this paper, the interface strength between thin films is evaluated on the basis of the fracture mechanics concept. The experiment on the interface between Si_3N_4/Cu films on a silicon substrate is conducted and the interface fracture toughness is determined from the stress intensity factor analyzed by using the BEM analysis.

FRACTURE MECHANICS CONCEPT

As shown in Fig. 1, a crack of length $2a$ is introduced in a homogeneous elastic body and a load P is applied on the axis normal to the crack. Theoretically, the distribution of stress in the vicinity of the crack tip is given by

$$\sigma_{ij} = \frac{K}{r^{0.5}} f(\theta) \qquad (1)$$

where r and θ are components of polar coordinates where the origin is located at the tip of the crack. $f(\theta)$ is a function of θ and K is the stress intensity factor depending on the load and the length of the crack. The stress takes infinity at the tip and the stress singular field appears in the vicinity of the crack tip. Equation 1 shows that the stress intensity factor K characterizes the stress singular field. The stress intensity factor is usually correlated with the applied load and is given by

$$K = \alpha(a)P. \qquad (2)$$

Here, $\alpha(a)$ is a function of the half-length of crack a which depends on the configuration of the specimen. If K is small, the crack does not propagate. K increases with P and the crack begins to propagate when K reaches K_c. It is verified by many investigations that the critical value K_c is independent on the configuration of the specimen and the length of the crack, though the critical load varies with the configuration of the specimen and the length of the crack [1]. Then, the K_c characterizes the fracture toughness of the cracked body. In other words, when K,

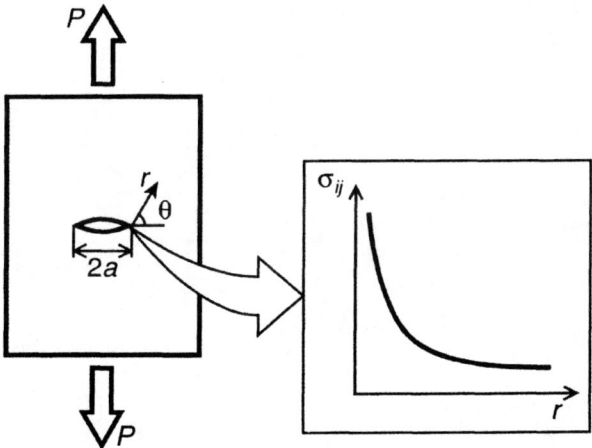

FIGURE 1. A schematic illustration of a cracked body under a load P and the stress singularity in the vicinity of the crack tip.

185

which characterizes the driving force to initiate the fracture, overcomes K_c, which characterizes the resistance of the material to the fracture, the fracture takes place at the crack tip. It implies that the criterion of the crack propagation can be evaluated by the stress intensity factor.

For K_c determination, the experiment requires continuous measurement of load (P) versus load-line displacement (δ). The critical load P_c is obtained from the P-δ curve. Substituting P_c into Eq. (2), the K_c is determined. Details of the method for measurement of fracture toughness are described in several standards such as ASTM E399 and E1820 [2].

EVALUATION PROCEDURE OF INTERFACE STRENGTH

The material tested is schematically illustrated in Fig. 2(a). Silicon nitride (Si_3N_4, passivation), copper (conductor metal), and tantalum nitride (barrier metal) are fabricated on a silicon substrate. The focus is put on the interface between Si_3N_4 and Cu in this study. Figure 2(b) shows the schematic illustration of the specimen and loading system [3,4]. Since the plastic deformation during the test is excluded, this test method is appropriate for the interface strength determination [3-5]. The load, P, is applied at the edge of the cantilever. Since stresses concentrates at the edge of the interfaces, the crack initiates at the edge. Experiments are carried out in an air at room temperature and the loading rate is constant during the test. Load versus load-line displacement is measured continuously during the test. Two specimens under the same condition are tested in order to check the repeatability.

The stress distribution along the interfaces is analyzed by the boundary element method (BEM) [6]. Considering the configuration of specimen and loading system, the analysis is carried out in a two-dimensional model. Figure 3 shows the mesh used for the analysis. The interface where the fracture takes place is carefully divided into a fine mesh. The elastic constants in the calculation are listed in Table 1.

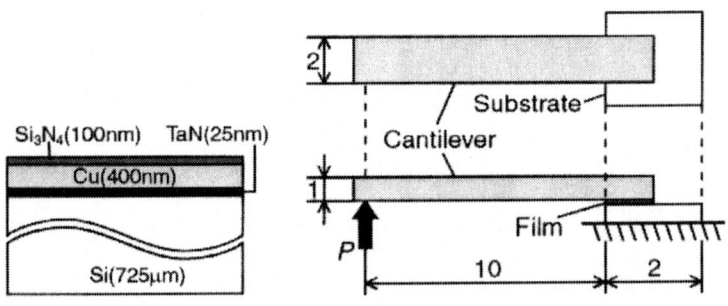

(a) Material tested   (b) Specimen and loading system

FIGURE 2. Schematic illustration of experiment for the fracture toughness at the edge of the interface.

TABLE 1. Elastic constants.

Material	Young's modulus E, GPa	Poisson ratio υ
Silicon nitride (Si₃N₄)	304	0.27
Copper (Cu)	129	0.34
Tantalum nitride (TaN)	350	0.35
Silicon (Si)	170	0.30
Epoxy	2.5	0.30
Steel (SUS304)	200	0.30

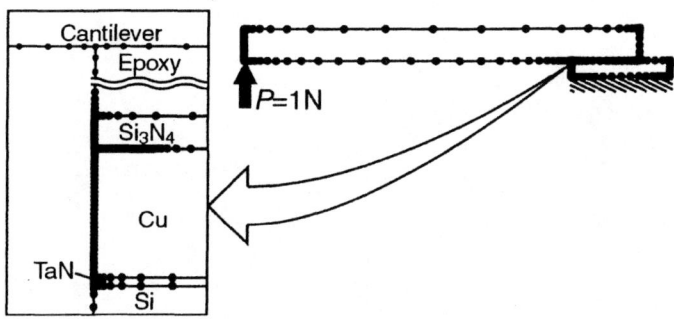

FIGURE 3. Mesh division for the BEM analysis.

RESULT AND DISCUSSION

Figure 4(a) plots the distribution of stress σ_{yy} along the Si₃N₄/Cu interface under $P=1$N. The horizontal axis indicates the distance from the left edge along the interface. The tensile stress concentrates at the left edge and the compressive one concentrates at the right edge because of the mismatch of the deformation. The crack initiates at the left edge. Figure 4(b) magnifies the distribution of Fig. 4(a) near the left edge. The singular stress field appears in the vicinity of the edge and is usually formulated by [7]

$$\sigma_{yy} = K_f \frac{1}{r^\lambda}. \tag{3}$$

Here, r is the distance from the edge and K_f indicates the stress intensity. The above equation is similar with Eq. (1) but the singular index of r is λ, while it is 0.5 for the crack. λ is constant depending on the elastic constants of materials on each side of the interface and the shape of the edge. Then, K_f characterizes the singular field of Eq. (3) and is similar to the K characterizing that of Eq. (1). K_f is directory correlated with the applied load P as

$$K_f = \alpha_f P, \tag{4}$$

where α_f is the proportional coefficient depending on the configuration of the specimen and can be obtained from the stress distribution such as shown in Fig. 5(b). If K_f is small, a crack does not initiate. The crack initiates when $K_f=K_{fc}$. As an analogy to the significance of K_c on the basis of the fracture mechanics concept, K_{fc} characterizes the resistance of the interface to the fracture (the interface fracture toughness). The independence of K_{fc} on the configuration of the specimen is reported elsewhere [3], which agrees with the analogy between K_{fc} and K_c.

Figure 5 shows the relationship between the load (P) and the displacement at the load point (δ) during the test. The crack initiates at the edge of the interface and the fracture instability takes place without the stable crack propagation in $P=P_c$. The observation of the fracture surface after the test by means of an optical microscopy

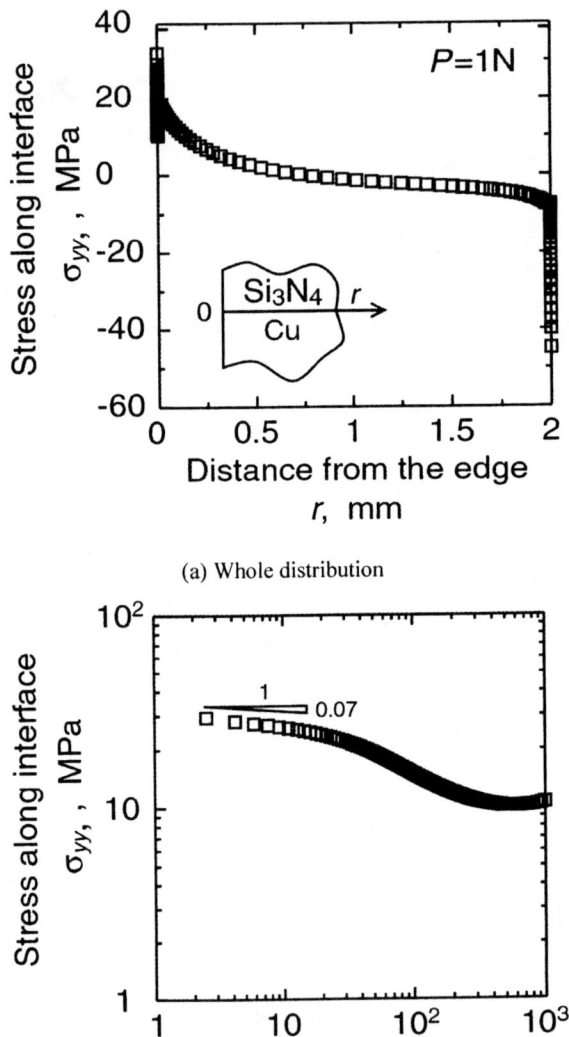

(a) Whole distribution

(b) Stress singularity in the vicinity of the edge

FIGURE 4. Distribution of stress σ_{yy} along the interface between Si_3N_4 and Cu.

reveals that the crack propagates along the Si$_3$N$_4$/Cu interface. For Si$_3$N$_4$/Cu interface, λ calculated from material properties of Si$_3$N$_4$ and Cu in Table 1 is 0.07. K_{fc} is determined by substituting the critical load P_c into Eq. (4) where α_f is obtained from the stress distribution in Fig. 4(b). Then, the criterion of the interface crack initiation at the edge is K_{fc}=20 MPa·m$^\lambda$ (λ=0.07).

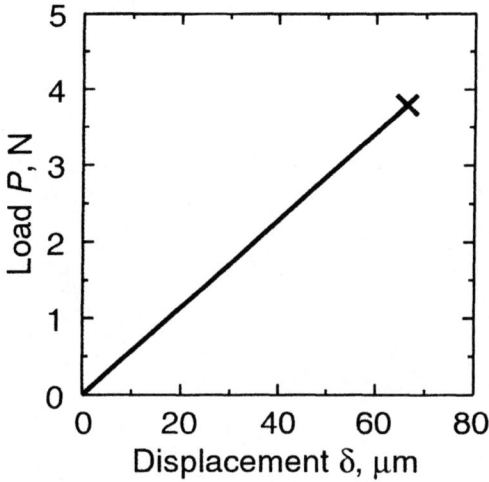

FIGURE 5. Relationship between load and load-line displacement.

ACKNOWLEDGMENTS

This investigation was supported by the Grant-in-aids for Encouragement Young Scientists (No.13750073).

REFERENCES

1. Broek D, *Elementary Engineering Fracture Mechanics*, Noordhoff International Publishing (1974).
2. Annual Book of ASTM Standards, 03.01.
3. Kitamura, T., Shibutani, T., and Ueno, T., *Trans. JSME*, **66**, pp. 1568-1573 (2000).
4. Kitamura, T., Shibutani, T., and Ueno, T., to be published.
5. Ma, Q., Fujimoto, H., Flinn, P., Jain, V., Adibi-Rizi, F., Moghadam, F., and Dauskardt, R. H., in *Materials Reliability in Microelectronics V*, MRS Symposium Proceedings 391, Pennsylvania: Materials Research Society, 1995, pp. 91-96.
6. Arai, M., Adachi, T., Nakayama, K., and Matsumoto, H., in *Japan & China International BEM Conference*, edited by Tanaka, M and Yao, M., Elsevier, 1995, pp.133-142.
7. Bogy, D. B., *J. Applied Mechanics*, **35**, pp.146-154 (1968).

PART IV: DEFORMATION AND STRESSES

Microdiffraction and Microfluorescence Studies of Electromigration-Induced Stresses and Composition Changes

G. S. Cargill III

Department of Materials Science and Engineering
Lehigh University, Bethlehem, PA 18015

Abstract. This paper reviews results obtained by x-ray microbeam energy dispersive Laue diffraction for electromigration-induced stresses in passivated pure Al conductor lines. Combined x-ray microdiffraction and microfluorescence experiments are described which provide new information on relationships between local Cu concentration and Al diffusivity during electromigration in Al(0.25 at.% Cu) conductor lines.

INTRODUCTION

Although stress has long been recognized as an important factor in electromigration, quantitative measurements of electromigration-induced stresses have been quite limited. In 1976 Blech and Herring [1] used x-ray topography to demonstrate qualitatively that a stress gradient developed during electromigration, and in 1977 Blech and Tai [2] used x-ray topography during electromigration at 340°C to determine the magnitude of stress gradients developed in 50 μm wide unpassivated aluminum conductor lines. They introduced the now well known equation

$$Z^* = \frac{\Omega}{e\rho j} \frac{\partial \sigma}{\partial x} \qquad (1)$$

to relate t the effective charge Z^* to the steady state stress gradient $\frac{\partial \sigma}{\partial x}$ for current densities j smaller than the threshold for continuing electromigration. The equation involves the atomic volume Ω, the stress normal to the grain boundaries σ, the distance x along the conductor line, the electronic charge e, the conductor line resistivity ρ, and the current density j. From their results, they concluded that $Z^* = -1.2$, with a minus sign to indicate that the direction of atom movement was the same as the direction of electron flow. The maximum electromigration-induced stress from their experiments was 120 MPa.

Fourteen years later, in 1991, Hemmert and Costa [3] reported stress measurements during electromigration at 200°C for 4 μm wide, SiO_2 passivated Al(4% Cu)

CP612, *Stress-Induced Phenomena in Metallization:* Sixth Int'l. Workshop, edited by S. P. Baker et al.
© 2002 American Institute of Physics 0-7354-0058-X/02/$19.00

conductor lines using npn transistors as local strain gauges. They estimated that stresses as large is 1 GPa were developed by electromigration.

In 1995 Ma et al. [4] reported strain measurements made by micro-Raman spectroscopy for 0.8 μm wide, SiO_2 passivated Al(0.5% Cu) conductor lines, with and without prior electromigration. They found nearly uniform hydrostatic stresses of about 570 ± 50 MPa along a conductor without electromigration, and widely varying hydrostatic stresses, ranging from about 180 MPa to 620 MPa, along a line after electromigration which had increased the line's resistance by 20%.

In 1996 we reported the first real time measurements of electromigration-induced strains using synchrotron-based x-ray microdiffraction [5]. In 1997 we reported microdiffraction results showing the evolution of a stress gradient during electromigration [6] in a passivated 10 μm wide pure Al line, and in 1998 we reported the extension of these measurements to steady state stress gradients, from which values of Z* (−1.6) and of the effective grain boundary diffusion coefficient D_{eff} [7] were obtained. Chiras and Clarke [8] have used optical piezospectroscopy to map stress in passivated 2 μm and 5 μm wide Al lines and determined Z* = −1.3±0.2.

More recently, we have reported x-ray microbeam fluorescence and diffraction studies of composition and stress evolution in Al(0.25 at.% Cu) conductor lines during electromigration [9-11]. From these measurements values of Z* and D_{eff} for Cu electromigration in Al(Cu) were obtained, as well as an estimate for the critical Cu concentration for substantial reduction of Al electromigration in Al(Cu). Microbeam x-ray topography studies of electromigration-induced stresses and composition changes have also been reported by Wang et al. [12, 13]. Our results from x-ray microbeam diffraction and fluorescence studies of electromigration [7, 9-11] are summarized in this paper. Most of this work has been carried out as graduate research by P.-C. Wang and H.-K. Kao. Other collaborators for parts of the work have been C.-K. Hu and I. C. Noyan.

INSTRUMENTATION AND SAMPLES FOR X-RAY MICROBEAM MEASUREMENTS

The x-ray microbeam measurements were made using beamlines X26C and X6A of the National Synchrotron Light Source (NSLS). The Al (111) interplanar spacing d_{111} along the film normal, Cu K_α fluorescence, Ti K_α fluorescence, and conductor line resistance were measured simultaneously. A novel 4-circle diffractometer used a Ge-solid state detector for diffraction measurements and a Si-pin diode detector for fluorescence measurements. The x-ray microbeam was formed by a tapered capillary with beam spot size of about 10 μm ×10 μm on the conductor line. The sample was translated to bring different positions into the x-ray beam to monitor the evolution of stress and Cu concentration profiles along the conductor line.

The samples used were either pure Al or Al(0.25 at. %Cu) prepared on thermally oxidized Si substrates by magnetron sputtering, e-beam lithography, and reactive ion etching (RIE). The conductor lines were 200 μm long, 10 μm wide, and 0.5 μm thick, with W bars at both ends and an underlayer of 100 Å Ti / 600 Å TiN. The W bars serve as diffusion barriers for Al and Cu, and they connect each end of the conductor line to Al(Cu) or Al wire-bonding pads. These samples were deposited at room temperature, patterned by RIE, and then annealed at 400°C. The 1.5 μm SiO$_2$ passivation layer was deposited at 350°C by plasma enhanced chemical vapor deposition (PECVD) and was removed from the top of the Al(Cu) wire-bonding pads by RIE. The polycrystalline Al and Al(Cu) films had <111> fiber texture with grain sizes expected to be on the order of the film thickness. The polycrystalline W bars had <110> fiber texture.

Procedures used for the x-ray microbeam strain and composition measurements have been described earlier [5, 14, 15]. Changes in Al d_{111} values were measured along the length of the conductor line during electromigration, using a symmetric reflection scattering geometry, by monitoring changes in the diffracted photon energy E_{111} at each measurement location. Also, values of the W (110) interplanar spacing d_{110} were measured at the W pads at each end of the conductor line to correct for drifts in angular or energy calibration during the experiment. About one hour was required for a complete set of measurements. The scattering angle 2Θ was about 26°, and the photon energy E_{hkl} was about 11.8 keV for Al (111) and about 12.3 keV for W (110).

Cu concentration was determined from the intensity of Cu K_α fluorescence excited from the sample within the volume of Al(Cu) illuminated by the x-ray microbeam, as measured by the Si-pin detector. The Ti K_α fluorescence intensity was used to normalize the Cu fluorescence intensity in each measurement to correct for changes in the incident x-ray flux with time. Background corrections for scattering from the Si substrate and from other sources were made by subtracting spectra measured when the conductor line was translated out of the incident x-ray beam, which then fell only on the substrate.

ELECTROMIGRATION-INDUCED STRAINS IN PURE ALUMINUM CONDUCTOR LINES

Before electromigration, the Al d_{111} values were nearly the same for all measurement locations, with

$$\frac{d^i - <d>}{<d>} \tag{2}$$

smaller than ±0.001, where d^i is the value of d_{111} at the ith location along the conductor line, and $<d>$=2.346Å is the average of the ten initial d^i values. The scatter in initial d-values is due in part to uncertainty in 2Θ values, because of finite incident beam divergence and finite detector aperture size [14]. More reliable are values of

$$\Delta d^i_{EM}(t) = d^i_{EM}(t) - d^i_{pre-EM} \tag{3}$$

where $d^i_{EM}(t)$ is the value of d^i at time t during electromigration, and d^i_{pre-EM} is the value d^i before the start of electromigration. $\Delta d^i_{EM}(t)$ represents the changes in Al d_{111} values at each location i and at each time t caused by electromigration.

For conductor lines with small thickness-to-width aspect ratios, thermal and electromigration-induced strains are expected to be equibiaxial, even if the lines are passivated, based on analytical modeling [16] and finite element calculations [17]. In these cases, the biaxial stress $\sigma_{//}$ can be calculated from the normal strain $\varepsilon_\perp$, Young's modulus Y, and Poisson's ratio v,

$$\sigma_{//} = \frac{-Y}{2v}\varepsilon_\perp \quad . \tag{4}$$

This equation has been used to calculate values of the electromigration-induced in-plane stress σ_{EM} from the measured electromigration-induced perpendicular strains $\Delta d^i_{EM}(t)/d^i_{pre-EM}$.

In this section we summarize the results of the electromigration measurements with a current density below the threshold current density for these passivated 200 μm long conductor lines for steady state electromigration [7]. Figure 1 shows the changes in d_{111} values at several locations along the Al line after nine hours of electromigration at 1.4×10^5 A/cm^2, as compared with those values measured before electromigration. Also shown in Fig. 1 is the scale of the electromigration-induced compressive stress σ_{EM} calculated from the $\Delta d^i_{EM}(t)$ values using Eq. (4) as described above. Generally, electromigration caused d_{111} values to decrease near the cathode end of the line and to increase near the anode end of the line. The measured Δd_{111} are the Poisson responses of the electromigration-induced in-plane film stresses: biaxial compressive stress increased the d_{111} value measured along the film normal, while biaxial tensile stress decreased the d_{111} value.

The atom flux J during electromigration can be expressed as [1, 2]

$$J = n\frac{D_{eff}}{kT}\left[Z^*ej\rho - b\frac{\partial\sigma_{EM}}{\partial x}\Omega\right] \tag{5}$$

where n is the atomic density, D_{eff} is the effective grain boundary diffusion coefficient, k is Boltzmann's constant, T is the absolute temperature, Z^* is the effective valence of the diffusing species, e is the electron charge, j is the current density in the line, ρ is the electrical resistivity of the conductor line, Ω is the atomic volume, $\partial\sigma_{EM}/\partial x$ is the electromigration-induced compressive stress gradient along the length of the conductor line, and b is a stress state-dependent coefficient, with $b=2/3$ for an equibiaxial stress state [7, 14], assuming no stress dependence of the diffusion coefficient D_{eff}. A linear stress gradient is expected to extend over the Al line with flux blocking boundaries at both ends, in agreement with results shown in Fig. 1.

For currents below the threshold value, $j_{th}=1.6\times10^5$ A/cm^2 [7], the stress gradient developed eventually counterbalances the electron wind and the net atom flux becomes zero.

By substituting $J=0$ in Eq. (5), the steady-state stress gradient is related to current density through Eq. (1) but with incorporation of a factor of $b=2/3$. The effective valence Z^* can be determined from the steady-state stress gradient ($\partial\sigma_{EM}/\partial x$) measured as a function of current density j below the threshold value. As shown in [7], the steady-state compressive stress gradient increased linearly with current density below the threshold current density j_{th}. Further increase in the applied current above the threshold value did not change the stress gradient. The value of $e\rho Z^*/b\Omega$ was determined from the slope of the linear fit to the data obtained below threshold current density, shown by the dashed line in Fig. 1. Taking for aluminum $\Omega=1.7\times10^{-23}$ cm^3 and $\rho=5.5\times10^{-6}$ ohm·cm, we obtained an effective valence $Z^*=-1.6$. More recent measurements by Chiras and Clarke [8] (-1.2) using piezospectroscopy and by Verbruggen [18] (-1.8) using piezoresistance measurements are close to the value from x-ray microdiffraction. Earlier estimates of Z^* for Al, obtained by less direct methods, ranged from 1 to 10, as discussed in [7].

Figure 2 shows the stress gradients measured as a function of time during current loading for current densities below the threshold value. There was no stress gradient before current loading. When a 1.0×10^5 A/cm^2 current was passed through the conductor line, the stress gradient built up to about 1.2 MPa/μm within three hours. The stress gradient relaxed back to zero within three hours after the current flow was halted, and it changed sign when the current was reversed.

The analytical model derived by Korhonen et al. [19] can be used in analyzing the stress evolution,

$$\frac{\partial\sigma_{EM}}{\partial t} = \frac{\partial}{\partial x}\left[\frac{D_{eff}B\Omega}{kT}\left(\frac{\partial\sigma_{EM}}{\partial x} - \frac{Z^*ej\rho}{b\Omega}\right)\right] \tag{6}$$

where B is the modulus for an aluminum film in an equi-biaxial stress state, which is about 0.75 times the Young's modulus [19]. $b=2/3$ has been introduced because of the equi-biaxial, rather than hydrostatic, stress state [7]. Taking $Z^*=-1.6$ as determined previously, and $B=50$ GPa for aluminum with a <111> fiber texture, the compressive stress $\sigma_{EM}(x,t)$ was calculated from Eq. (6) as a function of current density j for several different values of the effective grain boundary diffusion coefficient D_{eff}, assumed to be independent of stress, for Al in Al. Stress gradients were calculated as the slopes of linear fits to $\sigma_{EM}(x,t)$ versus x curves calculated with Eq. (6) for different times t. Although the electromigration-induced stress $\sigma_{EM}(x,t)$ did not depend linearly on position x during the very early stage after current changes, the calculated stresses were characterized by their average gradient for comparison with the experimental values. The results of these calculations are shown in Fig. 2 by lines for three different values of D_{eff}. Good agreement is obtained for the intermediate value, $D_{eff}=8.2\times10^{-11}$ cm^2/sec, as shown by the solid line in Fig. 2. This value agrees well with those determined over a range of temperatures by Verbruggen [18] from piezoresistance measurements.

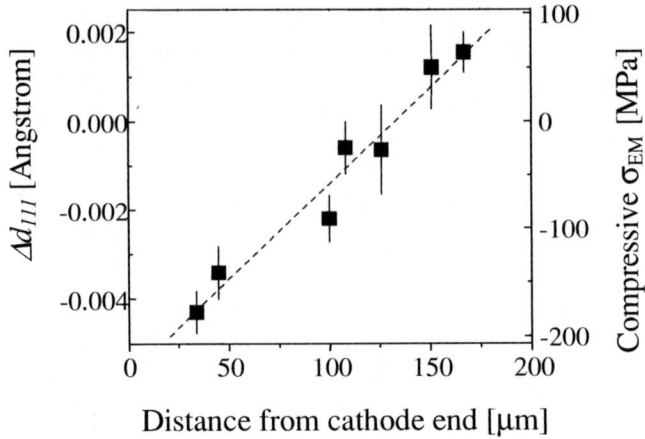

FIGURE 1. Electromigration-induced Al (111) plane spacing changes Δd_{111} measured at several positions along the aluminum conductor line after nine hours of 1.4×10^5 A/cm^2 current passage. The error bars indicate experimental uncertainties from counting statistics. The compressive stresses σ_{EM} from the Δd_{111} values based on the equi-biaxial stress model are shown on the right axis. A linear stress gradient of 1.8 MPa/μm was induced by electromigration, as shown by the linear fit dashed line. From [7].

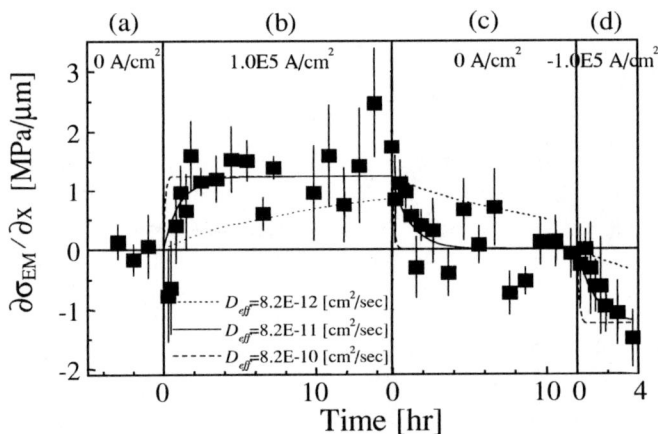

FIGURE 2. Evolution of the electromigration-induced compressive stress gradient (a) before current loading, (b) while passing 1.0×10^5 A/cm^2 through the aluminum line, (c) after halting the current flow, and (d) while passing 1.0×10^5 A/cm^2 in the opposite direction (current reversal). The curves are the stress gradient as a function of time calculated using the effective grain boundary diffusion coefficient $D_{eff} = 8.2 \times 10^{-12}$ cm^2/sec (dotted line), 8.2×10^{-11} cm^2/sec (solid line), and 8.2×10^{-10} cm^2/sec (dashed line). From [7].

ELECTROMIGRATION-INDUCED COMPOSITION AND STRESS CHANGES IN Al(0.25 At.% Cu) CONDUCTOR LINES

The role of Cu in reducing the rate of electromigration damage in Al(Cu) conductor lines has been studied for many years. Additions of a few percent of Cu increase electromigration lifetimes by one or two orders of magnitude [20]. Cu atoms tend to segregate to the Al grain boundaries and apparently reduce the grain boundary mobility of Al atoms. Most models for electromigration in Al(Cu) conductor lines predict that an incubation time is needed, after the start of current flow, before the onset of significant Al migration (see for example ref. [21]). This delay in Al migration is attributed to the Cu concentration having to drop below some threshold level within a current-dependent, critical length of the cathode end of the conductor line. This threshold concentration has been estimated to be ~0.1 at.% [21], but no direct measurements had been reported until our recent publication [11]. During electromigration in an Al(Cu) conductor line with blocking boundaries, a stress gradient is expected to develop in the part of the line where electromigration of Al occurs, but not in the part of the line where the Cu concentration remains above the threshold value which drastically reduces Al migration. This section reviews recent x-ray microbeam measurements of Cu concentration and stress changes during electromigration in passivated 10 μm wide Al(0.25 at.% Cu) conductor lines [9-11, 15].

The Cu concentration distribution along a conductor line was monitored in real-time during a series of electromigration tests while controlling the direction and the magnitude of the electron flow. Figure 3(a) shows the evolution of Cu concentration at four positions along the Al(Cu) line during electromigration with $j=0.3$ MA/cm^2 and T=310 °C. Before the start of electromigration Cu concentration was found to be uniform along the line. The Cu was quickly depleted near the cathode end of the line, dropping from 0.25 at. % to less than 0.02 at. % in less than four hours. Near the middle of the line the Cu concentration dropped more slowly, reaching 0.03 at. % after about 12 hours and remaining at this level. Nearer the anode the Cu concentration first increased to about 0.27 at. % and then after five hours slowly decreased to between 0.10 and 0.15 at. % after 10 hours. Nearest the anode end of the line, the Cu concentration increased to about 0.35 at. % during the first hour and remained at this level for 10 hours. The Cu concentration there then dropped to 0.25–0.30 at. % and remained at this level until the end of the electromigration test. The buildup in Cu concentration to 0.35 at. % at 184 μm from the cathode end during the first 10 hours results from the imbalance of the incoming and outgoing flux at that location when an Al$_2$Cu precipitate is growing at the anode end of the conductor line. Once a precipitate has been formed, the Cu concentration near the precipitate should stay near the equilibrium solubility (~0.43 at. %) for the electromigration temperature of 310°C, eventually resulting in the decrease of the Cu concentration at the 184μm location to about 0.25 at % after 12 hours.

From Fig. 3(a), Cu concentrations appear to reach steady state values by 19 hrs. Averaged values of C_{Cu} from 19 hrs to 36 hrs are shown as a semilog plot versus measurement location in Fig. 3(b). This type of behavior was first reported by Blech

[22], who explained it as the component of the Cu flux due to the electrical current density j

$$J_e = \frac{D_{Cu}^{eff} C_{Cu}}{kT} \left(Z_{Cu}^* e \rho j \right) \qquad (7)$$

being counterbalanced by the component of the Cu flux due to the Cu concentration gradient $\partial C_{Cu}/\partial x$

$$J_c = \frac{D_{Cu}^{eff} C_{Cu}}{kT} \left(-kT \frac{\partial \ell n C_{Cu}}{\partial x} \right) \qquad (8)$$

and the steady state Cu concentration profile being given by $J = J_e + J_c = 0$, or

$$Z_{Cu}^* = -\frac{kT}{e\rho j} \frac{\partial \ell n C_{Cu}}{\partial x} \qquad (9)$$

In Eqs. (7) and (8), D_{eff}^{Cu} is the effective grain boundary diffusivity for Cu in Al. Using Eq. (9) with the Cu concentration profile shown in Fig. 3(b), the effective charge Z_{Cu}^* was determined to be −8.6±0.6. This values falls among the Z_{Cu}^* values reported by Blech [22] (−4.1 to −14.9) and by Ho and Howard [23] (−16.8). The values found for Z_{Cu}^* are significantly larger than $Z_{Al}^* = -1.6$ for pure Al.

The effective diffusivity D_{Cu}^{eff} of Cu in Al was determined by modeling the evolution of Cu concentration at several locations along the conductor line as shown in Fig. 3(a), using D_{Cu}^{eff} as a fitting parameter. This was carried out for experimental data obtained at temperatures between 275°C and 325°C. Resulting values of $D_{Cu}^{eff}(T)$ are shown in Fig. 4 together with D(T) for Cu in single crystal Al [24] and with two values of D_{Cu}^{eff} from earlier electromigration studies [23, 25]. The activation energy obtained from the present data was 0.76±0.19 eV. The scatter of values at 300°C and 325°C reflect changes in electromigration kinetics which occurred during these experiments (see discussion in ref. [11]).

Changes in d_{111} values were measured, as well as Cu concentrations, during 38 hrs of electromigration at 10 locations for one conductor line [10]. Figure 5(a) shows strain measurement results after 9 hrs of electromigration with 7.5 mA, or current density of 1.5×10^5 A/cm^2, at 300°C. The decrease of the d_{111} values near the cathode end ($x/L=0$) of the line is due to the Al atoms removed from the grain boundaries by the electron flux, causing in-plane tensile stress or reduction of in-plane compressive stress. Farther down the conductor line, the accumulation of Al atoms within the grain boundaries creates compressive stress normal to the grain boundaries, causing d_{111} to increase due to the Poisson expansion along the film normal.

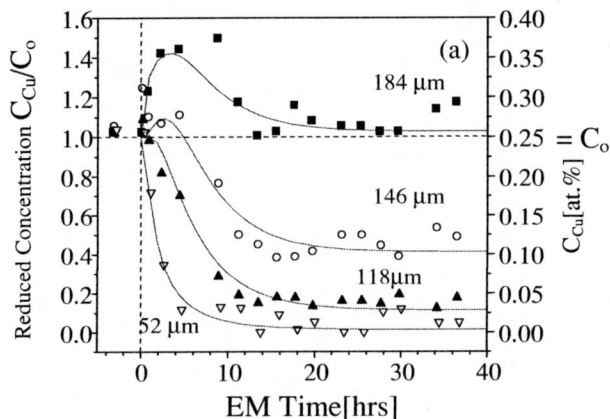

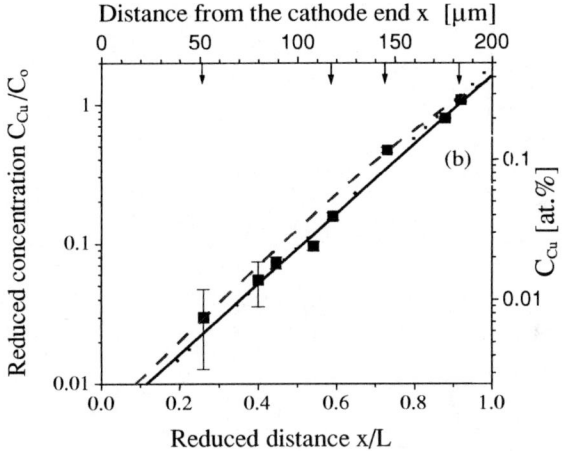

FIGURE 3. (a) Cu concentration versus electromigration time for four different measurement locations along an Al(Cu) conductor line with $j = 0.3$ MA/cm^2 and T = 310°C. Distances given are measured from the cathode end of the conductor line. C_o is the initial, uniform Cu concentration. Model simulation profiles are shown as solid curves. (b) Cu concentrations measured during electromigration at several positions along the conductor line averaged over the period of 19 hrs to 36 hrs. L is the conductor line length, 200 μm. The solid line is the weighted least square fit of the experimental data. The dotted line is the result from the model calculation for the averaged Cu concentration over the same period of time as the experimental data. The dashed curve shows the model-calculated result including an extra Cu flux term J_σ due to the redistribution of the Cu atoms without compensating Al backflow. The four arrows shown on the upper axis correspond to the measurement locations in (a). From [11].

In Fig. 5(a), a nearly linear strain distribution is observed along about 60% of the total conductor line length from the cathode end. The error bars in Fig. 5(a) are the standard deviation from the average of three measurements at each location. The changes in in-plane compressive stress $\Delta\sigma_{EM}$ shown on the right axis of Fig. 5(a) were calculated with Eq. (4) from changes in the d_{111} values. As shown by the linear fit dashed line, a stress gradient $\Delta\sigma_{EM}/\Delta x$ of about 3MPa/μm extending over 60% of the line length developed during electromigration.

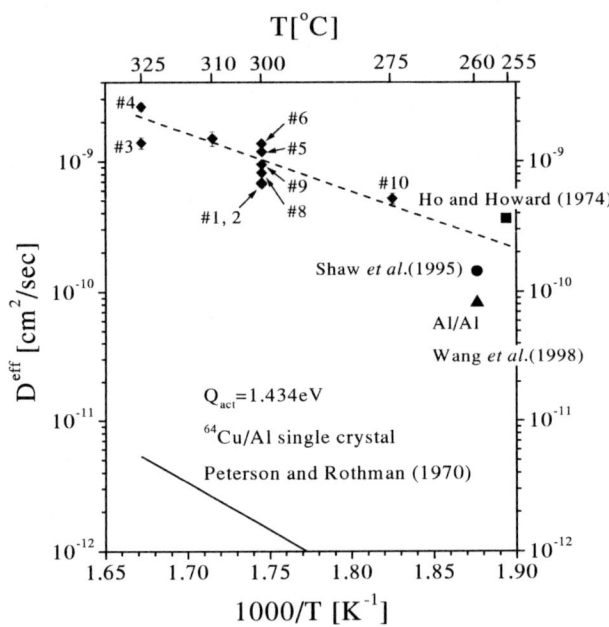

FIGURE 4. Arrhenius plot of effective grain boundary diffusivities. The dashed line is the least square fit of the Cu effective grain boundary diffusivities ∫ measured at different temperatures using x-ray microbeam fluorescence. The numbers, #1-#10, associated with ∫ data points indicate the order in which the measurements were made. The unnumbered ∫ data point is from measurements on a different sample. The solid line is the extrapolation of the Cu lattice diffusivity measured at higher temperature [24]. Also shown are previously reported effective diffusivities for Cu in Al(Cu) [23, 25] , ● and ◊ , and for Al in Al, ∫ [7]. From [11].

The absence of stress changes in the remaining 40% of the line length nearest the anode is presumably due to the Cu concentration in this region being above the threshold value which greatly slows down Al diffusion [9,11]. This threshold Cu concentration can be estimated from the Cu concentration profile along the line shown in Fig. 5(b). If we choose the location at 118 μm from the cathode end as the location

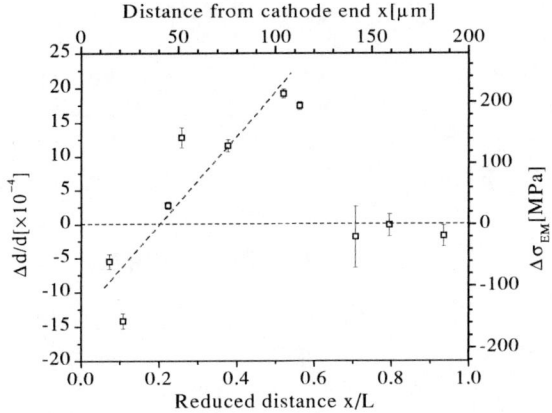

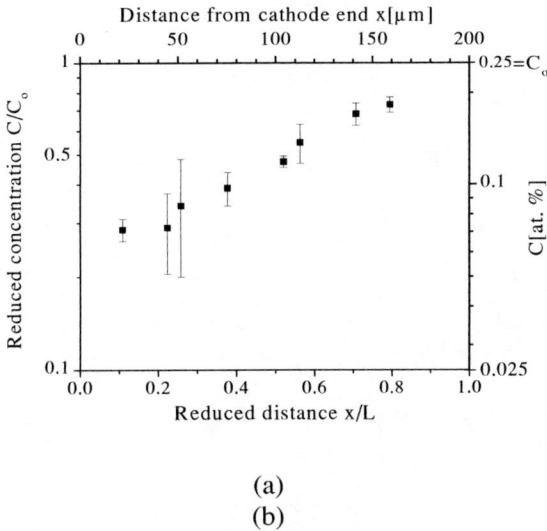

(a)

(b)

FIGURE 5. (a) Electromigration-induced Al(111) plane spacing changes, and (b) Cu concentration measured at different positions along the Al(Cu) conductor line after 9 hrs of electromigration. From [10].

where the Cu concentration is high enough to drastically reduce the Al grain boundary diffusion, the data in Fig. 5(b) indicate that the critical Cu concentration is ~0.15 at. %. The absence of significant changes in strain in the downstream 40% of the line length, although appreciable Cu enrichment has occurred in this region, also indicates that Cu electromigration by itself does not introduce significant compressive stresses. This observation supports the proposal by Shaw et al. [25] that Al and Cu movements are coupled during electromigration of Cu, so that downstream flow of Cu is balanced by upstream flow of Al.

CONCLUSIONS

X-ray microdiffraction and microfluorescence experiments have provided new information about electromigration, including Z^* and D_{eff} values for Al in pure Al and for Al and Cu in Al(0.25 at.% Cu), and the critical Cu concentration for drastic reduction of Al electromigration in Al(Cu) conductor lines.

ACKNOWLEDGMENTS

This work has been supported by NSF grants DMR-9796284 and DMR-9896002. The experiments have been carried out at the National Synchrotron Light Source, Brookhaven National Laboratory, which is supported by the Department of Energy. Most of the work described in this paper has been carried out as graduate research by P.-C. Wang and H.-K. Kao. Other collaborators for parts of the work have been C.-K. Hu, I. C. Noyan and K. J. Hwang. We also gratefully acknowledge valuable discussions with R. Rosenberg.

REFERENCES

1. I. A. Blech and C. Herring, Appl. Phys. Lett. **29**, 131 (1976).
2. I. A. Blech and K. L. Tai, Appl. Phys. Lett. **30**, 387 (1977).
3. R. S. Hemmert and M. Costa, IEEE Int. Reliability Phys. Symp. Proc. (IEEE, New York, 1991), p. 64.
4. Q. Ma, S. Chiras, D. R. Clarke, and Z. Suo, J. Appl. Phys. **78**, 1614 (1995).
5. P.-C. Wang, G. S. Cargill III, I. C. Noyan, E. G. Liniger, C.-K. Hu, and K. Y. Lee, MRS Symp. Proc. **427**, 35 (1996).
6. P.-C. Wang, G. S. Cargill III, I. C. Noyan, E. G. Liniger, C.-K. Hu, and K. Y. Lee, MRS Symp. Proc. **473**, 273 (1997).
7. P.-C. Wang, G. S. Cargill III, I. C. Noyan, and C.-K. Hu, Appl. Phys. Lett. **72**, 1296 (1998).
8. S. Chiras and D. R. Clarke, J. Appl. Phys. **88**, 6302 (2000).
9. H.-K. Kao, G. S. Cargill III, K. J. Hwang, A. C. Ho, P.-C. Wang, and C.-K. Hu, Mat. Res. Soc. Symp. Proc. **563**, 163 (1999).
10. H.-K. Kao, G. S. Cargill III, and C.-K. Hu, Mat. Res. Soc. Symp. Proc. **612**, D1.8.1 (2000).
11. H.-K. Kao, G. S. Cargill III, and C.-K. Hu, J. Appl. Phys. **89**, 2588 (2001).
12. P.-C. Wang, I. C. Noyan, S. K. Kaldor, J. L. Jordan-Sweet, E. G. Liniger, and C.-K. Hu, Appl. Phys. Letters **76**, 3726 (2000).
13. P.-C. Wang, I. C. Noyan, S. K. Kaldor, J. L. Jordan-Sweet, E. G. Liniger, and C.-K. Hu, Appl. Phys. Letters **78**, 2712 (2001).
14. P.-C. Wang, *Thermal and Electromigration Stress Distributions Measured by X-ray Microdiffraction*, D.E.S. thesis, Columbia University, 1997.
15. H.-K. Kao, *In-situ X-ray Microbeam Fluorescence and Strain Measurements on Al (0.25 at. % Cu) Conductor Lines During Electromigration*, D.E.S. thesis, Columbia University, 2000.
16. M. A. Korhonen, R. D. Black, and C.-Y. Li, J. Appl. Phys. **69**, 1748 (1991).
17. Sauter and W. D. Nix, IEEE Trans. Components Hybrids Manuf. Technol. **15**, 594 (1992).
18. A. Verbruggen, *Piezoresistance Electromigration Measurements*, Sixth Intern. Workshop on Stress-Induced Phenomena in Metallizations, July 2001.
19. M. A. Korhonen, P. Børgesen, K. N. Tu, and C.-Y. Li, J. Appl. Phys. **73**, 3790 (1993).
20. R. Rosenberg, J. Vac. Sci. Technol. **9**, 263 (1971).
21. M. A. Korhonen, T. Liu, D. D. Brown, and C.-Y. Li, MRS Symp. Proc. **391**, 411 (1995).
22. Blech, J. Appl. Phys. **48**, 473 (1977).
23. P. S. Ho and J. K. Howard, J. Appl. Phys. **45**, 3229 (1974).
24. N. L. Peterson and S. J. Rothman, Phys. Rev. B **1**, 3264 (1970).
25. T. M. Shaw, C.-K. Hu, and K. Y. Lee, Appl. Phys. Lett. **67**, 2296 (1995).

Dominant Inelastic Mechanisms In FCC Metallic Thin Films and Lines

Mauro J. Kobrinsky and Carl V. Thompson

Department of Materials Science and Engineering, Massachusetts Institute of Technology, Cambridge, MA 02139

Abstract. During fabrication, metallic thin films and lines typically develop high levels of stress that affect their performance and can lead to failure of devices based on them. A quantitative reliability assessment requires a thorough understanding of the mechanical behavior of these metallic structures that can only be attained if the dominant inelastic mechanisms are identified and characterized. In this paper, we discuss the inelastic behavior of Ag and Cu thin films, and damascene Cu lines. For the thin films, we identified the dominant inelastic mechanisms that generate plastic deformation during thermal cycling: diffusional creep at high temperatures (approximately above 0.3 of the melting temperature), and dislocation-mediated plasticity controlled by the thermally activated glide of dislocations through forest obstacles at low temperatures. As expected, the characteristic lengthscale for diffusional creep is determined by the microstructure (i.e. average grain size and film thickness). The characteristic lengthscale for dislocation plasticity is the length of the moving dislocation segments, which was found to depend on film thickness and to be on the order of 50-100 nm. Our experimental results indicate that the mechanical behavior of the lines is essentially identical to that of the films, and that it can also be understood in terms of diffusional creep and dislocation plasticity. Finally, a simple constitutive equation, based on the dislocation-mediated inelastic mechanism, is presented.

INTRODUCTION

A wide variety of advanced applications make use of the electric, magnetic, optical, and/or thermal properties of metallic structures with reduced dimensionality, in which at least one dimension is in the range 10 nm-1 µm. Two of the most common structures found in applications are continuous (or blanket) thin films and lines, which are usually deposited on much thicker substrates. In many cases, a capping layer is deposited on top of the films and lines to provide electrical insulation, and to prevent the degradation of the metal that could result from exposure to the atmosphere.

During processing, thin films and lines are subjected to temperature cycles that generate high levels of thermal stresses. Numerous failures of commercial devices are attributed to stress-related processes such as stress-induced cracking, electromigration-induced voiding, delamination, hillock formation, and fracture of the passivation layer.[1,2] Because the mechanical behavior of these metallic structures during thermal cycling is mostly inelastic, in order to predict the evolution of stresses, inelastic constitutive equations are required. Unfortunately, because the inelastic behavior of thin films and lines is different from that of their bulk counterparts, constitutive equations derived for bulk metals are inadequate for thin films and lines. The

CP612, *Stress-Induced Phenomena in Metallization:* Sixth Int'l. Workshop, edited by S. P. Baker et al.
© 2002 American Institute of Physics 0-7354-0058-X/02/$19.00

development of physically based constitutive equations for thin films and lines requires the identification and characterization of the dominant inelastic mechanisms.

A distinctive property of metallic structures with reduced dimensionality is that their dimensions are comparable to (and in many cases significantly smaller than) the characteristic lengthscales of the relevant inelastic mechanisms. Consequently, the mechanical behavior of these metallic structures is strongly affected, if not entirely determined, by their physical dimensions.

In this paper, we compare the mechanical behavior of Ag and Cu thin films, and damascene Cu lines, in the range of temperatures, T, $-50 < T < 500$ °C, with the goal of identifying the dominant inelastic mechanisms and their characteristic lengthscales. A more extensive description of most of the experimental results discussed here can be found in Refs. 3–6. Finally, we derive a simple physically based constitutive equation for dislocation plasticity in metallic thin films.

EXPERIMENTAL

Ag and Cu thin films were deposited on two types of substrates: (001)-oriented single-crystal Si wafers coated with 130 nm-thick thermally grown SiO_2 layers [Fig. 1(a)], and micromachined Si membranes. A more detailed description of the membranes can be found in Refs. 4 and 7. The films were deposited using electron beam evaporation under ultra high vacuum. The pressure during deposition was below 1.5×10^{-7} mbar. The continuity of the films after thermally cycling was confirmed using Scanning Electron Microscopy (SEM), Atomic Force Microscopy (AFM), and Transmission Electron Microscopy (TEM). Some samples were capped with SiO_x ($x \approx 2$) films using sputter deposition from SiO_2 targets.

The damascene Cu lines [Fig. 1(b)] consisted of a (001)-Si wafer (670 μm-thick), a $SiO_{x \approx 2}$ film (0.5 μm-thick), a Si_3N_4 layer (0.1 μm-thick), $SiO_{x \approx 2}$ trenches (0.5 μm-depth), a 500 Å-thick TaN film, a 0.1 μm-thick sputtered Cu layer, and an electroplated Cu film. A more detailed description of the fabrication process can be found in Refs. 5 and 8. Three different linewidths, w, were included in this study: 300, 500, and 800 nm. The depth of the trenches, h_l, was 500 nm in all samples. The distance between the centers of adjacent Cu lines was equal to $2w$.

The stresses in the films and lines during thermal cycling were obtained by measuring the curvature changes of the substrates using the laser scanning technique.[9] The films and the lines were thermally cycled in a reducing ambient provided by a flow of forming gas (20 % H_2, 80 % N_2) while the curvature of the samples was being measured. The heating and cooling rate was kept constant at 6 °C/min. In the case of the films, an equibiaxial stress state develops, and the magnitude of the biaxial stress can be calculated from the measured curvature using Stoney's equation.[10] In the case of the lines, a complex three-dimensional stress state is present during thermal cycling. There is no closed-form equation available to calculate the stresses *in* the Cu lines from the measured curvature changes for the general case in which inelastic deformations occur. Nevertheless, the volume-averaged stresses in the Cu-SiO_2 layer induced by the Cu lines can be calculated from the measured curvatures along and across the lines using a generalized Stoney's equation.[5] Since only the Cu in the lines was found to deform inelastically,[5] the volume-averaged stresses provide direct information on the inelastic behavior of the Cu lines.

In-situ TEM experiments using thin films deposited on micromachined Si membranes were used to investigate the characteristics of the dislocation motion during thermal cycling. More information about these experiments can be found in Refs. 4 and 6. The TEM samples were subjected to thermal cycles similar to the ones used for the curvature measurements.

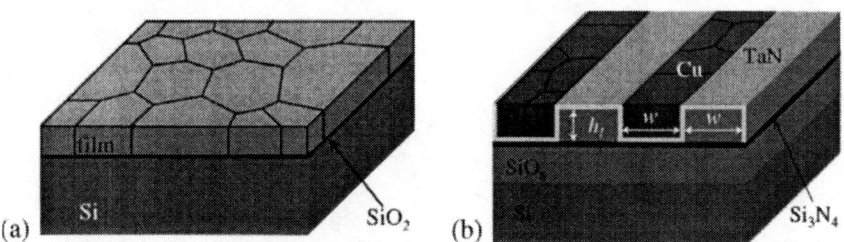

FIGURE 1. Schematic diagram of the samples. (a) Thin film on a substrate. (b) Damascene Cu lines, with linewidth w and height h_l.

Ag and Cu thin films have a strong (111) texture, and their grain structure is two dimensional in nature,[11] as shown schematically in Fig. 1(a). After annealing at 500 °C, the average in-plane grain size is about three times the film thickness.[11,12] No further grain structure evolution occurs after the first heating of the as-deposited films.[11] The average grain size of the Cu lines considered in this work is on the order of w, and, consequently, one grain spans the full trench width along most of the length of the lines.[13]

MECHANICAL BEHAVIOR DURING THERMAL CYCLING

Figure 2 shows a typical result for the evolution of the biaxial stress with temperature in an uncapped Ag film during thermal cycling between −50 and 500 °C. The in-plane biaxial strain rate, $\dot{\varepsilon}_{appl}$, imposed on the film by the substrate is given by:

$$\dot{\varepsilon}_{appl} = \left(\alpha_s - \alpha_f\right)\dot{T}, \tag{1}$$

where α_s and α_f are, respectively, the thermal expansion coefficients of the substrate and film, and $\dot{T}$ is the rate of temperature change. During the first heating from room temperature to 500 °C (not shown in Fig. 2), compressive stresses are generated in the film. After the first heating, cooling and heating produces a repeatable loop, which is shown in Fig. 2. During cooling from 500 °C to −50 °C, tensile stresses are present in the films. During heating from −50 °C, a linear elastic behavior is first observed. The slope of this linear regime is given by the applied strain times the equibiaxial elastic modulus of the film.[9] Figure 2 shows that a significant deviation from elastic behavior occurs before the stress changes from tensile to compressive, which has been attributed to a Bauschinger-type of behavior.[14]

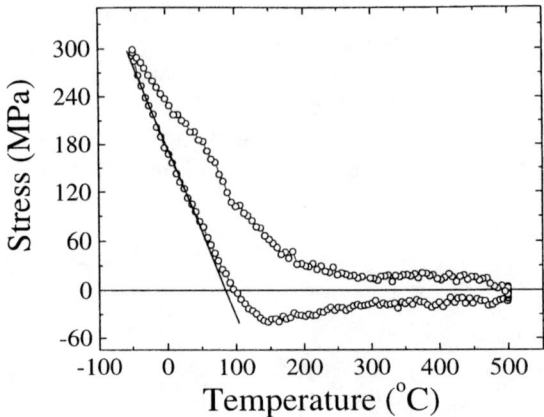

FIGURE 2. Biaxial stress in an uncapped 500 nm-thick Ag film on an oxidized Si substrate, as a function of temperature. The line in the plot is shown to facilitate the identification of the deviation from linear behavior.

It is interesting to compare the mechanical behavior of Ag thin films with that of Cu thin films and damascene Cu lines. Figs. 3 and 4 show, respectively, the stress evolution during thermal cycling for a 470 nm-thick Cu film, and for 500 nm-height damascene Cu lines. It can be observed that the three types of samples display a similar behavior, which seems to be representative of uncapped FCC metallic structures with reduced dimensionality. Figs. 2–4 show a relatively low and almost constant stress at high temperatures, and a rapid increase of the stress with decreasing temperature at low temperatures. Consequently, it is reasonable to conclude that the same inelastic mechanisms are dominant in these samples. This type of behavior is also exhibited by uncapped Au films.[15]

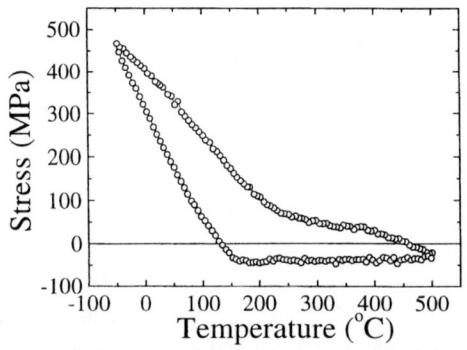

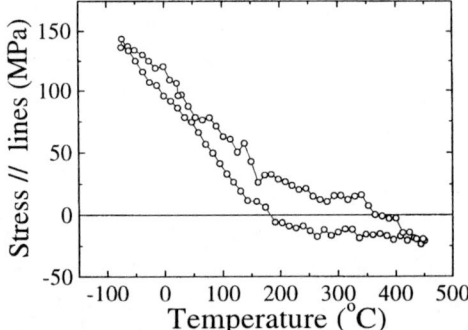

FIGURE 3. Biaxial stress in an uncapped 470 nm-thick Cu film on an oxidized Si substrate, as a function of temperature.

FIGURE 4. Cu-induced volume-averaged stress along damascene Cu lines during thermal cycling. The width and the height of the lines were 500 nm. The substrate is a Si wafer.

208

It is important to remark that the mechanical behavior shown in Figs. 2–4 is mostly inelastic, which is indicated by the fact that the stress-temperature curves are not a single straight line.

DOMINANT INELASTIC MECHANISMS

Despite the attention that the inelastic behavior of metallic thin films and lines has received in the past decades, it is still not well understood. The ultimate goal is to obtain constitutive equations based on the actual physical processes that generate inelastic strains in these metallic structures with reduced dimensionality. A conventional approach consists on proposing one or more inelastic mechanisms and then fitting the unknown parameters of the model to the experimental results, which usually results in good agreement between model and experimental data (see, for example, Refs. 16–18). Unfortunately, models based on different inelastic mechanisms can fit the same experimental results with similar accuracy. Consequently, the dominant inelastic mechanisms are difficult to identify using this approach.

To identify the dominant inelastic mechanisms that generate inelastic strains, we focus our attention on the stress evolution during cooling (after annealing at 500 °C), which allows us to avoid unnecessary complexities arising from microstructure evolution during the first heating, and from Bauschinger-type of effects during heating from low temperatures.[14] Figure 5 shows typical result for the stress evolution during cooling for uncapped Ag films with three different thicknesses.

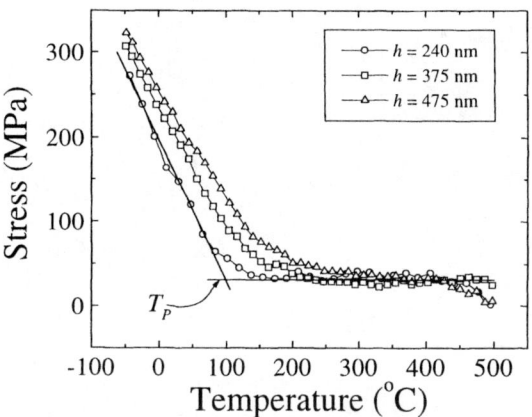

FIGURE 5. Biaxial stress in uncapped Ag films with different thicknesses, as a function of temperature, during cooling from 500 °C. The figure illustrates the definition of the transition temperature T_P (shown only for the thinnest film), and its dependence on film thickness, h.

From the results of Fig. 5, it is possible to identify two different temperature regimes, in which different inelastic mechanisms are dominant.[3] At high temperatures, the stress is approximately constant and relatively low, on the order of a few tens of MPa, while at low temperatures, the stress rapidly increases with decreasing temperature. A transition temperature, T_P, between these two regimes can be operationally defined as shown in Fig. 5. In the following two sections, we will first discuss the high-temperature regime, and then will focus on the low-temperature regime.

High-Temperature Inelastic Regime

At high temperatures (above T_P), because of the fine microstructure of the films and lines (the average in-plane grain sizes are typically below a μm), diffusional creep is expected to be important. Figure 6 schematically shows how diffusional processes can relax stress in thin films and lines.

In the past, a number of diffusional creep models that provide the strain rate as a function of temperature and stress have been reported (see, for example, Refs. 16, 19, 20). Although the models were developed under different assumptions (i.e. occurrence or absence of sliding at the film/substrate interface, grain boundary diffusion rate-controlled, surface diffusion rate-controlled), a common feature is that the predicted strain rate depends on h^{-3}, where h is film thickness, and on temperature in an Arrhenius manner. Unfortunately, because the experimentally determined stress evolution during thermal cycling is almost featureless at high temperatures (i.e. almost a flat line), it is difficult to prove that diffusional creep is actually taking place in these metallic structures by fitting the models to the experimental results. Instead, we focused on the thickness dependence of the transition temperature, and found that it decreases with decreasing film thickness, as shown in Fig. 5.[3]

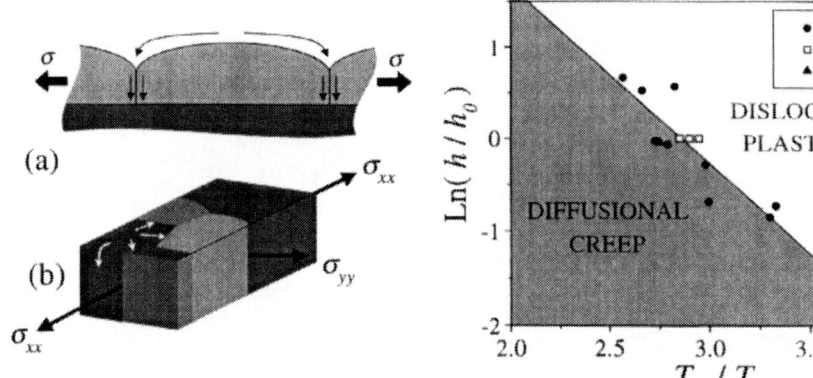

FIGURE 6. Schematic illustration of the diffusional creep process. The thin arrows indicate flow of matter. (a) In thin films. (b) In lines.

FIGURE 7. Log(h) as a function of T_P^{-1} for Ag and Cu films, and for damascene Cu lines. h is thickness of the films or the height of the lines. The transition temperatures, T_p, were normalized using the absolute melting temperatures of the materials, T_M. $h_0 = 500$ nm.

The activation energy for this process can be found by plotting the log of the thickness as a function of the inverse of the transition temperature[3] (see Fig. 7). Figure 7 shows experimental results for Ag and Cu films, as well as for damascene Cu lines. All the data fall along the same line because the transition temperature was normalized by the melting temperature to account for differences in activation energies for diffusion, which, in metals, typically scale with the melting temperature.[21] For the case of Ag films, we found a value of 0.6 eV for the activation energy, which is consistent with diffusional creep processes.[3] Additional evidence for the occurrence of diffusional creep can be found by depositing an adherent capping layer on the films or lines to suppress or hinder this diffusional process by removing the presence of a free surface. It has been found that when a TaN capping layer is deposited on top of damascene Cu lines, the stress plateau at high temperatures disappears,[5] providing compelling evidence for the occurrence of diffusional creep in these metallic structures at high temperatures.

Low-Temperature Inelastic Regime

Being a thermally activated process, the rate of plastic strain generated by diffusional creep rapidly decreases with temperature, and at low temperatures (below T_P) cannot accommodate the applied (constant) strain rate given by Eq. 1. As a consequence, the stress rapidly increases with decreasing temperature. However, the behavior of the films and lines is not elastic, indicating that another inelastic mechanism is generating a measurable amount of plastic deformations.

We identified the dominant inelastic mechanism at low temperatures using in-situ TEM, and by calculating the activation volume from stress relaxation experiments.[4] In-situ TEM experiments showed a profuse jerky-type of dislocation motion,[4,6,22] as illustrated in Fig. 8, which is indicative of a thermally activated motion of dislocations through obstacles.[23] Experimental evidence suggests that the obstacles are forest dislocations.[4] It was found that dislocations are pinned at different points, and do not move most of the time, until a segment of dislocation of length 2λ (see Fig. 8) overcomes a pinning point with the aid of thermal activation, and moves forward. Figure 9 shows the distribution of lengths of moving dislocation segments, $f_h(\lambda)$, measured in an uncapped Ag film at room temperature, after thermal cycling.

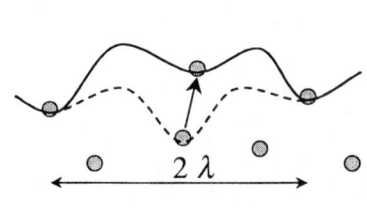

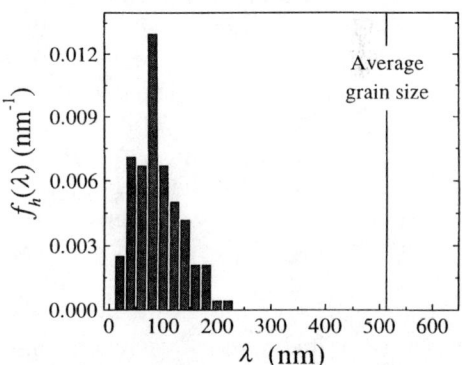

FIGURE 8. Schematic illustration of jerky-type of dislocation motion. The points represent obstacles (forest dislocations); the dashed and full lines represent, respectively, the initial and final positions of the dislocation.

FIGURE 9. Distribution of lengths of moving dislocation segments, $f_h(\lambda)$, measured for an uncapped Ag film at room temperature, after thermal cycling. The vertical line indicates the in-plane average grain size.

Activation volumes provide information about the *average* (over the entire film) lengthscale over which the dislocation-motion events take place. The measured values were found to be on the order of 100 b^3, where b is the length of the Burgers vector. These values are consistent with thermally activated glide of dislocations through forest dislocation obstacles, and are in agreement with the results obtained from the in-situ TEM experiments.[4] It is relevant to mention that the activation volume and the distributions of λ are related to each other.[4,6,23]

It is important to emphasize that Fig. 9 shows that the measured λ's are significantly smaller than the average in-plane grain size. Because the stress that thin films and lines can sustain depends on λ^{-1}, the results of Fig. 9 explain their high strength.

SIZE EFFECTS

The dependence of the diffusional creep mechanism on film thickness, h, and in-plane average grain size, g, is well understood.[16,19,20] For metallic thin films, if grain boundary diffusion is the rate controlling process, the strain rate is proportional to $g^{-1}h^{-2}$.[16,19,20] On the other hand, if surface diffusion is rate controlling, the strain rate is proportional to $g^{-2}h^{-1}$.[16] However, because the in-plane average grain size is usually proportional to the film thickness, the inelastic strain rate scales with h^{-3} in both cases. In the case of the damascene Cu lines considered in this study, the average grain size is proportional to the linewidth. Because the linewidth was varied over a small range of values, and the height of the lines was kept constant, it was not possible to determine conclusively the size dependence of diffusional creep for the lines (i.e. $w^{-1}h^{-2}$ or $w^{-2}h^{-1}$). Finally, in the case of the lines, there is an additional size effect at high temperatures, which results from the fact that one grain spans the full trench width along most of the length of the lines.[5] Because of this characteristic, there is an absence of grain boundaries aligned parallel to the lines. Since the relaxation of the stress component normal to the lines requires diffusion of matter from the surface into grain boundaries that are not perpendicular to the length of the line, the bamboo type of microstructure of the Cu lines hinders the stress relaxation across the lines. It is important to emphasize that this conclusion does not assume grain-boundary diffusion to be the rate controlling process. Consequently, diffusional creep is more effective relaxing stresses along the length of the lines than across the lines.[5]

At low temperatures, a common approach to investigate thickness effects has been to plot the room temperature stress in the films after thermal cycling, σ_{RT}, as a function of film thickness. Early investigations, mostly performed on Al, found that the stress approximately depends on the inverse of the thickness.[9,24,25] In a sense, it can be said that the strength of the films increases with decreasing film thickness. However, for the case of uncapped Ag[3] and Au[15] films, it has been found that σ_{RT} does not monotonically increase with decreasing film thickness, but instead reaches a maximum value and then decreases with decreasing film thickness, as shown in Fig. 10. The departure of the behavior of the thinner uncapped films from a monotonic increase of σ_{RT} with decreasing h can be understood by considering the fact that diffusional creep becomes more important with decreasing film thickness. At room temperature, the fraction of applied strain that has been accommodated by diffusional creep (during cooling from 500 °C) increases with decreasing h, and is substantial for the thinner uncapped films. This also explains why this decrease of σ_{RT} with decreasing h is not observed in capped Ag films, and in Al films, which have a natural oxide layer that acts as a capping layer.

The increase of σ_{RT} with decreasing film thickness in the case of thick uncapped films and capped films indicates that the low-temperature dislocation plasticity mechanism depends on film thickness. We have found that the distributions of λ depend on film thickness, and that the mode of the distributions, λ_m (i.e. the most probable value of λ), increases with increasing film thickness,[6] as shown in Fig. 11. A similar conclusion was obtained by investigating the thickness dependence of the activation volume.[6]

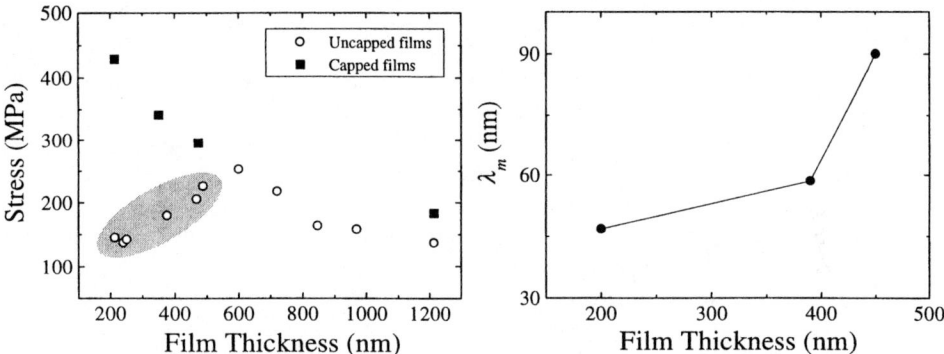

FIGURE 10. Room temperature stress in Ag thin films after thermal cycling, σ_{RT}, as a function of film thickness, for capped and uncapped films. The shaded area encloses the cases that deviate from the monotonic increase of σ_{RT} with film thickness.

FIGURE 11. λ_m as a function of film thickness, obtained from in-situ TEM experiments on uncapped Ag films. The measurements were performed at temperatures of approximately 100 °C, after cooling from 400 °C.

Because the stress that films can sustain increases with decreasing λ_m, the increase with thickness of the characteristic lengthscale for dislocation plasticity (λ_m) explains why the "strength" of thin films decreases with increasing h. However, the physical origin of the thickness dependence of λ_m has still to be understood. The thickness dependence of λ_m suggests that the forest dislocation density increases with decreasing film thickness, which can be rationalized in different ways. A possible explanation could be that larger dislocation densities are necessary near grain boundaries to satisfy compatibility of strains. This effect becomes more important with decreasing grain size.[6,26] Another possible explanation is based on the observation that, in experiments[27] and simulations,[28] dislocations nucleate at grain boundaries and/or grain boundary triple junctions (where three grains meet). This effect also becomes more important with decreasing grain size. However, more work is necessary to fully answer this important question.

Finally, it is relevant to mention that in the case of damascene Cu lines, we have observed that, at low temperatures, the amount of inelastic deformations decreases with decreasing linewidth, which is possibly caused by the additional constraint imposed on dislocation motion by the presence of trenches with widths comparable to the measured values of λ.[5]

DEVELOPMENT OF A CONSTITUTIVE EQUATION FOR DISLOCATION-MEDIATED PLASTICITY

The high-temperature diffusional creep mechanism has been modeled by many authors, and is relatively well understood. On the contrary, at low temperatures, a physically based constitutive equation is still needed. In this section, we will present a constitutive equation for thin films, which will be developed in more detail elsewhere.[29]

With the goal of developing a simple model based on experimental results, the following postulates, which are inspired by experimental results, were proposed:

1. Inelastic deformations are generated by the thermally activated glide of segments of dislocations through forest dislocation obstacles.
2. The dislocation density reaches a saturation value during the first heating, and then remains approximately constant.
3. Most of the inelastic events occur in the "interior" of the grains.

Postulate 2 is based on in-situ TEM observations, which shows only a small change in the dislocation density during cooling, after annealing at 500 °C. Postulate 3 is suggested by the fact that the length of the moving dislocation segments (and the distances that they move forward) is significantly smaller than the film thickness and average in-plane grain size. Consequently, postulate 3 is equivalent to neglecting the direct influence of image forces (arising from the presence of the surface, grain boundaries, and the film/substrate interface) on the moving dislocation segments. The model is illustrated in Fig. 12, which shows two dislocations moving in a slip plane through forest dislocation obstacles. In addition, to simplify the problem further, let us replace the apparently random distribution of pinning points in the slip plane with a periodic array of obstacles such that the distance between pinning points along the dislocation line is λ_m. Under these conditions, the inelastic biaxial strain rate, $\dot{\varepsilon}_{pl}$, is given by:[26,29]

$$\dot{\varepsilon}_{pl} = \dot{\varepsilon}_o \exp\left[-\frac{\Delta G^*(\tau, \lambda_m)}{kT}\right],$$ (2)

where T is temperature, k is the Boltzmann's constant, τ is the shear stress in the slip plane, $\dot{\varepsilon}_o$ is a constant, and ΔG^* is the Gibbs free energy of activation for the dislocation-dislocation cutting process, which depends on the shear stress acting on the moving dislocations and on the distance between the pinning points.[23] ΔG^* can be calculated from the force-distance curves, which provide the force that dislocations "feel" as they move towards each other. Unfortunately, the force-distance curves for FCC metallic films such as Cu and Ag are not well known. Based on experimental results for Cu,[30] it is reasonable to use a linear force-distance curve, which gives:[26]

$$\Delta G^*(\tau, \lambda) = \hat{K} b \left(1 - \frac{\tau b \lambda_m}{\hat{K}} \right)^2,$$ (3)

where $\hat{K}$ is the maximum force that the obstacles can sustain, and is approximately 0.32 μb^2, where μ is the shear modulus of the metal.[30,31] A more rigorous model should take into account the fact that there is a distribution of strengths of obstacles[32] and a distribution of lengths of moving dislocation segments.[26] Equations (2) and (3) provide the rate of generation of inelastic strain as a function of temperature and stress. It is important to notice that size effects are taken into account in Eq. (3) by incorporating the thickness-dependent obstacle spacing, λ_m, which can be approximately obtained from Fig. 11 and from the results reported in Ref. 6.

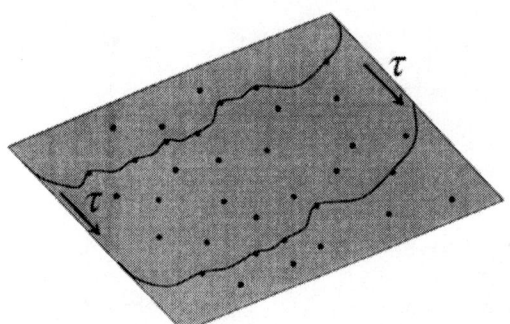

FIGURE 12. Schematic illustration of the proposed model for dislocation-mediated plasticity in thin films. The figure shows two dislocations gliding in the slip plane (gray surface) through an array of forest dislocation obstacles, under an applied shear stress τ.

CONCLUSIONS

We found that Ag and Cu films, and damascene Cu lines, display a similar mechanical behavior, which is expected to be representative of FCC metallic thin films and lines. Two different inelastic mechanisms were identified: diffusional creep at high temperatures (approximately above 0.3 of the absolute melting temperature), and dislocation plasticity controlled by thermally activated glide of dislocations through forest dislocation obstacles at low temperatures. A simple inelastic constitutive equation for the low-temperature dislocation plasticity was derived. The model takes into account the thermal nature of the dislocation-mediated inelastic mechanism in thin films and lines, and incorporates the thickness-dependent characteristic lengthscale of dislocation-mediated plasticity.

ACKNOWLEDGMENTS

The authors would like to thank Steve Seel for useful discussions. This work was supported by the NSF through Contract No. DMR-9710139.

REFERENCES

1. Vaidya, S., Sheng, T., and Sinha, A. K., *Appl. Phys. Lett.*, **36**, 464 (1980).
2. Thompson, C. V. and Lloyd, J. R., ed., *"Materials Reliability in Microelectronics II"*, Materials Res. Soc. Proceedings, San Francisco, 1992.
3. Kobrinsky, M. J. and Thompson, C. V., *Appl. Phys. Lett.* **73**, 2429 (1998).
4. Kobrinsky, M. J., and Thompson, C. V., *Acta mater* **48**, 625 (2000).
5. Kobrinsky, M. J., Thompson, C. V., and Gross, M. E., *J. Appl. Phys.* **89**, 91 (2001).
6. Kobrinsky, M. J., Dehm, G., Thompson, C. V., and Arzt, E., *Acta mater.* **49**, 3597 (2001).
7. Blanchet, A., Masters thesis, Department of Materials Science and Engineering, Massachusetts Institute of Technology, Cambridge, MA (1998).
8. Lingk, C., and Gross, M. E., *J. Appl. Phys.* **84**, 5547 (1998).
9. Nix, W. D., *Metall. Trans.*, **20A**, 2217 (1989).
10. Stoney, G. G., *Proc. R. Soc. Lond.* **A82**, 172 (1909).
11. Carel, R., Ph.D. thesis, Department of Materials Science and Engineering, Massachusetts Institute of Technology, Cambridge, MA (1995).
12. Thompson, C. V., *Annu. Rev. Mater. Sci.* **30**, 159 (2000).
13. Gross, M. E., Lingk, C., Siegrist, T., Coleman, E., Brown, W. L., Ueno, K., Tsuchiya, Y., Itoh, N., Ritzdorf, T., Turner, J., Gibbons, K., Klawuhn, E., Biberger, M., Lai, W. Y. C., Miner, J. F., Wu, G., Zhang, F., MRS Conference Proceedings **514**, 1998, p. 293.
14. Baker, S. P., Keller, R. M., and Arzt, E., in *Thin Films-Stresses and Mechanical Properties VII*, MRS Conference Proceedings, Boston 1997, p. 605.
15. Leung, O. S., Munkholm, A., Brennan, S., and Nix, W. D., *J. Appl. Phys.* **88**, 1389 (2000).
16. Thouless, M. D., *Acta Metall. Mater.* **41**, 1057 (1993).
17. Flinn, P. A., Gardner, D. S., and Nix, W. D., *IEEE Trans. on Electron Devices* **34(3),** 689 (1987).
18. Keller, R.-M, Baker, S. P., and Arzt, E., *Acta mater.* **47**, 415 (1999).
19. Turlo, J., Ph.D. thesis, Department of Materials Science and Engineering, Stanford University, Stanford, CA (1992).
20. Gao, H., Zhang, L., Nix, W. D., Thompson, C. V., and Arzt, E., *Acta mater.* **47**, 2865 (1999).
21. Brown, A. M., and Ashby, M. F., *Acta Metall.* **28**, 1085 (1980).
22. Dehm, G. and Arzt, E., *Appl. Phys. Lett.* **77**, 1126 (2000).
23. Argon, A. S., "Mechanical Properties of Single-Phase Crystalline Media: Deformation at Low Temperatures", in *Physical Metallurgy*, edited by Cahn, R. W. and Haasen, P., Amsterdam: North-Holland, 1996.
24. Doerner, M. F., Gardner, D. S., and Nix, W. D., *J. Mater. Res.* **1**, 845 (1986).
25. Venkatraman, R. and Bravman, J. C., *J. Mater. Res.* **7**, 2040 (1992).
26. Kobrinsky, M. J., Ph.D. thesis, Department of Materials Science and Engineering. Massachusetts Institute of Technology, Cambridge, MA (2001).
27. Owusu-Boahen, K., and King, A. H., *Acta mater.* **49**, 237 (2001).
28. Caro, A., Van Swygenhoven, H., Derlet, P., Scherrer, P., Farkas, D., Caturla, M. J., and Diaz de la Rubia, T., to be published in the Proc. of the Fall MRS Conference, Boston, 2000.
29. Kobrinsky, M. J. and Thompson, C. V., "A Low-Temperature Physically Based Constitutive Equation for Polycrystalline Metallic Thin Films", *in preparation*.
30. Argon, A. S. and East, G. H., in *Proceedings of the 5th International Conference on the Strength of Metals and Alloys*, edited by Haasen, P., Gerold, V., and Kostorz, G., Vol. 1, 1979, p. 9.
31. Frost, H. J., and Ashby, M. F., *Deformation-mechanism maps: the plasticity and creep of metals and ceramics*, 1st ed., Pergamon Press, 1982.
32. Shenoy, V. B., Kukta, R. V., and Phillips, R., Phys. Rev. Lett. **84**, 1491 (2000).

High Resolution Microdiffraction Studies Using Synchrotron Radiation

R. Spolenak[1,*], N. Tamura[2], B. C. Valek[3], A.A. MacDowell[2], R. S. Celestre[2], H.A. Padmore[2], W.L. Brown[1], T. Marieb[5], B.W. Batterman[2,4], and J.R. Patel[2,4]

[1] Agere Systems, formerly of Bell Laboratories, Lucent Technologies, Murray Hill, NJ
[2] ALS/LBL,1 Cyclotron Road, Berkeley, CA
[3] Dept. of Mat. Sci. & Eng., Stanford University, Stanford, CA
[4] SSRL/SLAC, Stanford University, Stanford, CA
[5] Intel Corp., Portland, OR
* currently Lehigh University, Bethlehem, PA

Abstract. The advent of third generation synchrotron light sources in combination with x-ray focusing devices such as Kirkpatrick-Baez mirrors make Laue diffraction on a submicron length scale possible. Analysis of Laue images enables us to determine the deviatoric part of the 3D strain tensor to an accuracy of $2x10^{-4}$ in strain with a spatial resolution comparable to the grain size in our thin films. In this paper the application of x-ray microdiffraction to the temperature dependence of the mechanical behavior of a sputtered blanket Cu film and of electroplated damascene Cu lines will be presented. Microdiffraction reveals very large variations in the strain of a film or line from grain to grain. When the strain is averaged over a macroscopic region the results are in good agreement with direct macroscopic stress measurements. However, the strain variations are so large that in some cases in which the average stress is tensile there are some grains actually under compression. The full implications of these observations are still being considered, but it is clear that the mechanical properties of thin film materials are now accessible with new visibility.

INTRODUCTION

In the past years several theories have been developed to describe thin film plasticity [1-4]. They discuss the influence of film thickness and grain size as well as the average grain orientation (texture) on the onset of plastic yielding. All of those models, however, are based on describing average phenomena. Experimental studies have been reported that investigate local phenomena [5,6] such as local dislocation density, dislocation-dislocation and dislocation-interface interaction and local glide planes by TEM studies. Some groups are even measuring the local stresses at the high spatial resolution given by the TEM using convergent beam electron diffraction (CBED) [7-9]. Due to the limitation of statistics in TEM data (the number of grains measured) the insight gained can be incorporated into material responses only through macroscopic models.

The relatively new technique of X-ray microdiffraction on a submicron length scale as developed by [10-14], provides a unique opportunity for studying local effects, but

CP612, *Stress-Induced Phenomena in Metallization:* Sixth Int'l. Workshop, edited by S. P. Baker et al.
© 2002 American Institute of Physics 0-7354-0058-X/02/$19.00

on a coarser spacial scale than TEM. Since the focal spot of the X-rays is comparable to or smaller than the grain size investigated and since the rate of data acquisition is high enough to measure a relative large number of grains (~100), local phenomena can be measured and also correlated to each other, i.e. the interaction of grains with one another can be examined.

In this paper the technique itself will first be described. We will then present the results of a temperature cycle on thin Cu films and lines on a Si substrate. Mechanical stresses are introduced as a consequence of the mismatch in thermal expansion coefficient between the substrate and the metal. Varying geometries and microstructures have different effects on the onset of plastic deformation on one hand and on the complexity of the 3D stress state on the other. This set of experiments has been chosen because the technique allows for the *in-situ* observation of changes in stress at the same location, i.e. within the same individual grains and their neighbors.

EXPERIMENTAL

Description Of Beamline 7.3.3 At The ALS

These experiments were all carried out on beamline 7.3.3 at the Advanced Light Source (ALS) at Lawrence Berkeley National Laboratory. The X-ray synchrotron beam from a bending magnet source is focused via a pair of bendable Kirkpatrick-Baez mirrors to a submicron spot ($0.7x0.8$ μm^2 FWHM). A 4-crystal Si (111) monochromator is used to switch between white and monochromatic beams illuminating the same area on the sample. In this paper, however, only the white beam capabilities will be utilized.

The sample is mounted in a 45° reflective geometry. We are interested in aluminum and copper materials, which have a grain size the order of the x-ray beam size. Single crystal Laue diffraction takes place and sets of Bragg reflections are produced for each grain or sub-grain that is illuminated. These diffraction spots are detected by a large area CCD detector ($9x9$ cm^2 active area from Bruker Inc.). The number of reflections collected per diffracting grain depends on the solid angle covered by the detector and the energy range available from the synchrotron source. In this case the detector is usually placed at a distance of 30 mm from the sample and the energy range of the primary beam is 6 to 13 keV. For Cu this leads to 8-12 reflections. For Al with its larger lattice constant the number is even larger. More reflections can also be found for crystals with lower symmetry. As X-rays are highly penetrating, the entire thickness of a thin film can be examined. Even passivating layers over the film of interest do not interfere with the measurements and no special sample preparation is necessary. To investigate a large region of a sample, the sample is stepped in x and y under the focused x-ray spot.

Once a "frame" (the Laue spot pattern for one sample position, which takes about 1 second for our copper samples) has been taken, the set of Laue spots for each grain within the illuminated area can be automatically indexed thus determining its orientation. By comparing the predicted spot positions for a perfect crystal (with the symmetry of the material investigated) and the actual positions, shear distortions of the crystal can be quantified. This yields 5 components of the 3D strain tensor, i.e. the deviatoric part. The actual lattice parameter of a particular lattice plane can be determined by scanning the energy utilizing the above mentioned monochromator. This provides the dilatational component of the complete strain tensor. Details of this method are described elsewhere [15]. The strain tensor is subsequently converted into a stress tensor by application of the stiffness tensor for the material investigated.

Maps in stress or orientation are produced from the sequential frames of the x-y sample scan. This sample translation can be carried out with a precision of 0.1 μm with a piezo-driven stage. In addition to scanning the sample, the stage allows for *in-situ* heating and the application of electrical current to the sample.

A solid state detector coupled with a multichannel analyzer allows for the parallel collection of fluorescence signals, which can be used for chemical analysis and sample positioning. A software package developed at the ALS (X-MAS for X-Ray Microdiffraction Analysis Software) by one of the authors [16] is used for data collection, Laue pattern indexing, strain refinement and monochromatic beam scans. For polycrystals, an orientation map "smoothing" algorithm also allows for the automatic determination of grain boundaries by fitting the intensity profile of each individual crystal grain in the stepping map and intersecting the resulting normalized profiles.

Sample Preparation

The first sample is a 1.5 μm film of sputtered Cu deposited on 200 nm of CVD silicon nitride as a barrier layer on a substrate silicon wafer In order to choose and find an area of interest (in this case, (100) oriented grains surrounded by (111) oriented grains) the sample has been imaged as well as marked (by sputtering of material) by a focused ion beam (FIB) system.

For studies of the influence of geometrical constraints on material deformation in temperature cycling, damascene electroplated Cu lines with several different widths have been prepared. As these are also designed for electromigration experiments so-called 'Blech' structures have been produced. Cu segments of varying length are connected by Ta links. For the preparation of the structures on top of the underlying Ta links, silicon oxide has been patterned into trenches. These trenches were coated with a Ta barrier layer and a sputtered Cu seed layer. Then the structures were electroplated after which the excess Cu was removed by a chemical mechanical polishing step. After an ammonia plasma clean the samples were passivated by 200 nm of plasma enhanced CVD- silicon nitride. A schematic cross-section of the layers can be seen in Fig. 1.

All samples were annealed at 400 °C for 30 minutes in forming gas to stabilize the microstructure. For the damascene lines this step happened before the CMP process.

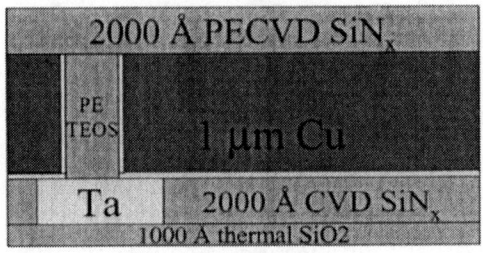

Si

FIGURE 1. Schematic of the stack structure for the electroplated Cu damascene lines.

Experimental Procedure

The general experimental concept has been outlined above. In the case of an *in-situ* thermal cycle some additional steps have to be taken. First, the sample is mounted using a high temperature epoxy or cement to minimize sample drift. Rough alignment is carried out by a long focal length optical microscope. This, however is just barely sufficient to align the beam roughly with the area of interest. Subsequently, a fluorescence scan was performed in order to accurately locate the structures. In the case of a continuous film, markers as described above are needed. For both kinds of samples Cu K_α fluorescence radiation was measured . The authors are aware that making index marks in a sample with an FIB can lead to Ga contamination and a change in mechanical properties. However, we believe that for the qualitative concepts reported here these contamination effects are not important.

After the area of interest has been found, Laue images are collected by scanning the sample relative to the beam. Upon completion of the map the sample is heated to the next temperature and both fluorescence and Laue scans are repeated. The procedure is applied repeatedly to the maximum temperature and back down.

Although our main interest is in examining the local variations in stress, the average of our measurements and its comparison with macroscopic measurements is valuable. The main axes of the stress tensor are shown below split into a deviatoric and a hydrostatic part.

$$
\begin{pmatrix} \sigma_{xx} & 0 & 0 \\ 0 & \sigma_{yy} & 0 \\ 0 & 0 & \sigma_{zz} \end{pmatrix} = \begin{pmatrix} \sigma'_{xx} & 0 & 0 \\ 0 & \sigma'_{yy} & 0 \\ 0 & 0 & \sigma'_{zz} \end{pmatrix} + \begin{pmatrix} \sigma_h & 0 & 0 \\ 0 & \sigma_h & 0 \\ 0 & 0 & \sigma_h \end{pmatrix} \tag{1}
$$

where the dash (') denotes the deviatoric components and the h subscript the hydrostatic stress. By definition, the trace of the deviatoric stress tensor has to be zero. If we now average all the data, we can assume that the average stress out-of-plane σ_{zz} has to be zero to satisfy the boundary condition of a free surface. Furthermore, in-plane the stress state has to be biaxial. The Laue measurements provide only the deviatoric part of the tensor, but the average biaxial stress can be determined in the following way:

Since

$$\sigma_{zz} = 0$$
$$\sigma_{xx} = \sigma_{yy} = \sigma_b$$

(2)

apply to a blanket thin film, combining the two equations, the average biaxial stress σ_b can be defined as:

$$\sigma_b = \frac{\sigma'_{xx} + \sigma'_{yy}}{2} - \sigma'_{zz}$$

(3)

This analysis is used in the following to evaluate the average data of the temperature cycles. The biaxial stress, however, can only be determined when the geometrical constraint does not violate Eq. 2, as it clearly does in the case of narrow damascene lines. The averaging as used in the results sections is done by creating a histogram of stress states (see Fig. 4) of all available data points. The median of the distribution is then obtained by plotting the percentile of stress states as a function of the stress state and applying a sigmoidal fit. This avoids having a few high stress values in the histogram unreasonably influence the result.

RESULTS

Sputtered Cu Thin Film

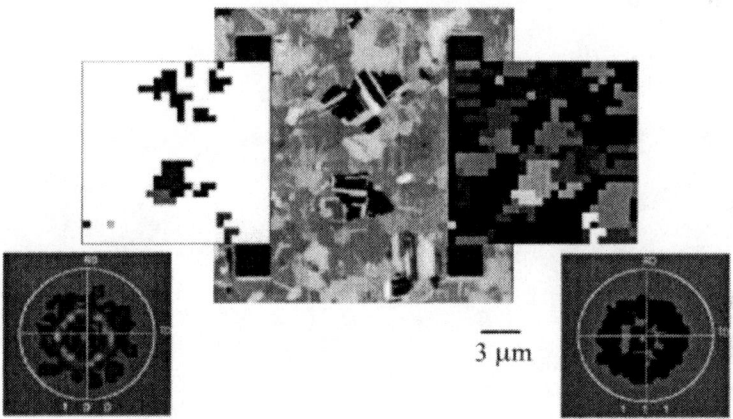

3 μm

FIGURE 2. FIB picture of the 18 μm x 18 μm area investigated (Black areas show (100) oriented grains), two microdiffraction orientation maps and a (100) and (111) pole figure showing the two texture components of the film.

The first experiment investigates the stress state in a sputtered blanket Cu film. Fig. 2 shows the high resolution microstructure through the channeling contrast in an FIB. Two large (100) grains can be identified. The identifying properties are the dark channeling contrast and the 90^0 angle between the twin planes ((111) planes) when they intersect the surface. In the case of (111) oriented grains out of plane this angle would be 60°. To the left and the right of the FIB micrograph in Fig. 2 are two orientation maps as determined by X-ray microdiffraction.

The left map shows the angle between the <100> direction and the surface where black is zero degrees. Both (100) grains can clearly be identified. The pixel size in the orientation map is 0.5x0.5 µm². On the right, the in-plane orientation is shown which reveals more about the grains around the (100) grains (grayscale corresponds to angle between (100) direction and vertical axis). On the bottom left and right one finds a (100) and (111) pole figure, respectively. The pole figures, which include all of the microdiffraction data points, clearly demonstrate the two texture components of the film.

When heated above room temperature the stress state in the film starts to change in a non homogeneous way. Fig. 3 shows the change from room temperature to 125 °C. The square symbols show the averaged microdiffraction results obtained by the method described above. The round symbols show the average (111) component and the triangles the average (100) component. The closeness of the average to the (111) component is indicative of the dominant (111) makeup of the film. The thermoelastic slopes of the microdiffraction results are compared to literature values for a (111) and (100) texture. These are 3.6 MPa/K and 1.44 MPa/K, respectively [19]. The (100) component shows a smaller thermoelastic slope as well as a smaller yield stress in compression than the (111) component. The slope of the (111) component agrees well with the literature value. The slope of the (100) component is slightly higher than expected. One should bear in mind, however, that the (100) results only originate from two big grains rather than a representative set of grains as for the (111) texture.

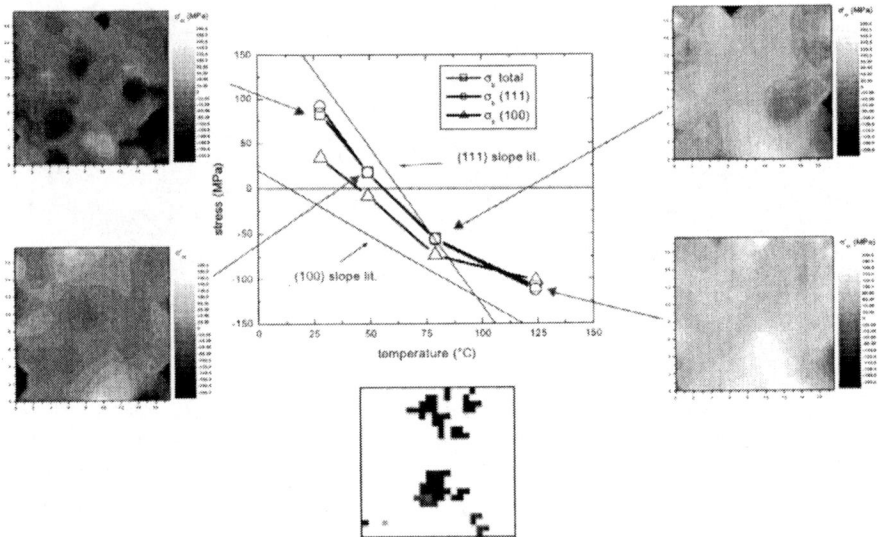

FIGURE 3. Average biaxial stress as a function of temperature in a 1.5 µm thin sputtered Cu film. Included are also the thermoelastic slopes for a (111) and (100) texture. The insets show the local deviatoric stress out of plane (white: 200 MPa, black: -200 MPa). The bottom map shows the orientation of the grains pinpointing two big (100) oriented grains in the middle.

222

Below the stress-temperature plot is a texture map showing the location of the two (100) grains. Around the stress-temperature plot are maps that show the local stress distribution at each of the four measurement temperatures. The maps show the deviatoric stress component out of plane. If this component is positive, it implies a compressive stress in plane and vice versa if it is negative. It is interesting to note that at the beginning of the experiment (at 30 °C) one can see regions where the stress state is compressive in-plane (the regions of lightest contrast) even though the median of the whole film is tensile. At higher temperatures the stress variations become smaller as the in-plane stress becomes dominantly compressive. They seem to even out, rather than just being transposed elastically. At the highest temperatures, in the compressive regime, the highest stresses develop at the bottom of the maps in a (111) oriented region, finally also extending into the lower (100) grain. The regions of the highest compressive stress do not seem to correlate with any obvious features of the initial (30 °C) stress distribution.

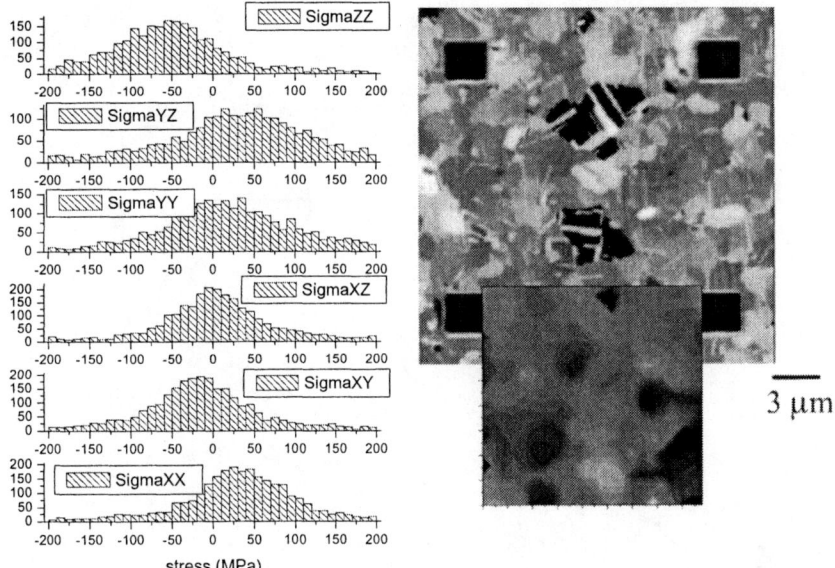

FIGURE 4. Histograms of deviatoric stress components in the analyzed area at room temperature. (The numbers on the ordinate are the number of data points). The FIB image shows the detailed microstructure and the stress map shows the deviatoric stress out of plane (zz) (black: -200 MPa, white: 200 MPa). The white area indicates a compressive stress state in-plane.

Fig. 4 shows the distribution of stresses in detail. All the different components of the deviatoric stress tensor are on the left. The cross components are zero on average. An important feature to notice is that the distributions are very broad. For the out-of-plane component σ_{zz}, for instance, several data points are even on the positive side, which implies an in-plane compressive stress state even though the median of the distribution is tensile. This can also be seen in the stress map, where the white area indicates an in-plane compressive grain. The three dark areas are strongly tensile grains. All of these have a (111) out of plane orientation. The two large (100) grains do not stand out as unique and they do not even seem to have the same stress although both are slightly in-plane tensile.

Damascene Electroplated Cu lines

So far only an on-average biaxially stressed blanket sample has been looked at. In lines rather than blanket films the scenario is quite different. Several properties are changed in the transition from a one dimensional constraint (blanket film) to a two dimensional constraint (line). Fig. 5 shows the average stress components in a line that is 5 μm wide and 1 μm high. As one can see from the similarity of the stress component across (y) and the component along the line (x), the stress state is still approximately biaxial. For a 5um wide, 1um thick line we have not yet entered a line aspect ratio in which the side walls play a major role. From the perspective of yield stress this is not too surprising as grain boundaries already act as obstacles and one would only expect a significant change when the line width becomes the order of the grain size. Because of the similarity of the x and y stress components we have calculated the in-plane biaxial stress as a function of temperature. It is plotted in Fig 5 and shows the tensile-to-compressive and back to tensile hysteresis loop that is familiar from wafer curvature measurements.

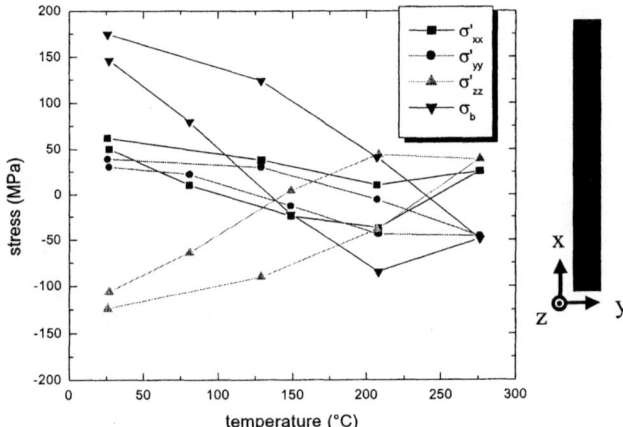

FIGURE 5. Plot of the average deviatoric stress components and the biaxial stress as a function of temperature in a 5 μm wide damascene line.

As the line width is further reduced the stress component across the line (σ'_{yy}) starts to show the influence of the additional geometrical constraint. The major part of plasticity happens along the line rather than across it. The overall stress levels increase and the area within the hysteresis loop decreases. The stress state changes from a biaxial stress state in-plane to a triaxial one.

The results for a case in which the aspect ratio is higher than one (0.8 um wide and 1 μm thick) are shown in Fig. 6. One sees first of all that the stress components across and along the line change signs. Secondly, the hysteresis virtually disappears over this temperature range. That means that the yield stress of the line is never reached in compression. Finally, the overall stress levels are increased. It is also interesting to observe that one could again call it a biaxial stress state since the two stress components across the line and out of plane have nearly the same values. The observation seems reasonable as the aspect ratio of the line is close to one. However, it still does not mean that the absolute biaxial stress can be computed, as the boundary condition of a free surface does not apply.

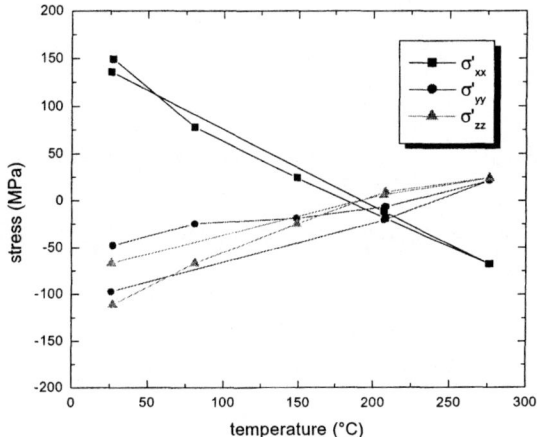

FIGURE 6. Plot of the average deviatoric stress components as a function of temperature in a 0.8 μm wide damascene line.

Looking at stress on a grain to grain basis one can see that even if the average looks smooth as shown in Figs. 5 and 6 above, the local variations are large. The continuous line in Fig. 7 shows the out-of-plane orientation of the 0.8 μm wide line. As one can see upon comparison to the square symbols that show the out-of-plane stress at room temperature, grain orientation and stress levels are definitely correlated. Local variations especially close to grain boundaries can give up to double the average stress value. When heated to 81 °C, within the precision of our measurements, the grain structure has not changed. (Since the line had been pre-annealed to 400 °C no change would be expected). At this higher temperature σ_{zz} increases, going from dominantly compressive to dominantly tensile, but also the differences in stress decrease. This is comparable to the effect one has seen for the blanket sputtered Cu film where stress levels became more uniform before new variations were introduced between different grains in the compressive regime. The stress map at the bottom indicates that stresses not only vary along the line but across the narrow line as well.

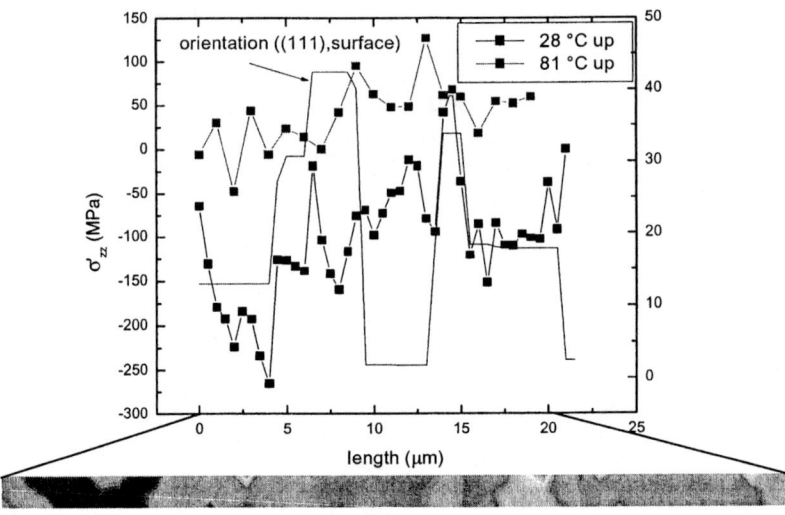

FIGURE 7. Plot of the deviatoric stress normal to the sample surface as a function of space in a 0.8 μm wide 20 μm long damascene line at two different temperatures. The continuous line shows the orientation of the grains in this bamboo line. On the bottom one sees a 2D map of the stress in the line at room temperature showing variations across the line.

DISCUSSION

By averaging microdiffraction data the same stress properties reported from macroscopic measurements are obtained, as, of course they should be. For increasingly constrained geometries (line width to line thickness ratio decreasing) the following trends are observed: a) The area of the hysteresis in stress vs. temperature plots, and thus the amount of plasticity, decreases with decreasing line width. b) The compressive yield stress of copper under increased geometrical constraint goes up. c) The sign of the deviatoric components of stress in plane changes and thus moves the maximum resolved shear stresses to different glide systems. This change in sign provides the potential for altering the macroscopic yield stress by texture control as described by [4,17].

The real merit of microdiffraction studies is not, however, in averaging their results, but in the access they give to examination of highly local variations. One of the most striking features we have observed is the wide distribution of stresses that are found even in a blanket film. Even at room temperature the deviatoric stress map out of plane is positive in local regions when that component of stress in the film as a whole is strongly negative. This implies that there are small regions of compressive (negative) stress components in-plane. In macroscopic investigations of stress vs. temperature [18] it has frequently been found that the stress begins to deviate from the thermoelastic straight line even before the average stress state becomes compressive. This phenomenon has usually been attributed to a 'Bauschinger' type behavior [18], where a dislocation pile-up causes the back flow of dislocations even before the sign of the shear stress acting on them is reversed.

The observation that locally compressive regimes do exist may add new insight in the explanation of this phenomenon. It is straight forward that already-compressive regions may start yielding much earlier then the majority component when a film is heated up, causing an early deviation from the perfectly elastic behavior in the on-average still tensile film.

An important question that arises is how such a broad stress distribution with some regions having stress of opposite sign to the average is created in a prior temperature cycle. Presumably all the deviatoric components disappear at high temperatures when diffusion is active and shear stresses are quickly relieved. So it is when the film is subsequently cooled that the distribution in stresses has to be established. In Cu there are several factors that can lead to this distribution. The first is a distribution of grain sizes, which should lead to a distribution in yield stresses. The second is the elastic anisotropy of Cu, through which a distribution in grain orientations will lead to a distribution in stresses. The last but probably not least is the fact that in stabilized metallic films the grain size is usually of the order of the film thickness. In conjuction with the continuum mechanical effect that tractions at the grain boundary have an effect towards the center of a grain to the distance of the order of the thickness, it becomes clear that grain to grain next neighbor interactions must have an important effect on the distribution of stresses as well.

An indication of this can be seen in the blanket sputtered Cu film and in the narrowest Cu line. Even when macroscopically the film is behaving elastically, differences in stress that should be conserved are smoothed out when the temperature is increased. High compressive stresses then develop at different locations. Another indication of this phenomenon is that the thermoelastic line for the (100) component (Fig. 3) does not agree with the literature data. This can be attributed to the stress state in the two (100) grains that were used for averaging not being biaxial. Hence the biaxial modulus does not apply and the local stress state relative to the crystallographic orientation of the grain has to be taken into account. An explanation for the local stress state not being biaxial can be the influence of the surrounding grains as well as the presence of crystallographic twins.

OUTLOOK

One of the missing details in understanding polycrystalline plasticity is the role of grain to grain interaction. X-ray microdiffraction offers a unique tool to investigate this phenomenon. It should be clear from the data presented above and the associated comments on it that there are many details of the relationship between stress and microstructure that we do not yet understand. We have tantalizing data which we are attempting to analyze with that objective, but simple correlations do not stand out boldly. There is no question that new insight will begin to evolve as these techniques are more widely applied.

Synchrotron microdiffraction provides a sufficiently high spacial resolution and a fast enough data acquisition rate to enable investigation of enough grains to be statistically relevant, and still perform precise single crystal diffraction. Microdiffraction retains the tremendous advantage of x-ray diffraction in not having to resort to complicated sample preparation or to compromise the structure of interest by removing passivating overlayers or alter underlayers. It has the one disadvantage that so far it is possible only at synchrotron sources which currently limits the number of researchers and experiments that can exploit its strength.

ACKNOWLEDGMENTS

The authors acknowledge the members of the Silicon Fabrication Research Lab at Agere Systems for help in sample fabrication. The Advanced Light Source is supported by the Director, Office of Science, Office of Basic Energy Sciences, Materials Sciences Division, of the U.S. Department of Energy under Contract No. DE-AC03-76SF00098 at Lawrence Berkeley National Laboratory. We thank Intel Corp. for their funding support for the end station on beam line 7.3.3.

REFERENCES

1. L. B. Freund, J. of Appl. Mech. **54**, 553 (1987).
2. W. D. Nix, Metall. Trans. A **20**, 2217 (1989).
3. C. V. Thompson, J. Mater. Res. **8** (2), 237-238 (1993).
4. J. E. Sanchez Jr. and E. Arzt, Scripta Metall. et Mater. **27**, 285-290 (1992).
5. G. Dehm and E. Arzt, Appl. Phys. Lett. **77**, 1126 (2000).
6. M. J. Kobrinsky, and C. V. Thompson, Acta mater. **48**, 625-633 (2000).
7. S. Kraemer, J. Mayer, C. Witt, A. Weickenmeier, and M. Ruehle, Ultramicroscopy **81**, in press (2000).
8. H. J. Maier, R. R. Keller, H. Renner, H. Mughrabi, and A. Preston, Philosophical Magazine **A74**, 23-46 (1996).
9. M. E. Kassner, M.-T. Perez Prado, K. S. Vecchio, and M. A. Wall, Acta mater. **48**, 4247-4254 (2000).
10. J.S. Chung, N. Tamura, G.E. Ice, B.C Larson, J.D. Budai, W. Lowe, In Materials Reliability in Microelectronics IX, , Mat. Res. Soc. Symp. Proc., Edited by C.A. Volkert, A.H. Verbruggen, D. Brown, **563** (1999) 169-174
11. N. Tamura, J.-S. Chung, G.E. Ice, B.C Larson, J.D. Budai, J.Z. Tischler, M. Yoon, E.L. Williams and W.P. Lowe, In Materials Reliability in Microelectronics IX, , Mat. Res. Soc. Symp. Proc., Edited by C.A. Volkert, A.H. Verbruggen, D. Brown, **563** (1999) 175-180.
12. B.C. Larson, N. Tamura, J.-S. Chung, G.E. Ice, J.D. Budai, J.Z. Tischler, W. Yang, H. Weiland, W.P. Lowe, , Mat. Res. Soc. Symp. Proc., Edited by S.R. Stock, S.M. Mini, and D.L. Perry, **590** (2000) 247-252.
13. N. Tamura, B. C. Valek, R. Spolenak, A. A. MacDowell, R. S. Celestre, H.A.Padmore, W. L. Brown, T. Marieb, J. C. Bravman, B. W. Batterman and J. R. Patel, Mat. Res. Soc. Symp. Proc., Edited by G.S. Oehrlein, K. Maex, Y.-C. Joo, S. Ogawa and J.T. Wetzel, **612** (2000) D8.8.1-D8.8.6
14. R. Spolenak, D.L. Barr, M.E. Gross, K. Evans-Lutherodt, W.L. Brown, N. Tamura, A.A. MacDowell, R.S. Celestre, H.A.Padmore, J.R. Patel, B.C. Valek, J.C. Bravman, P. Flinn, T. Marieb, R.R. Keller, B.W. Batterman, Mat. Res. Soc. Symp. Proc., Edited by G.S. Oehrlein, K. Maex, Y.-C. Joo, S. Ogawa and J.T. Wetzel, **612** (2000) D.10.3.1-D10.3.7.
15. A.A.MacDowell, R.S.Celestre, N.Tamura, R.Spolenak, B.C. Valek, W.L.Brown, J.C.Bravman, H.A.Padmore, B.W.Batterman & J.R.Patel, SRI Conference Proc., Berlin, Nuclear Instruments and Methods in Physics Research A (2000) in press
16. N. Tamura (2001)
17. R. Spolenak, C. A. Volkert, K. M. Takahashi, S. A. Fiorillo, J. F. Miner, and W. L. Brown, Mat. Res. Soc. Proc **594**. (1999).
18. R.-M. Keller, S. P. Baker, and E. Arzt, J. Mater. Res. **13** (5), 1307-1317 (1998).
19. Landolt-Börnstein, *Zahlenwerte und Funktionen aus Physik und Chemie*, ed. H. H. Borchers, K.-H. Hellwege, K. L. Schäfer, E. Schmidt, Springer Verlag, Berlin, Vol 4, 691-692, (1964)

GRAIN STRUCTURE EVOLUTION DURING ANNEALING OF ELECTROPLATED COPPER

S. H. Brongersma, E. Kerr, I. Vervoort, and K. Maex.

IMEC, Kapeldreef 75, B-3001 Leuven, Belgium

Abstract The recrystallization of electroplated Copper proceeds in two distinctly different phases. Firstly, secondary grain growth occurs with an activation energy of ~ 0.92 eV. Impurities, incorporated in the layer during plating are pushed into the grain boundaries during this process and Carbon then readily diffuses out to the surface. However, Hydrogen and Sulfur are less mobile and accumulate in the boundaries, thereby stabilizing the structure. Only during an extended anneal at elevated temperature is a desorption of these elements observed and does a further grain enlargement commence. This is accompanied by a strong increase in the stress driven (200) volume fraction as a result of grain boundary volume elimination.

INTRODUCTION

Electroplating of Copper in damascene structures exhibits superior filling properties where a traditional conformal fill is no longer sufficient [1]. This has made it the most popular method used for Cu deposition in the microelectronics industry today. The various additives in the plating bath result in a 'bottom-up' or 'super-filling' of large aspect ratio structures and a small as-deposited grain size. Unfortunately this renders the grain structure unstable as the resulting high density of grain boundary and defect energy results in a spontaneous recrystallization at room temperature. During this event the sheet-resistance R_S typically drops ~ 20 % to a near bulk value. Additionally, changes are also seen in grain orientation and stress, although these observables have proven themselves to depend critically on the plating chemistry [2,3].

CP612, *Stress-Induced Phenomena in Metallization:* Sixth Int'l. Workshop, edited by S. P. Baker et al.

EXPERIMENTAL DETAILS

All sample preparation and experiments described in this study are on 8" Si(100) wafers with a native oxide. First, a 30 nm TaN diffusion barrier and a 150 nm Cu seed layer are plasma vapor deposited in an Endura tool from Applied Materials. The subsequent electrochemical deposition (ECD) of 1μm Cu films takes place in an Equinox tool from Semitool *Inc.*, using the Nanoplate2001^{TM} plating chemistry from Shipley *Inc.*. The chemistry contains some proprietary additives, amongst others. brightener, designed to improve the filling of deep trenches with Cu during the ECD process. The concentration of brightener [B] is determined using a Cyclic Voltametric Stripping technique (QL-10) and was kept constant at 3 ml/l in this study. Sheet resistance measurements with a KLA-Tencor RS75 reflect the average of 49 points distributed evenly over the wafer surface while the experiments on transition times at elevated temperatures were performed in a Sum 11k probe station from Cascade in combination with a HP 4156.

RESULTS AND DISCUSSION

The time evolution of the sheet-resistance is strongly dependent on several parameters, including layer thickness, brightener concentration, and plating current I_P. Figure 1 shows how the transformation time t_R, defined as the time needed for half the decrease to occur, increases considerably with decreasing I_P. At the same time the size of the change in R_S diminishes, but as indicated by the Arrhenius plot in fig 2, the activation energy of ~ 0.92 eV is the same for all these layers. This value is consistent

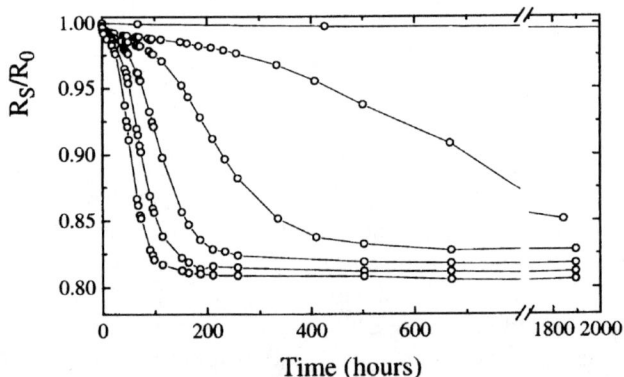

FIGURE 1. Plot of normalized sheet resistance versus time as measured at room temperature for 1 μm Copper layers plated on a 200 mm wafer at 6.0, 4.5, 3.0, 1.5, 0.75, and 0.3 A from left to right respectively.

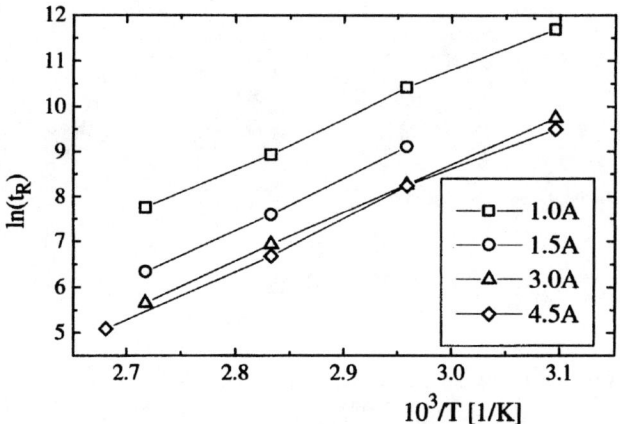

FIGURE 2. Arrhenius plot for 4 values of the plating current, showing that the activation energy is the same for all conditions, as expected for grain boundary self diffusion.

with the one quoted previously by Gupta [4] for grain boundary self diffusion in very pure Copper. At the same time impurities, incorporated in the layer during plating, are forced into grain boundaries during secondary grain growth, as shown by Hau-Riege and Thompson [5] in a TEM study, and then diffuse out to the surface through the untransformed matrix. As we published previously [6], there is indeed a substantial migration of Carbon during the recrystallization as shown through a combined desorption-LEIS (low energy ion scattering) study. However, our simulations on grain growth [7] clearly show that a stagnation of grain growth occurs at a specific size a_0 (equal to ~ 2 μm), which was attributed to a critical concentration of impurities in the boundaries that inhibits further growth.

The focussed ion beam images in fig 3 show the grain structure at three different times after plating and indicate the fairly small grain size after a stabilization of the sheet resistance. Thus the question remains which impurity could be responsible for such a stagnation and whether it can be removed in order to allow further growth.

A further high temperature anneal whilst monitoring the sample surface with LEIS and using a gas analyzer to check for desorption shows that both Sulfur and Hydrogen were still in the layer after the structure had stabilized. Hydrogen in fact stays in the Cu layer up to 400 °C, which is well above the temperature expected for Copper and as approximately 40 monolayers of Hydrogen were recorded there were clearly enough impurities in the layer to saturate all the grain boundaries and block grain boundary motion. However, whether Hydrogen or Sulfur is truly responsible for this remains to be investigated. Also it is unclear whether these two elements remain in the sample in a bonded state, which would provide an explanation for the high desorption temperature.

The resulting grain structure after release of the boundaries, as shown in fig 4, has indeed increased considerably in size. The accompanying grain boundary volume

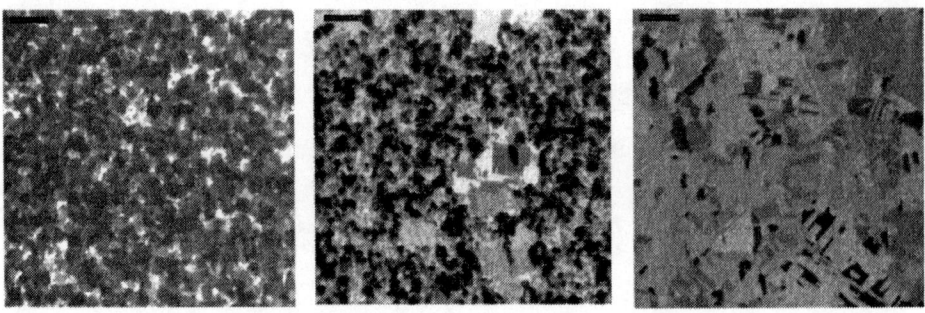

FIGURE 3. Three FIB images taken just after plating, during the recrystallization, and after the sheet resistance has stabilized from left to right respectively. Each image shows an area of ~ 10x10 μm^2.

elimination results in a strong increase in the stress driven (200) component (from 16 to 37 %), as determined from XRD polar figures described previously [8]. Thus, even though the final values of the sheet resistance observed in fig 1 appear stable, they are found to decrease further to a full 20 % of the as-plated sheet resistance for all the layers after the high temperature anneal. Thus a lower plating current results in a less complete first phase of the transformation, but this is compensated by a larger decrease of R_S during the continuation of grain growth after the impurities that initially stabilized the structure are removed.

FIGURE 4. Top view FIB image showing that considerable grain growth occurs during extensive annealing, even though the sheet resistance appeared stable after the first phase of the recrystallization.

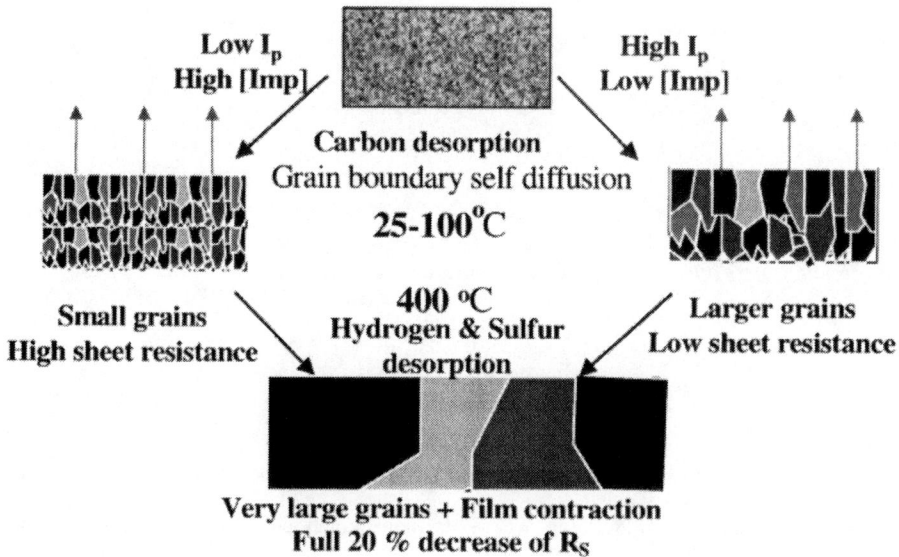

Low I$_p$
High [Imp]

Carbon desorption
Grain boundary self diffusion
25-100°C

High I$_p$
Low [Imp]

Small grains
High sheet resistance

400 °C
Hydrogen & Sulfur
desorption

Larger grains
Low sheet resistance

Very large grains + Film contraction
Full 20 % decrease of R$_S$

FIGURE 5. Overview of the influence of impurities during two phases of grain growth.

An overview of the various stages of growth is given in fig 5. First an initial deposit is formed with an impurity concentration that increases with decreasing plating current [8,9]. During secondary grain growth these impurities are pushed into the grain boundaries and may then either diffuse out to the surface or accumulate in the boundary. In this case Carbon diffuses out while Sulfur and Hydrogen remain, causing the existence of a limit to final grain size. Thus the transition time changes because the number of nucleation events (used in the meaning of onset of growth for a secondary grain) needed to transform the entire matrix depends on the final size of each individual grain. Also, depending on grain size the final stable sheet resistance after this first stage will vary.

During a second stage the impurities initially inhibiting growth, namely Hydrogen and/or Sulfur, are removed from the layer and grain growth continues. This leads to very large grains observed in fig 4, a significant increase in the (200) volume fraction, and a further reduction of the sheet resistance after which the value for all layers has decreased by the full 20%.

It is important to note that the existence of two separate stages is induced by a growth inhibiting impurity that was originally incorporated into the layer during the plating process. This indicates that for another chemistry, which leads to less or no Sulfur incorporation, this behavior may not be present. In such a case [3] the grain boundary volume reduction would already occur in the first (and then only) stage.

Build up of tensile stress and an increase in (200) fraction should then even be observed at T_R.

CONCLUSIONS

From the above it is clear that impurities play an important role in the dynamics of grain growth during the recrystallization and that the details may depend strongly on small changes in plating conditions and chemistry. In our case it induces two stages that are separated by a Sulfur and/or Hydrogen induced stagnation of growth.

REFERENCES

1   D.C. Edelstein, Proc. SPIE Conf, Multilevel Interconnect Technol. II **3508**, 8 (1998).
2   D. Walther, M.E. Gross, K. Evans-Lutterodt, W.L. Brown, M. Oh, S. Merchant, and P. Naresh, Met. Res Soc. Proc. **612**, in press.
3   H. Lee, S.D. Lopatin, and S.S. Wong, Proc. of the International Interconnect Technology Conference 2000, p.114. (ISBN 0-7803-6327-2).
4   D. Gupta, Mat. Res. Soc. Proc. **337**, 209 (1994).
5   S.P. Hau-Riege and C.V. Thompson, Appl. Phys. Lett **76**, 309 (2000).
6   S.H. Brongersma, E. Kerr, I. Vervoort, and K. Maex, Proceeding of the Advanced Metallization Conference 2000, accepted.
7   S.H. Brongersma, E. Richard, I. Vervoort, and K. Maex, Proceedings of the International Interconnect Conference 2000, p.31. (ISBN 0-7803-6327-2).
8   S.H. Brongersma, E. Kerr, I. Vervoort, A. Saerens, and K. Maex, Submitted to Journal of Materials Research.
9   M.A. Gribelyuk, S.G. Malhotra, P.S. Locke, P. DeHaven, J. Fluegel, C. Parks, A.H. Simon, and R. Murphy, Proc. of the International Interconnect Technology Conference 2000, p.188. (ISBN 0-7803-6327-2).

Stress Induced Metallurgical Effects in Ti/TiN/AlCu/TiN Metal Stacks

Klaus Koller, Martina Hommel*[)], Stefan Hummelt, Heinrich Koerner

Infineon Technologies AG, Wireless Products, WS TI T,
Otto-Hahn-Ring 6; D-81739 Munich; Germany
*[)] Infineon Technologies AG, Corporate Frontend CFE PT RM
Otto-Hahn-Ring 6; D-81739 Munich; Germany

Abstract. Integrated circuits with aluminum metallization for products with high current densities need a metal stack with liner and antireflective coating (ARC) which can fulfill several requirements (e.g. low sheet resistance, high reliability, smooth surface, good adhesion, thermal stability, etc.). In this work different multilayer metal stacks are investigated and several phenomena which can be observed after thermal annealing of Ti/TiN/AlCu/TiN stacks are described and discussed. Metallurgical, electrical and mechanical properties of different layer combinations are investigated after thermal annealing and stress tests are done to compare the electromigration and life time behaviour of each metal stack. For all investigated metal stacks it is shown that an interface reaction between Ti and aluminum will form $TiAl_3$ phase. Even with very thick TiN layers on top of titanium or with only TiN liner the phase formation occurred. Explanations and models for the formation of different phenomena (hillocks, depressions and elevations), are discussed. The origin of each phenomena is stress related and assisted either by the liner material and/or the ARC layer. A qualitative model which explains the different observed layer reactions is discussed.

INTRODUCTION

The requirements on metallization and interconnect systems in advanced SiGe BiCMOS technology increase with decreasing feature sizes. Especially for advanced SiGe bipolar technologies, the current densities through the transistors and thus in the metallization are getting more demanding (≈ 0.3 MA/cm²) than e.g. in CMOS technologies with comparable feature sizes (critical dimension 0.4 µm for metal lines).

Especially for high-speed communication devices, the lifetime requirements will be high (≥ 15 years). Additional to that, passive devices (like MIM capacitors) need to be integrated into the metallization schemes and create additional requirements on e.g. surface roughness, stress behaviour and the stability of metal stacks during manufacturing and under operational or stress conditions.

State of the art AlCu metal stacks typically contain bottom layers (liner) and top layers (ARC) to fulfil a lot of different manufacturing and reliability aspects. The liner, which consists in most cases of Ti or Ti/TiN, is required for optimum adhesion and reliability behaviour. The antireflective coating (ARC) layer is necessary to achieve fine pitch lithography requirements.

During the BEOL manufacturing process the wafers go through several thermal processes like PECVD depositions, spin on glass (SOG) annealing and forming gas

CP612, *Stress-Induced Phenomena in Metallization:* Sixth Int'l. Workshop, edited by S. P. Baker et al.
© 2002 American Institute of Physics 0-7354-0058-X/02/$19.00

anneal. These temperature activations cause several interface reactions and lead to substantial mechanical stress. The reason for the creation of stress is different coefficients of thermal expansion of the materials used. The aluminum itself expands 8 times more than silicon and the diffusion barrier layers based on titanium and/or titanium nitride have a three times higher expansion coefficient than silicon.

Temperature annealing processes used in the BEOL go up to 430°C and create stress related surface artefacts in the different metal stacks. Hillock formation [1] is one main observation which can be explained by material diffusion along grain boundaries. This phenomena is well known in IC aluminum metallization processes. But with increasing miniaturisation of structures additional phenomena are observed [2]. Besides hillock formation two other striking observations are "depressions" (deepenings in the metal surface) and "elevations" (smooth large elevations up to 20 microns in diameter and 200 nm in height). These phenomena can be very easily observed by means of optical microscopy. They can only be seen on patterned wafers and were up to now not very well studied. With this work we try to find out the origin of the artefacts which occur in Ti/TiN/AlCu/TiN metal stacks.

Sample Preparation and Experiments

Different metal stacks were prepared using patterned and unpatterned wafers. The different metal stack split groups were insitu sputtered in an Applied Materials ENDURA PVD system. The investigated multilayer metal stacks consist of (Ti) : (TiN) / AlCu / (Ti) : TiN layer combinations of different thicknesses and combinations. The titanium or titanium nitride and the AlCu (0.5 at. % Cu) layer were deposited at a temperature of 150°C. Only titanium nitride was used as an ARC layer on top of the metal stack in most of the cases.

Patterned wafers for electromigration investigations were prepared with a two metal level, multilayer interconnect. Metal one (with an AlCu thickness of 400 nm, line widths 0.4 µm) was sputtered onto a BPSG oxide layer. Deposition of a PECVD oxide and SOG and chemical mechanical polishing (CMP) followed for intermetal dielectric planarization after patterning of the metal one layer. After via etch and tungsten via-plug process the same metal two stack was sputtered (AlCu thickness 600 nm, line widths 0.5 µm) and patterned. Metal two was passivated with PECVD oxide (200 nm) and PECVD nitride (300 nm). Finally, a forming gas anneal (5% H_2 in N_2) was carried out at a temperature of 430°C.

The sheet resistance and mechanical characterisation of the different metal stacks were done on unpatterned, unpassivated wafers after the metal sputtering process. These wafers were annealed up to 6 times in forming gas atmosphere after the first characterization. The first annealing process was applied to simulate the temperature stress which occurs during the typical interconnect manufacturing process (e.g. PECVD oxide deposition at 390°C and SOG anneal after spinning at 400°C for 1 hour). The following annealing procedures were done in order to study the temperature and material stability of the stacks with respect to interface reactions or hillock growth.

The main combinations for the material interaction studies are listed in table 1.

TABLE 1. Layer combinations.

	Titanium (nm)	Titanium nitride (nm)	AlCu (nm)	Titanium nitride or Ti / TiN (nm)
1. group	5 to 80	0	400 or 600	0 / 32
2. group	10 to 40	10 to 50	400 or 600	0 / 32
3. group	10 to 20	0 to 50	400 or 600	0 / 32
4. group	0	10 to 50	400 or 600	0 / 32
5. group	10 to 40	10 to 50	400 or 600	8 / 32

A) Al grain growth on different liners

For the evaluation of the grain growth on different liner materials, wafers with only a few atomic layers of AlCu were prepared (adjusted AlCu sputter parameters, less RF-power). There is a clear difference in grain growth within the investigated liner materials. Titanium liner w/o TiN show a homogenous layer of aluminum (figure 1a). This corresponds to a "Frank-van-der-Merwe-Growth" which gives a single crystal layer with a wetting angle of zero.

For liners with a thin TiN layer above the titanium, the aluminum layer deposits inhomogeneously (figure 1b/c). The aluminum forms small islands and shows a mixed growth which corresponds to "Stranski-Krastanov". AlCu sputtered on TiN liner without titanium underneath shows a crystalline aluminum growth (figure 1d). The islands are separated and at this point no formation of a continuous layer occurred. This material growth modus is described in the "Volmer-Weber" model.

 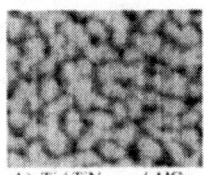 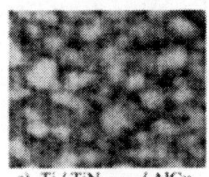 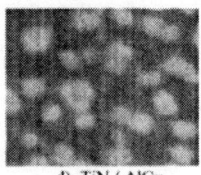

a) Ti / AlCu   b) Ti / TiN$_{<20 nm}$ / AlCu   c) Ti / TiN$_{>20 nm}$ / AlCu   d) TiN / AlCu

FIGURE 1. SEM picture of AlCu seed layer a) on Ti b/c) on Ti and TiN d) on TiN only.

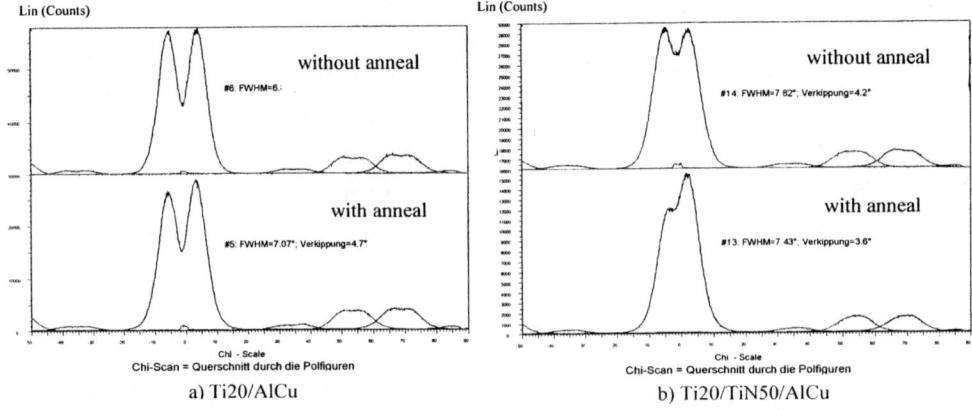

Lin (Counts)

Lin (Counts)

without anneal

#6. FWHM=6.

with anneal

#5: FWHM=7.07°; Verkippung=4.7°

Chi - Scale
Chi-Scan = Querschnitt durch die Polfiguren

a) Ti20/AlCu

without anneal

#14: FWHM=7.82°; Verkippung=4.2°

with anneal

#13: FWHM=7.43°; Verkippung=3.6°

Chi - Scale
Chi-Scan = Querschnitt durch die Polfiguren

b) Ti20/TiN50/AlCu

FIGURE 2. Texture of a) Ti 20/AlCu without anneal with anneal b) Ti20/TiN50/AlCu

B) Texture measurements

Texture measurements were done with a Siemens D 5000 goniometer. Samples with different $Ti_x/TiN_y/AlCu$ layers with (FG 430°C) and without anneal were measured. There was no significant difference between these two groups for all liner thicknesses. The AlCu exhibited a prefered <111> crystallographic orientation (fig. 2).

C) Resistance measurements

The resistivity measurements were done with a four-point-method by "Valdes" using a Prometrix Omnimap RS30.

Resistivity measurements were done with different titanium liner thicknesses after each annealing step (one step: 430°C, 30 min in FG). The largest resistance increase typically was observed after the first annealing step due to $TiAl_3$ formation because of the higher resistivity of $TiAl_3$ (40 $\mu\Omega$*cm [6]) compared to AlCu (2.7 $\mu\Omega$*cm). The resistance increase with each following anneal was less pronounced. The higher the Ti thickness, the larger was the sheet resistance increase (figure 3).

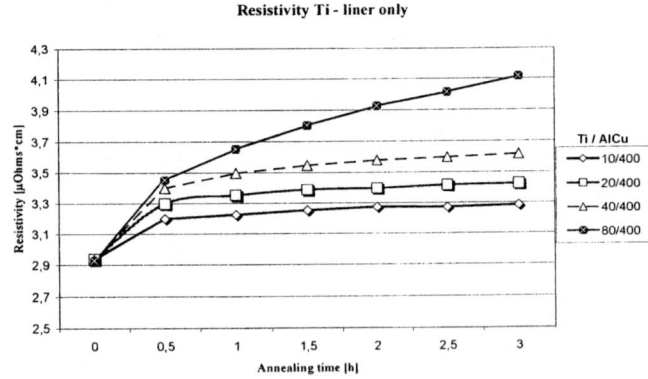

Resistivity Ti - liner only

Ti / AlCu
10/400
20/400
40/400
80/400

FIGURE 3. Sheet resistance behaviour for different titanium liner thicknesses after annealing steps

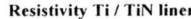

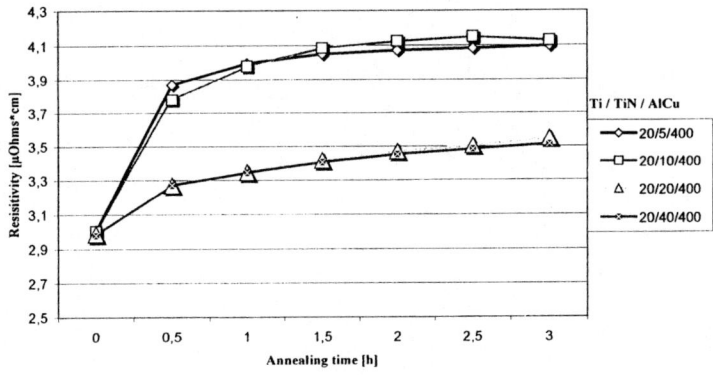

FIGURE 4. Resistivity as a function of annealing time

The resistivity of combinations of Ti/TiN liners shows a significant dependence on the TiN layer thickness. TiN thicknesses less than 20 nm on top of Ti could not suppress the TiAl$_3$ formation, leading to a significant resistance increase after the first annealing step. However, liner thicknesses greater than 20 nm reduced the TiAl$_3$ formation significantly as can be revealed in the sheet resistance progress after annealing procedures. Furthermore, such liners show no saturation of the sheet resistance even after 3h total anneal time. It is assumed that a steady titanium diffusion through the TiN layer occured during annealing steps and thus the sheet resistance increased continously due to ongoing phase formation (figure 4).

D) Mechanical stress measurements

For a relative comparison, the wafer bow was measured before and after deposition as well after each annealing step with a Tencor MX 203 tool. This is an easy method which uses a capacity measurement at four points across the wafer. Corrugated surface effects could not be detected with this measurement method.

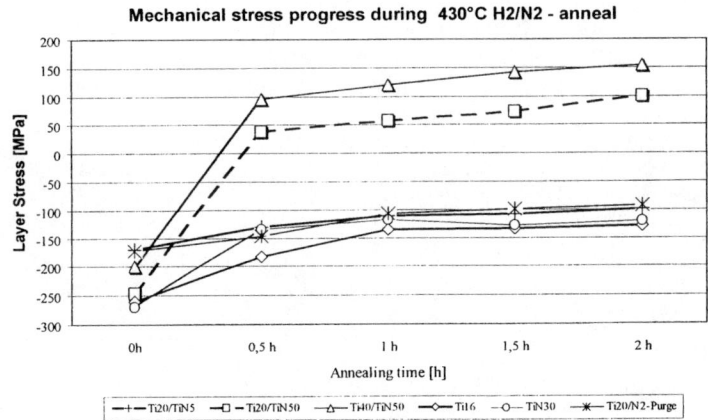

FIGURE 5. Stress measurements of different liner stacks w/o AlCu as a function of annealing time

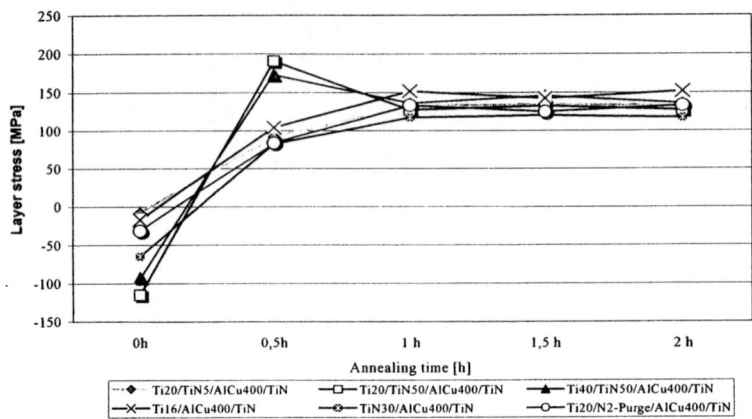

FIGURE 6. Stress measurements of different liner stacks with AlCu as a function of annealing time

The mechanical stress measurements were done with two groups, one with (figure 6) and one w/o aluminum (figure 5) on top of the Tix/TiNy liner combinations, respectively. In both cases, the liner material mainly influences the mechanical stress during the annealing process. The aluminum which has a compressive stress after deposition reduces or compensates the tensile stress of the liner. The absolute amount of the stress difference is in the order of the yield point of the bulk aluminum ($\approx$ 80MPa).

After the first annealing step the liner stacks w/o TiN and with TiN thicknesses of less than 20 nm compensate the compressive stress due to $TiAl_3$ formation and a tensile stress results. For liners with thick TiN layers or only TiN, the stress is independent of the presence of an aluminum layer. However, the stress is not saturating after the first anneal step because of ongoing $TiAl_3$ phase formation and aluminium relaxation (TiN thickness 50 nm). At the interface of these stacks, the liner material continuously forms $TiAl_3$ and AlN.

E) Surface roughness and reflectivity measurements

Two measurement methods were used for a relative comparison of the surface roughness: the reflectivity using Tencor Prometrix UV-150 (for g-line and i-line wavelength) and the determination via Atomic Force Microscope (AFM).

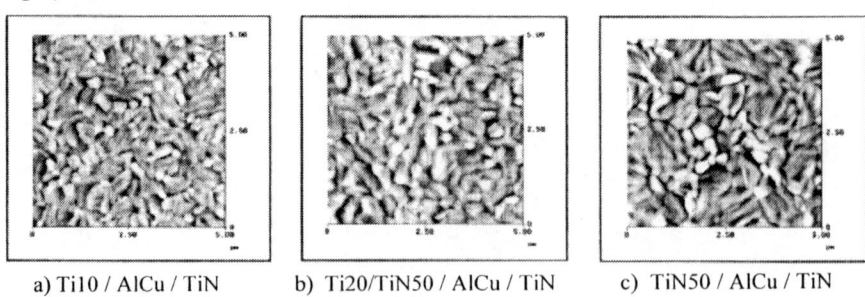

a) Ti10 / AlCu / TiN   b) Ti20/TiN50 / AlCu / TiN   c) TiN50 / AlCu / TiN

FIGURE 7. AFM pictures of different metal stacks

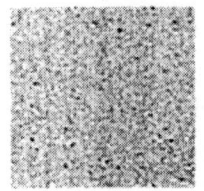 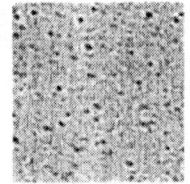 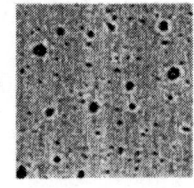 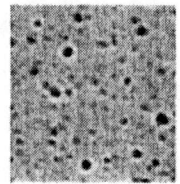

a) Ti10 / AlCu / TiN   b) Ti20/TiN<20nm   c) Ti20/TiN>20nm   d) TiN50 / AlCu / TiN

FIGURE 8. Microscope pictures of different metal stacks

As seen in the electrical measurements there is a good correlation between the main groups. The metal stacks which form the TiAl$_3$ phase show a higher reflectivity after the first annealing step than those with thick TiN liners. This can be explained by less hillock growth (hillock formation = less reflectivity) due to phase formation and stress relaxation. This will only change after titanium has completely reacted with aluminum.

AFM measurements show a different aluminium structure in dependence of the liner composition. A comparison of three different liner materials is shown in figure 7. Figure 7a) shows a liner with Titanium with Nitrogen purge after deposition . The grain size is much smaller and more inhomogenous than with TiN only (figure 7c). The surface height difference between stacks with Ti only compared to the liner with TiN (figure 7b) and c) was up to 35 % which means a larger surface roughness especially for the TiN liner only.

After the annealing process, these both main groups can also be easily distinguished because of the different number of hillocks formed (figure 8 a to d).

Results and Discussion

One of the main impacts of the liner material on aluminum multilayer metallization during thermal processing is relaxation of stress, which has been built up because of the different coefficients of thermal expansion [3] of the materials (see table 2).

A) "Hillock" formation

The liner and the deposited metal stack on silicon oxide over silicon bulk material exhibit a compressive stress and a convex wafer bow. During thermal annealing, the aluminum layer tends to expand. Compressive stress builts up in the aluminum layer because the liner materials (Ti and/or TiN) underneath prevent the expansion or relaxation. This compressive stress generates a vertical aluminum flux along the grain boundaries to the surface because the material mobility at the aluminum surface is much higher than at the interface.

TABLE 2. Coefficients of thermal expansion for various materials				
	Silicon/Silicon oxide	Aluminum	Titanium	Titanium nitride
Coefficient α	$3 * 10^{-6}$ K^{-1}	$23.6 * 10^{-6}$ K^{-1}	$8,4 * 10^{-6}$ K^{-1}	$9,6 * 10^{-6}$ K^{-1}

241

Annealing of the aluminum forms a bamboo grain structure (line widths <0.5 μm). Aluminum atoms will diffuse preferentially along the vertical grain boundaries towards the surface (especially at triple points) and the accumulated material will form hillocks.

The amount of hillock formation in the aluminum film was investigated as a function of the various metal liners. In case of thin TiN layers less hillock formation was observed. If no or less hillock growth occurred only two explanations are possible: a) there has been no compressive stress or b) no diffusion of Al could occur. It is well known that Al lines wth less than 1 μm width, as used in this study, always form a bamboo structure. Therefore, vertical diffusion paths have been present and option b) is not a suitable explanation for the suppressed hillock growth. Therefore, only the absence of layer stress can be the appropriate reason for less or no hillock formation.

In the case of thin (≤ 20 nm) TiN layers between Al and bottom Ti, TEM analysis (figure 9) show a large amount of $TiAl_3$ phase formation. The $TiAl_3$ phase has a higher density and leads to a volume shrinkage of 6% compared to pure aluminum. Thus, the compressive stress is compensated by this interfacial reaction and no hillock growth occurs. In addition, the electrical resistance measurements also indicate the $TiAl_3$ phase formation. It forms during the first annealing step, which leads to a significant increase of the resistance. The following annealing procedures have a minor or no impact on sheet resistance.

The diffusion of titanium through thin TiN layers (≤ 20 nm) occurs easily. At TiN - thicknesses above this threshold, Ti diffusion is slowed down. Therefore, there is only a less pronounced increase of resistance observed and the resistance does not change with such metal stacks after each thermal anneal procedure. In addition, the formation of thin AlN interlayers was measured with ESI (Electron Spectroscopy Imaging) at the interface of TiN to aluminum which will also slow down or hinder the Ti diffusion. In such cases which do not lead to significant $TiAl_3$ formation, there is no relaxation of the compressive stress in the Al layer and the stress enhances the Al diffusion through vertical grain boundaries and thus supports the excessive formation of hillocks. Extensive hillock formation will also take place if there is only a very thin Ti-layer beneath the TiN or the Al, which will lead to an insufficient formation of the stress releaving $TiAl_3$ phase. If sufficient formation of $TiAl_3$ phase is achieved (either by no or only thin TiN layers on top of a sufficiently thick Ti-layer), a relative smooth aluminum surface with a minimum number of hillocks will be achieved.

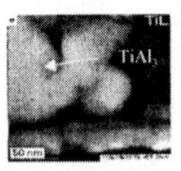
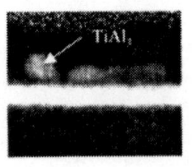
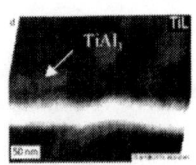

a) Ti10 / AlCu / TiN   b) Ti20/TiN<20nm   c) Ti20/TiN>20nm   d) TiN50 / AlCu / TiN

FIGURE 9. TEM analysis of different metal stacks

B) Formation of "depressions"

Depressions are mostly observed on patterned wafers because the mechanical stress is much higher at patterned metal lines (figure 10a). The phenomena is especially observed at rectangular bends of metal lines where the location of the depression was exactly at the edge of the structure with an angle of 45°. Depressions are preferentially forming in these locations because there is tensile stress on both length sides of the metal lines. Stress in directions vertical and diagonal to the metal line direction relaxes easily because there is no or only less surrounding material. In addition to that, the $TiAl_3$ phase is forming which is associated with volume contraction. The whole stress force is acting at the boundary of the material yield point and therefore the aluminum will be deformed.

Unpatterned wafers do not exhibit depressions, but show grooves. The reason for the groove formation (0.5 to 1.0 microns wide) is supposed to be the $TiAl_3$ formation with its volume shrinkage.

Tensile stress builds up during the cooling phase after annealing. It is caused by the phase formation and it will compensate the compressive stress which formed because of the different coefficients of thermal expansion.

C) Formation of "elevations"

The occurrence of elevations is very similar to the hillock formation. They arise only on metal stacks which have a TiN ARC layer on top and a thick TiN liner beneath. Furthermore, they are only observed on patterned wafers.

A very similar explanation as for the growth of hillocks is used to understand the formation. Because of the compressive stress in the stack, which is formed due to the different expansion coefficients of the layers, material transportation along the grain boundaries towards the surface will take place. In this case, a TiN ARC layer will hinder or prohibit a simple hillock creation. Therefore an additional force will build up against the direction of material transport and it will lead to a lateral Al diffusion along the interface of the TiN ARC and the AlCu. At spots where the TiN ARC is discontinues or weak, the material will be accumulated and "elevations" will form. The

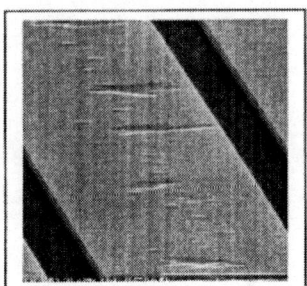 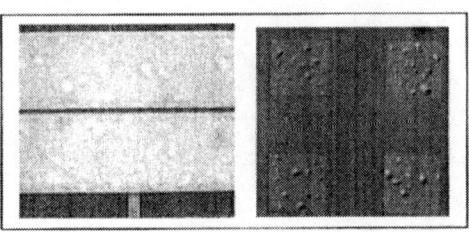

FIGURE 10. a) SEM picture of "depressions" b) microscope picture of "elevations"

assumption of a large lateral material flux at the interface is based on the high mobility of aluminium atoms at temperatures at 430°C.

Existing models which describe the material transportation under induced stress were described by Nabarro/Herring (1) and by von Coble (2) [3]. These models describe the material transport from the bulk to the aluminum surface (diffusion mechanism).

$$\frac{d\varepsilon_{NH}}{dt} = A \cdot \frac{D_V \sigma \Omega}{d^2 kT} \qquad (1)$$

$$\frac{d\varepsilon_C}{dt} = A' \cdot \frac{\delta D_B \sigma \Omega}{d^3 kT} \qquad (2)$$

ε - dilatation	Ω - atomic volume
$A^{()}$ - coefficent	d - grain diameter
D_V - volume diffusion coefficient	k - Boltzmann constant
D_B - grain boundary diffusion coefficent	T - absolute temperature
σ - compressive tension	δ - thickness of grain boundary

Electromigration results

Electromigration tests were performed in both directions of current flow using a via chain test structure with different metal line lengths (10 µm up to 300 µm). In the "via upstream" test (current flow from metal one through the via to metal two, figure 11) metal stacks with thick TiN liners (≥ 20 nm) over titanium exhibited the highest time to failure.

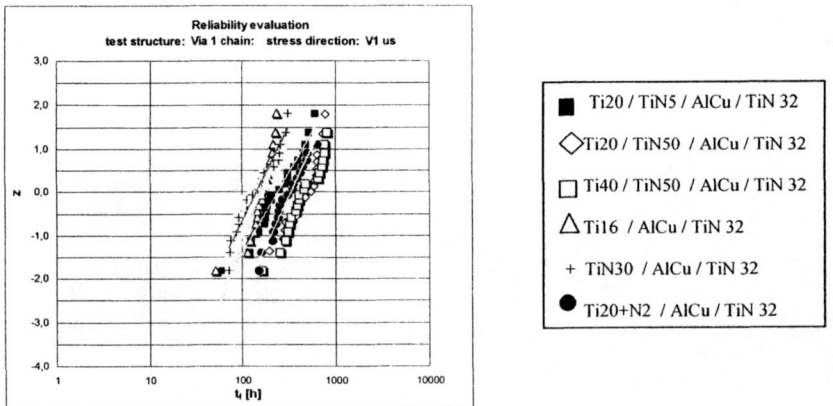

FIGURE 11. Lifetime and cumulative failure probability during electromigration stress (current flow via upstream)

This is explained by the absence of a pronounced TiAl$_3$ phase, which would provide additional interfaces and diffusion paths for lateral Al material diffusion. Metal stacks which form a thick TiAl$_3$ phases provide such additional diffusion paths and finally exhibit enhanced void formation above the via.

One specific group with only a titanium liner, which received a nitrogen purge directly after the sputtering process [5], showed a significantly better time to fail. In this case the TiAl$_3$ phase was formed as well but the lifetime of this specific sample in this stress mode seems to indicate a different failure mechanism and will require further analysis.

The group with TiN liner only showed significant shorter times to fail. This was caused by weak adhesion as confirmed with TEM analysis.

The different stacks showed only negligible differences with respect to lifetime when the current flow during electromigration stress was in the opposite direction (via downstream). The Ti liner with nitrogen purge and the TiN only liner achieved the highest and the lowest time to failure, respectively. Due to weak adhesion, the sample with TiN only liner exhibited the worst results for both stress directions.

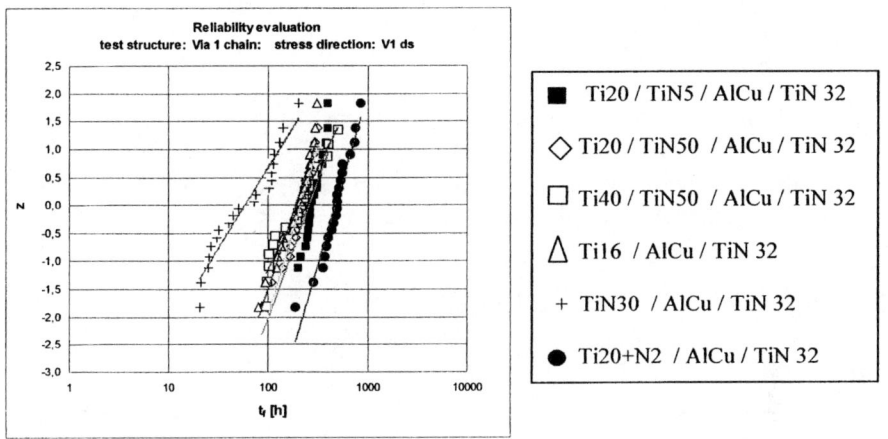

FIGURE 12. Lifetime and cumulative failure probability during electromigration stress (current flow via downstream)

245

CONCLUSION

In this work several phenomena which can be observed after thermal annealing of Ti/TiN/AlCu/TiN metal stacks were described and discussed.

For all investigated metal stacks it was shown that an interface reaction between Ti and aluminum will form $TiAl_3$ phase. Even with very thick TiN layers on top of titanium or with TiN liner only the phase formation occurred. Therefore the TiN liner between titanium and aluminum did not prohibit the diffusion of titanium into aluminum. A significant impact on the amount of diffusion was obtained with TiN layer thicknesses ≥ 20 nm. Explanations and models for the formation of different phenomena (hillocks, depressions and elevations), which have been observed with various metal stacks after annealing have been discussed.

The origin of each phenomena is stress related and assisted either by the liner material and/or the ARC layer. Each type of the artefacts can be found and depends on the chosen materials and the applied process (e.g. temperature budget).

Electromigration results of specific metal stack combinations were obtained with different stress modes. A superiour metal stack with respect to electromigration lifetime within the investigated metal stack repertoire could not be found. This may be due to the fact that several other aspects (e.g. patterning, vias, interfaces between lines and vias, electromigration stress modes etc) exhibit a certain influence and have to be taken into account.

ACKNOWLEDGEMENTS

The authors would like to thank Mrs. Jobst for texture analysis and Mrs. Zeitler for preparing the test material and inline measurements.

REFERENCES

1. F.Y. Gènin, W.J. Siekhaus; "Experimental study to validate a model of hillocks formation in aluminum thin films.", in *J. Appl. Phys.* **97(7)**, April 1996, pp. 3560-3566
2. D. K. Ferry; M. N. Kozicki; G. B. Raupp, "Some Fundamental Issues of Metallization in VLSI", in *Tempe Center for Solid State Electronics,* Arizona State Univ Research.; Army Research Office, Research Triangle Park, NC
3. Hermann Römpp, *Römpp-Chemistry-Lexikon Thieme Verlag,* 9th edition, **1995**, Vol. 1 pp. 573
4. S. Schmidbauer, S. Sinler, M.U. Lehr, J. Klotzsche, J. Hahn, *"Novel metallization scheme using ntirogen passivated Ti liner for AlCu based metallization"*; Proceedings of the SPIE Symposium **1999**
5. L.M. Gignac, et al, "Thin film interaction between Ti, TiN and Al during high temperature Al reflow of 1 Gb DRAM interconnects"; IBM-TJ Watson Research Center
6. J.A. Prybyla, R.R. Kola, R. Hull, D.J. Eaglesham, H.A. Huggins, "Microstructure and electromigration properties of submicron Al(0.5%Cu) lines; Mat. Res. Symp. Proc. **Vol. 317. 1994,** Materials Research Society

247

R

Ramanath, G., 10
Rauser, D., 86
Rosenberg, R. R., 33, 177

S

Saito, S., 49
Saka, M., 74
Sakaue, H., 94
Sasagawa, K., 74
Schmidt, H., 133
Schwaiger, R., 119
Sekiguchi, A., 169
Shaw, T. M., 33, 177
Shibutani, T., 184
Shin, C.-S., 10
Shingubara, S., 94
Shiratori, M., 184
Spolenak, R., 217
Sullivan, T. D., 33
Suzuki, M., 49

T

Tachibana, A., 105
Takahagi, T., 94
Tamura, N., 217
Thompson, C. V., 205

U

Ueno, K., 49

V

v. Glasow, A., 157
Valek, B. C., 217
Vervoort, I., 229
Volkert, C. A., 119

W

Wachnik, R. A., 33
Wada, M., 169
Wang, P.-C., 33
Weihnacht, M., 133
Wendrock, H., 86
Wetzig, K., 86, 133

Y

Yu, Q., 184

Z

Zitzelsberger, A. E., 157

SUBJECT INDEX